D1754799

Edited by
Michael Feig

Modeling Solvent Environments

Related Titles

Matta, C. F. (Ed.)

Quantum Biochemistry

Electronic Structure and Biological Activity

2010
ISBN: 978-3-527-32322-7

Helms, V.

Principles of Computational Cell Biology

From Protein Complexes to Cellular Networks

2008
ISBN: 978-3-527-31555-0

Comba, P., Hambley, T. W., Martin, B.

Molecular Modeling of Inorganic Compounds

2009
ISBN: 978-3-527-31799-8

Höltje, H.-D., Sippl, W., Rognan, D., Folkers, G.

Molecular Modeling

Basic Principles and Applications

Third, Revised and Expanded Edition
2008
ISBN: 978-3-527-31568-0

Reiher, M., Wolf, A.

Relativistic Quantum Chemistry

The Fundamental Theory of Molecular Science

2009
ISBN: 978-3-527-31292-4

Edited by Michael Feig

Modeling Solvent Environments

Applications to Simulations of Biomolecules

WILEY-VCH

WILEY-VCH Verlag GmbH & Co. KGaA

The Editor

Prof. Michael Feig
Michigan State University
Department of Biochemistry & Molecular Biology
Department of Chemistry
218 Biochemistry Building
East Lansing, MI 48824
USA

All books published by Wiley-VCH are carefully produced. Nevertheless, authors, editors, and publisher do not warrant the information contained in these books, including this book, to be free of errors. Readers are advised to keep in mind that statements, data, illustrations, procedural details or other items may inadvertently be inaccurate.

Library of Congress Card No.: applied for

British Library Cataloguing-in-Publication Data
A catalogue record for this book is available from the British Library.

Bibliographic information published by the Deutsche Nationalbibliothek
The Deutsche Nationalbibliothek lists this publication in the Deutsche Nationalbibliografie; detailed bibliographic data are available on the Internet at <http://dnb.d-nb.de>.

© 2010 WILEY-VCH Verlag GmbH & Co. KGaA, Weinheim

All rights reserved (including those of translation into other languages). No part of this book may be reproduced in any form – by photoprinting, microfilm, or any other means – nor transmitted or translated into a machine language without written permission from the publishers. Registered names, trademarks, etc. used in this book, even when not specifically marked as such, are not to be considered unprotected by law.

Composition Toppan Best-set Premedia Limited, Hong Kong
Printing and Bookbinding betz-druck GmbH, Darmstadt
Cover Design Schulz Grafik-Design, Fußgönheim

Printed in the Federal Republic of Germany
Printed on acid-free paper

ISBN: 978-3-527-32421-7

Contents

Preface *XIII*
List of Contributors *XV*

1 Biomolecular Solvation in Theory and Experiment *1*
Michael Feig
1.1 Introduction *1*
1.2 Theoretical Views of Solvation *2*
1.2.1 Equilibrium Thermodynamics of Solvation *2*
1.2.2 Radial Distribution Functions *3*
1.2.3 Integral Equation Formalisms *4*
1.2.4 Kirkwood–Buff Theory *5*
1.2.5 Kinetic Effects of Solvation *5*
1.3 Computer Simulation Methods in the Study of Solvation *6*
1.3.1 Molecular Dynamics and Monte Carlo Simulations *6*
1.3.2 Water Models *7*
1.3.3 Solvent Structure and Dynamics from Simulations *8*
1.3.4 Free Energy Simulations *10*
1.4 Experimental Methods in the Study of Solvation *11*
1.4.1 X-Ray/Neutron Diffraction and Scattering *11*
1.4.2 Nuclear Magnetic Relaxation *12*
1.4.3 Optical Spectroscopy *13*
1.4.4 Dielectric Dispersion *13*
1.5 Hydration of Proteins *13*
1.5.1 Protein Folding and Peptide Conformations in Aqueous Solvent *14*
1.5.2 Molecular Properties of Water Near Protein Surfaces *14*
1.5.3 Water Molecules at Protein–Ligand and Protein–Protein Interfaces *16*
1.6 Hydration of Nucleic acids *16*
1.7 Non-Aqueous Solvation *18*
1.7.1 Alcohols *18*
1.7.2 Urea *18*
1.7.3 Glycerol *19*
1.8 Summary *19*
References *20*

Modeling Solvent Environments. Applications to Simulations of Biomolecules. Edited by Michael Feig
Copyright © 2010 WILEY-VCH Verlag GmbH & Co. KGaA, Weinheim
ISBN: 978-3-527-32421-7

2		**Model-Free "Solvent Modeling" in Chemistry and Biochemistry Based on the Statistical Mechanics of Liquids** *31*
		Norio Yoshida, Yasuomi Kiyota, Yasuhiro Ikuta, Takashi Imai, and Fumio Hirata
	2.1	Introduction *31*
	2.2	Outline of the RISM and 3D-RISM theories *33*
	2.3	Partial Molar Volume of Proteins *38*
	2.4	Detecting Water Molecules Trapped Inside Protein *39*
	2.5	Selective Ion Binding by Protein *42*
	2.6	Water Molecules Identified as a Substrate for Enzymatic Hydrolysis of Cellulose *45*
	2.7	CO Escape Pathway in Myoglobin *47*
	2.7.1	Effect of Protein Structure on the Distribution of Xe *47*
	2.7.2	Partial Molar Volume Change Through the CO Escape Pathway of Myoglobin *49*
	2.8	Perspective *51*
		References *52*
3		**Developing Force Fields From the Microscopic Structure of Solutions: The Kirkwood–Buff Approach** *55*
		Samantha Weerasinghe, Moon Bae Gee, Myungshim Kang, Nikolaos Bentenitis, and Paul E. Smith
	3.1	Introduction *55*
	3.2	Biomolecular Force Fields *56*
	3.3	Examples of Problems with Current Force Fields *58*
	3.4	Kirkwood–Buff Theory *59*
	3.5	Applications of Kirkwood–Buff Theory *60*
	3.6	The General KBFF Approach *62*
	3.7	Technical Aspects of the KBFF Approach *64*
	3.8	Results for Urea and Water Binary Solutions *65*
	3.9	Preferential Interactions of Urea *68*
	3.10	Conclusions and Future Directions *71*
		Acknowledgments *72*
		References *73*
4		**Osmolyte Influence on Protein Stability: Perspectives of Theory and Experiment** *77*
		Char Hu, Jörg Rösgen, and B. Montgomery Pettitt
	4.1	Introduction *77*
	4.2	Denaturing Osmolytes *78*
	4.2.1	Does Urea Weaken Water Structure? *78*
	4.2.2	Effect of Urea on Hydrophobic Interactions *80*
	4.2.3	Direct Interaction of Urea with Proteins *82*
	4.3	Protecting Osmolytes *83*

4.3.1	Do Protecting Osmolytes Increase Water Structure?	83
4.3.2	Effect of Protecting Osmolytes on Hydrophobic Interactions	86
4.4	Mixed Osmolytes	87
4.5	Conclusions	88
	Acknowledgments	88
	References	88

5 Modeling Aqueous Solvent Effects through Local Properties of Water 93
Sergio A. Hassan and Ernest L. Mehler

5.1	The Role of Water and Cosolutes on Macromolecular Thermodynamics	93
5.2	Forces Induced by Water in Aqueous Solutions	95
5.2.1	Interactions in Water-Accessible Regions of Proteins	96
5.2.2	Insight from Atomistic Dynamic Simulations	96
5.2.3	Bulk and Non-Bulk Contributions of the Water Forces	99
5.2.4	Effects of Salts on the Intermolecular Interactions	103
5.3	Continuum Representation of Water	107
5.3.1	Dielectric and Structural Response of Water	108
5.3.2	Electrostatic and Liquid-Structure Forces on Solutes and Cosolutes	111
5.4	Modeling Water Effects on Proteins and Nucleic Acids	112
5.4.1	Calculation of Water Forces in Solutions of Arbitrary Compositions	113
5.4.1.1	Electrostatic Forces	114
5.4.1.2	Liquid-Structure Forces	116
5.4.2	Calculation of pK_a in Proteins	119
	Acknowledgments	122
	References	123

6 Continuum Electrostatics Solvent Modeling with the Generalized Born Model 127
Alexey Onufriev

6.1	Introduction: the Implicit Solvent Framework	127
6.1.1	Key Approximations of the Implicit Solvent Framework	128
6.1.2	The Poisson–Boltzmann model	129
6.2	The Generalized Born Model	131
6.2.1	Theoretical Foundation of the GB Model	131
6.2.2	Computing the Effective Born Radii	135
6.2.2.1	The Integral Approaches	135
6.2.2.2	Representations of the Dielectric Boundary	138
6.2.3	Accounting for Salt Effects	141
6.2.4	The Non-Polar Contribution	142
6.2.5	GB for Non-Aqueous Solvents	143

6.3	Applications of the GB Model	145
6.3.1	Protein Folding and Design	145
6.3.2	"Large-Scale" Motions in Macromolecules	145
6.3.3	Peptides and Proteins in the Membrane Environment	146
6.3.4	pK Prediction and Constant pH Simulations	146
6.3.5	Other Uses	147
6.4	Some Practical Considerations	147
6.4.1	The Accuracy/Speed Tradeoffs	147
6.4.2	GB Computational Time Relative to Explicit Solvent	148
6.4.3	Enhancement of Conformational Sampling	152
6.5	Limitations of the GB Model	154
6.6	Conclusions and Outlook	157
	Acknowledgments	159
	References	159
7	**Implicit Solvent Force-Field Optimization**	**167**
	Jianhan Chen, Wonpil Im, and Charles L. Brooks III	
7.1	Introduction	167
7.2	Theoretical Foundations of Implicit Solvent	169
7.2.1	General Principles of Implicit Solvent	169
7.2.2	Continuum Electrostatics	169
7.2.3	Treatment of Non-Polar Solvation	171
7.3	Optimization of Implicit Solvent Force Fields	171
7.3.1	Solvation Free Energies of Small Molecules	172
7.3.2	Potentials of Mean Force of Pairwise Interactions	173
7.3.3	Conformational Equilibria of Model Peptides	175
7.3.4	Folding Simulations of Small Proteins	179
7.3.5	Optimization Based on Other Experimental Measurables	182
7.4	Concluding Remarks and Outlook	182
	Acknowledgments	184
	References	184
8	**Modeling Protein Solubility in Implicit Solvent**	**191**
	Harianto Tjong and Huan-Xiang Zhou	
8.1	Introduction	191
8.2	The Models	192
8.2.1	Transfer Free Energy	192
8.2.2	Electrostatic Free Energy	193
8.3	Applications	193
8.3.1	Transfer Free Energy from Octanol to Water	193
8.3.2	Crystalline Phase as a Dielectric Medium	195
8.3.3	Protein Solubility: pH Dependence	197
8.3.4	Protein Solubility: Effects of Mutations	201
8.4	Summary and Outlook	203
	References	204

9 Fast Analytical Continuum Treatments of Solvation   209
François Marchand and Amedeo Caflisch
9.1 Introduction   209
9.2 The SASA Implicit Solvent Model: A Fast Surface Area Model   210
9.2.1 Description of the Model   211
9.2.2 Applications of the SASA Implicit Solvent Model   213
9.2.2.1 Reversible Folding of Structured Peptides   213
9.2.2.2 Peptide Aggregation   213
9.2.2.3 Other Applications   215
9.2.3 Limitations of the SASA Implicit Solvent Model   216
9.3 The FACTS Implicit Solvent Model: A Fast Generalized Born Approach   217
9.3.1 Description of the Model   219
9.3.1.1 Atomic (or Self) Electrostatic Solvation Energy   219
9.3.1.2 Total Electrostatic Solvation Energy   221
9.3.1.3 Atomic Solvent-Accessible Surface Area   223
9.3.1.4 Total Solvation Free Energy in the FACTS Model   223
9.3.2 Parameterization of FACTS   223
9.3.3 Validation and Applications of FACTS   224
9.3.3.1 Potential of Mean Forces of Side-Chain Dimers   224
9.3.3.2 Atomic Fluctuations   224
9.3.3.3 Peptide Aggregation   226
9.3.3.4 Scalar and Parallel Performance   226
9.4 Conclusions   227
Acknowledgments   228
References   228

10 On the Development of State-Specific Coarse-Grained Potentials of Water   233
Hyung Min Cho and Jhih-Wei Chu
10.1 Introduction   233
10.2 Methods of Computing Coarse-Grained Potentials of Liquid Water   235
10.2.1 Multi-Scale Coarse Graining (MS-CG) Method with Force Matching (FM)   235
10.2.2 The Iterative-YBG Method   236
10.2.3 Numerical Issues of the Iterative-YBG Method   237
10.3 Structural Properties and the "Representability" Problem of Coarse-Grained Liquid Water Models   238
10.3.1 Coarse-Grained Water Model Computed from All-Atom Water Potentials   239
10.3.2 Anisotropic Structural Property of Liquid Water   242
10.3.3 Properties of CG Water by Using a Self-Consistent Force Matching Force Field and Pressure Constraint   244

10.4	Conclusions	247
	Acknowledgment	248
	References	248

11 Molecular Dynamics Simulations of Biomolecules in a Polarizable Coarse-Grained Solvent 251
Tap Ha-Duong, Nathalie Basdevant, and Daniel Borgis

11.1	Introduction	251
11.2	Theory	253
11.2.1	A Non-Local Formulation of the Dielectric Continuum Theory in Terms of Polarization Density	253
11.2.2	A Local Polarization Density Free-Energy Functional	254
11.2.3	Particular Case of Solutes with Uniform Dielectric Constant	256
11.2.4	Electrostatics on Particles	257
11.2.5	A Phenomenological Dipolar Saturation	258
11.3	Applications: Solvation of All-Atom Models of Biomolecules	259
11.3.1	Parameters and Simulation Conditions	259
11.3.2	Stability of Small Proteins in the PPP Solvent	260
11.3.3	Hydration Properties of Nucleic Acids	264
11.4	Conclusion and Prospects	266
	References	268

12 Modeling Electrostatic Polarization in Biological Solvents 273
Sandeep Patel

12.1	Introduction	273
12.2	Current Approaches for Modeling Electrostatic Polarization in Classical Force Fields	275
12.2.1	Charge Equilibration Models: Electronic Polarization and Charge Dynamics	276
12.3	Parameterization of Charge Equilibration Models	278
12.3.1	Establishing Atomic Hardness and Electronegativity Parameters	278
12.3.2	Polarizability	279
12.3.3	Fitting and Force-Field Refinement	281
12.4	Applications of Charge Equilibration Models for Biological Solvents	281
12.4.1	Water	281
12.4.2	Non-Aqueous Solvents: Charge Equilibration Models of Alkanes	283
12.4.2.1	Bulk Liquid Properties	286
12.4.2.2	Local Chain Dynamics	288
12.5	Toward Modeling of Membrane Ion Channel Systems: Molecular Dynamics Simulations of DMPC–Water and DPPC–Water Bilayer Systems	289
12.5.1	Fully Polarizable DMPC Bilayers in TIP4P-FQ Solvent: Application of Charge Equilibration Models for Molecular Dynamics Simulations	290

12.5.2	Component Atomic and Electron Density Profiles 294
12.5.3	Lipid Chain Dynamics 296
12.5.4	Charge Distributions 296
12.5.5	Dipole Moment Variation 299
12.5.6	Dielectric Permittivity Profiles 300
12.6	Conclusions and Future Directions 302
	Acknowledgments 303
	References 303

Subject Index 309

Preface

Computer simulations of biological macromolecules have become an integral part of modern biological science, and in many respects the development of the relevant methodology is turning into a mature field. However, there continues to be considerable interest in solvent models. On the one hand, the accurate modeling of the solvent environment is essential for reflecting the key role that solvent plays in determining the structure and dynamics of biomolecules. On the other hand, the computational cost associated with modeling the solvent is often much larger than the cost of simulating a given biomolecular solute. This is one of the main reasons why modern simulations of biomolecules are still mostly limited to single molecules, and why biological time scales can often not be reached even with significant computer resources.

The *de facto* standard of modeling solvent in biomolecular simulations is to include explicit solvent molecules, mostly using venerable three-site water models. This approach emphasizes accuracy over efficiency and is generally successful in generating stable dynamics trajectories of solvated biomolecules. Standard explicit solvent models typically reproduce many properties of pure water under ambient conditions, but have difficulties describing co-solvent effects or the temperature dependence of solvation. However, the use of a fully atomistic explicit solvent is computationally expensive and limits the scalability of molecular simulations to large system sizes. As a result, many efforts have focused on developing models that provide improved accuracy without significantly increasing computational costs or that are more efficient and scale more favorably with an increase in system size without compromising accuracy. There are exciting recent developments in both respects. Kirkwood–Buff type force fields and continued advances in reference interaction-site model (RISM) type approaches are addressing deficiencies with modeling co-solvent effects, while polarizable force fields are entering the main stream. At the same time, efficient implicit and coarse-grained solvent models have passed the initial conceptual stages and are turning into viable alternatives for an explicit solvent. The main goal of this book is to collect the most promising recent developments into a single resource and to provide a broad view of what modern solvent models can offer beyond the well-trodden fixed-charge atomistic explicit solvent models. The emphasis on breadth over depth means that many types and variants of solvent models that have been proposed in the past could not be fully reflected here for lack of space. The book is more of a snapshot of the currently most popular approaches rather than a complete account of present and past developments.

This book is targeted at an audience with an understanding of molecular modeling methodology as well as an appreciation for the role of solvent in biomolecular structure and dynamics. Chapter 1 briefly reviews the basic concepts related to modeling and solvation, and also discusses what is known about solvation from an experimental point of view. Readers very familiar with those subjects may wish to skip this first chapter. The other chapters introduce specific approaches for developing solvent models, beginning with more fundamental ideas based on statistical mechanics, and moving on to implicit solvent, coarse-grained solvent, and recent advances in the development of polarizable models.

It is my hope that this book will be of broad interest to practitioners of molecular modeling tools, to enable them not only to better understand the choices that have become available for modeling solvent in biomolecular simulations, but also to spur the further development of new models by contrasting the strengths and weaknesses of the most recent advances.

Michigan State University *Michael Feig*
December 2009

List of Contributors

Nikolaos Bentenitis
Southwestern University
Department of Chemistry and
Biochemistry
Georgetown, TX 78626
USA

Nathalie Basdevant
Université d'Evry-Val-d'Essonne
Laboratoire Analyse et
Modélisation pour la Biologie
et l'Environnement
rue du Père André Jarlan
91025 Evry Cedex
France

Charles L. Brooks III
University of Michigan
Department of Chemistry and
Biophysics Program
Ann Arbor, MI 48109
USA

Daniel Borgis
Ecole Normale Supérieure
UMR CNRS Pasteur
Département de Chimie
24 rue Lhomond
75231 Paris Cedex 05
France

Jianhan Chen
Kansas State University
Department of Biochemistry
34 Chalmers Hall
Manhattan, KS 66506
USA

Amedeo Caflisch
University of Zürich
Department of Biochemistry
Winterthurerstrasse 190
8057 Zürich
Switzerland

Hyung Min Cho
University of California, Berkeley
Department of Chemical Engineering
101A Gilman Hall
Berkeley, CA 94720
USA

Jhih-Wei Chu
University of California, Berkeley
Department of Chemical Engineering
101A Gilman Hall
Berkeley, CA 94720
USA

List of Contributors

Michael Feig
Michigan State University
Department of Biochemistry and
Molecular Biology
Biochemistry Building 218
and Department of Chemistry
East Lansing, MI 48824
USA

Moon Bae Gee
Kansas State University
Department of Chemistry
213 CBC Building
Manhattan, KS 66506
USA

Fumio Hirata
Institute for Molecular Science
Department of Theoretical and
Computational Molecular Science
Okazaki National Research Institutes
and
The Graduate University for
Advanced Studies Sokendai
38 Nishigo-Naka
Myodaiji
Okazaki 444-8585
Japan

Sergio A. Hassan
National Institutes of Health
U.S. Department of Health and
Human Services
Center for Molecular Modeling
DCB/CIT
12 South Drive
Bethesda, MD 20892
USA

Char Hu
Baylor College of Medicine
Graduate Program in Structural and
Computational Biology and Molecular
Biophysics
Houston, TX 77000
USA

Tap Ha-Duong
Université d'Evry-Val-d'Essonne
Laboratoire Analyse et Modélisation
pour la Biologie et l'Environnement
rue du Père André Jarlan
91025 Evry Cedex
France

Yasuhiro Ikuta
Institute for Molecular Science
Department of Theoretical and
Computational Molecular Science
Okazaki National Research Institutes
and
The Graduate University for
Advanced Studies Sokendai
38 Nishigo-Naka
Myodaiji
Okazaki 444-8585
Japan

Takashi Imai
RIKEN
Computational Science Research
Program
Wako
Saitama 351-0198
Japan

Wonpil Im
University of Kansas
Department of Molecular Biosciences
and Center for Bioinformatics
Lawrence, KS 66047
USA

Yasuomi Kiyota
Institute for Molecular Science
Department of Theoretical and
Computational Molecular Science
Okazaki National Research Institutes
and
The Graduate University for
Advanced Studies Sokendai
38 Nishigo-Naka
Myodaiji
Okazaki 444-8585
Japan

Myungshim Kang
Kansas State University
Department of Chemistry
213 CBC Building
Manhattan, KS 66506
USA

François Marchand
University of Zürich
Department of Biochemistry
Winterthurerstrasse 190
8057 Zürich
Switzerland

Ernest L. Mehler
Cornell University
Weill Medical College
Department of Physiology and
Biophysics
New York, NY 10006
USA

Alexey Onufriev
Virginia Polytechnic Institute and
State University
Departments of Computer Science
and Physics
2160C Torgersen Hall
Blacksburg, VA 24061
USA

B. Montgomery Pettitt
University of Houston
Department of Chemistry and
Institute for Molecular Design
232 Fleming Building
Houston, TX 77204-5003
USA

Sandeep Patel
University of Delaware
Department of Chemistry and
Biochemistry
238 Brown Laboratory
Newark, DE 19716
USA

Jörg Rösgen
University of Texas Medical Branch
Department of Human Biological
Chemistry and Genetics
Galveston, TX 77555-1052
USA

Paul E. Smith
Kansas State University
Department of Chemistry
213 CBC Building
Manhattan, KS 66506
USA

Harianto Tjong
Florida State University
Department of Physics
Institute of Molecular Biophysics
Tallahassee, FL 32306-4380
USA

Samantha Weerasinghe
University of Colombo
Department of Chemistry
Colombo 00300
Sri Lanka

Norio Yoshida
Institute for Molecular Science
Department of Theoretical and
Computational Molecular Science
Okazaki National Research Institutes
and
The Graduate University for
Advanced Studies Sokendai
38 Nishigo-Naka
Myodaiji
Okazaki 444-8585
Japan

Huan-Xiang Zhou
Florida State University
Department of Physics
Kasha Laboratory of Biophysics
Tallahassee, FL 32306
USA

1
Biomolecular Solvation in Theory and Experiment
Michael Feig

1.1
Introduction

The modern understanding of biological macromolecules is unthinkable without a consideration of the solvent environment [1]. Solvent is crucial for the functioning of biological systems both directly, by actively participating in biological processes, and indirectly, by stabilizing biologically active conformations of proteins and nucleic acids. The aqueous milieu of virtually all living environments singles out water as the most important biomolecular solvent. Water is an extremely versatile solvent owing to its small size, and its inherently polar and polarizable nature. Its ability to act as both a hydrogen bond acceptor and donor allows it to form a variety of structures that can easily adapt to changes in environmental conditions and in particular to the presence of biomolecules and other co-solutes.

The interaction of water with biomolecules is a complex subject that has been studied for many decades. While the study of biomolecular solvation remains the focus of current research efforts, there are a number of major themes that have crystallized from past studies:

1) Solute–water interactions profoundly affect the conformational sampling of peptides [1]. In particular, protein folding to a unique native structure that can carry out biological function relies on the presence of water and in particular the hydrophobic effect [2–4] and, in peptides, the interaction with solvent determines secondary structure preferences [5]. Sufficient hydration is also necessary to maintain the biologically active B-form of DNA [6, 7].
2) Water is a critical participant in many biochemical reactions, where it can act as either an acid or a base and where it can rapidly transfer protons through the Grotthuss mechanism [8].
3) Solvent plays a key role in complex formation [9–11], ligand recognition [12–15], and DNA sequence recognition by DNA binding proteins [16].
4) The dynamic properties of water generally smooth solute energy landscapes [17] and, in particular, water has been found to act as a lubricant during protein folding [18].

Modeling Solvent Environments. Applications to Simulations of Biomolecules. Edited by Michael Feig
Copyright © 2010 WILEY-VCH Verlag GmbH & Co. KGaA, Weinheim
ISBN: 978-3-527-32421-7

5) Solvent interactions in biological environments are modulated further by the presence of co-solvents such as ions and a variety of small molecules [19], while entirely non-aqueous environments do not necessarily abrogate biological function for natively water-soluble proteins [20, 21]. Such environments are interesting in the context of industrial process biochemistry, where they may confer certain practical advantages [22]. Furthermore, non-aqueous biological environments exist in the form of lipid bilayers, where hydrophobic biomolecules can be sequestered from the aqueous solvent environment.

While a comprehensive review of biomolecular solvation in a single chapter is not possible, this chapter does attempt to provide a cursory overview of only those aspects that are most important to the subject of this book, the development of computational solvent models. For more in-depth discussions, the reader is referred to the relevant literature. This chapter is divided as follows: First, solvation is discussed from a thermodynamic and statistical mechanics perspective, followed by an overview of computational and experimental methods that are commonly used to study solvation around biomolecules. Next, the current knowledge about water–protein and water–nucleic acid interactions is summarized. Finally, the effects of non-aqueous solvents are described briefly.

1.2 Theoretical Views of Solvation

1.2.1 Equilibrium Thermodynamics of Solvation

The energetics of a given biomolecule is commonly separated into bonded and non-bonded interactions [23]. Bonded interactions stem from covalent bonding and effectively restrain local molecular geometries according to molecular bonding orbitals. The remaining non-bonded interactions add long-range charge–charge interaction, van der Waals type dispersion, and strong hard-sphere-like repulsion to avoid electron density overlap of non-covalently interacting atoms. The presence of solvent introduces solvent–solute and solvent–solvent terms in addition to the solute–solute non-bonded terms that are present in vacuum as well. The total free energy of a given biomolecule is therefore given as follows:

$$\Delta G_{\text{total}} = \Delta H_{\text{solute-solute}}^{\text{bonded}} + \Delta\Delta H_{\text{solvent-solvent}}^{\text{bonded}} \\ + \Delta H_{\text{solute-solute}}^{\text{non-bonded}} + \Delta H_{\text{solute-solvent}}^{\text{non-bonded}} + \Delta\Delta H_{\text{solvent-solvent}}^{\text{non-bonded}} \\ - T\Delta S_{\text{solute}} - T\Delta\Delta S_{\text{solvent}} \qquad (1.1)$$

where $\Delta\Delta H_{\text{solvent-solvent}}^{\text{bonded}}$, $\Delta\Delta H_{\text{solvent-solvent}}^{\text{non-bonded}}$, and $\Delta\Delta S_{\text{solvent}}$ denote the changes in solvent–solvent enthalpies and solvent entropy from bulk values.

As a good first approximation, the bonded terms in solute and solvent are largely unaffected by solute–solvent interactions, so that it is the balance of the non-bonded enthalpic terms and entropic terms that determines the conformational preference of biomolecules in a given solvent environment. With the solvation free energy

$$\Delta G_{\text{solvation}} = \Delta H_{\text{solute-solvent}}^{\text{non-bonded}} + \Delta\Delta H_{\text{solvent-solvent}}^{\text{non-bonded}} - T\Delta\Delta S_{\text{solvent}} \quad (1.2)$$

the biomolecular free energy then becomes

$$\Delta G_{\text{total}} = \Delta H_{\text{solute-solute}} - T\Delta S_{\text{solute}} + \Delta G_{\text{solvation}} \quad (1.3)$$

For proteins in aqueous solvent, the formation of well-defined, compact native structures results in favorable solute–solute enthalpies but decreased solute entropy. At the same time, native protein structures generally have unfavorable free energies of solvation compared to fully extended structures because favorable direct interactions between water and the peptide backbone are traded for intramolecular hydrogen bonding in the formation of secondary structure elements. Successful folding in aqueous solvent then requires that the folded structure retains sufficiently favorable interactions between the solvent and amino acid side chains. This results in the well-known hydrophobic effect, where hydrophobic amino acids that do not contribute significantly to $\Delta H_{\text{solute-solvent}}^{\text{non-bonded}}$ are sequestered into the core of the folded structure while a large part of the protein surface consists of polar and charged residues that interact favorably with water. Hydrophobic surface residues are further discouraged energetically because the lack of hydrogen bond donors or acceptors forces the surrounding water to adopt a more ordered, clathrate-like structure that is both enthalpically and entropically less favorable than bulk water.

1.2.2
Radial Distribution Functions

The thermodynamic view of solvation can be connected to the molecular level with the aid of statistical mechanics by using correlation functions that capture the molecular organization of the liquid solvent. Typically, pairwise radial distribution functions (RDFs) $g_{ij}(r)$ are used to describe the density distribution of an atomic or molecular species i as the function of distance from another species j. Using RDFs it is then possible to calculate thermodynamic quantities. For example, the total interaction energy between N species i in a monatomic system that is only subject to pairwise interactions can be calculated according to

$$U_{ii} = \frac{N\rho_i}{2} \int_0^\infty 4\pi r^2 g_{ii}(r) U(r) \, dr \quad (1.4)$$

where ρ_i is the density of species i and $U(r)$ is the pairwise interaction energy as the function of distance. This approach can be extended to systems with non-pairwise interactions by including higher-order correlation functions. Pairwise (or higher-order) distribution functions can be extracted directly from computer simulations (see Section 1.3.4) or from experiments.

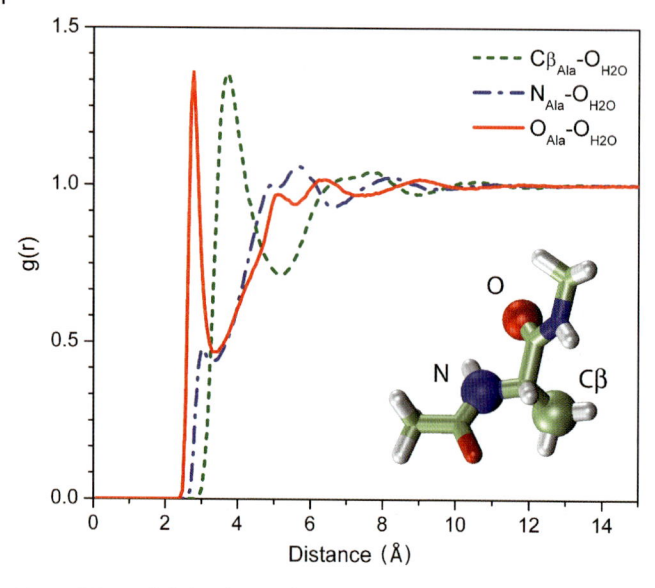

Figure 1.1 Radial distribution functions between the oxygen atom of water and selected sites on alanine dipeptide from 100 ns molecular dynamics simulation of blocked alanine dipeptide in explicit solvent.

As an example, Figure 1.1 shows RDFs between the oxygen atoms of water and the N, O, and C_β atoms of blocked alanine dipeptide from a molecular dynamics simulation in explicit solvent. For distances up to about 10 , the water oxygen density varies significantly from that of bulk solvent and also between different alanine dipeptide reference sites, indicative of the specific solvent structure around this solute.

1.2.3
Integral Equation Formalisms

The pairwise distribution function can be obtained alternatively by solving the Ornstein–Zernike integral equation, which decomposes the pairwise distribution function $h(r_{12}) = g(r_{12}) - 1$ between molecules 1 and 2 into a direct correlation function $c(r_{12})$ and an indirect contribution due to the presence of a third molecule:

$$h(r_{12}) = c(r_{12}) + \rho \int dr_3\, c(r_{13}) h(r_{23}) \tag{1.5}$$

Solution of the Ornstein–Zernike integral equation requires an expression for the direct correlation function $c(r_{12})$, which can be approximated under certain assumptions. In the reference interaction-site model (RISM), this approach is extended to heterogeneous systems, and in particular solute–solvent interactions, by introducing site–site correlation functions between solute and solvent interaction sites

[24–27] as in Figure 1.1. Using the pairwise site–site correlation functions $c(r)$ and $h(r)$, free energies, enthalpies, and entropies of solvation can then be calculated directly [28, 29]. This approach has been used to understand the solvation of peptides [30–32] and co-solvent effects [33], and more recently to predict ligand binding sites [34]. It is described in more detail in Chapter 2.

1.2.4
Kirkwood–Buff Theory

Another route for connecting thermodynamics with the molecular nature of solutions is through Kirkwood–Buff (KB) theory [35, 36] (see Chapter 3). In this theory, thermodynamic quantities are related to the so-called KB integrals,

$$G_{ij} = \int_0^\infty [g_{ij}(r) - 1] 4\pi r^2 \, dr \tag{1.6}$$

In KB theory, RDFs are required as input to calculate G_{ij} and from those to derive thermodynamic quantities. The inversion of KB theory also allows G_{ij} values to be determined from thermodynamic measurements [37], but RDFs and the corresponding information about liquid structure cannot be extracted from KB theory. In contrast, the integral equation approach based on the Ornstein–Zernike equation allows the determination of pair correlation functions based on a given interaction potential.

Like integral equation theory, KB theory has also been applied to analyze the solvation of peptides [38] and in particular to investigate co-solvent effects [39–42]. Recently, KB theory has been used in the parameterization of force fields [43, 44].

1.2.5
Kinetic Effects of Solvation

Solute–solvent interactions also play a critical role in kinetic processes of the solute, such as chemical reactions and conformational transitions. The solvent environment may alter the free energy of the transition states through specific solute–solvent interactions. Examples of such interactions are found in enzyme active sites, where the presence of water molecules has been proposed to lower the transition state to accelerate the kinetics of enzyme reactions [45–47]. In contrast, less-specific stochastic collisions between solute and solvent atoms provide both frictional drag forces and the activation energy necessary to overcome kinetic barriers. A decrease in solvent viscosity has been found to accelerate protein folding kinetic rates [48], and even with friction, the mere presence of stochastic collisions between solute and solvent leads to an enhancement of barrier crossings [49]. As a result, transition state barriers are effectively lowered and, more generally, the energy landscape is smoothed. Over short time scales, such collisions are non-equilibrium processes, with the rate of collisions linked to the dynamics of solvent molecules at the biomolecular surface. The stochastic effect of solvent on kinetics is captured by Grote–Hynes theory [50] and has been applied successfully in the context of variational transition-state theory [51] to explain the effect of solvent on reaction rates [52].

1.3
Computer Simulation Methods in the Study of Solvation

1.3.1
Molecular Dynamics and Monte Carlo Simulations

While the main subject of this book is the development of solvent models that can describe the effect of solvent on a given biological macromolecule in computer simulations, this chapter reviews how computer simulations can be used to investigate biomolecular solvation properties.

Typically, such simulations employ either molecular dynamics or Monte Carlo techniques, and rely on a classical model for the solute with an explicit representation of the solvent molecules. Both molecular dynamics and Monte Carlo simulations require an intra- and intermolecular interaction potential $V(\mathbf{r})$ commonly known as the "force field" [53]. In Monte Carlo simulations, the potential is used directly to accept or reject moves according to the Metropolis probability [54]

$$P = \min\left(1, e^{-[V(\mathbf{r}_2) - V(\mathbf{r}_1)]/kT}\right) \quad (1.7)$$

In molecular dynamics simulations, the gradient of the potential is used to describe the dynamics of a given system according to Newton's equation of motion,

$$m_i \frac{d^2 \mathbf{r}_i}{dt^2} = \mathbf{F}_i(\mathbf{r}) = -\nabla V(\mathbf{r}) \quad (1.8)$$

This equation is then integrated in a stepwise fashion to obtain a trajectory.

The force field $V(\mathbf{r})$ commonly consists of bonded terms for covalent bonds, angles, and torsions along with non-bonded electrostatic and Lennard-Jones terms [53]:

$$\begin{aligned}V(\mathbf{r}) = & \sum_{i=1}^{N_{\text{bonds}}} k_{\text{bond},i} (d - d_0)^2 + \sum_{i=1}^{N_{\text{angles}}} k_{\text{angle},i} (\theta - \theta_0)^2 \\ & + \sum_{i=1}^{N_{\text{torsions}}} k_{\text{torsion},i} [1 + \cos(n\phi - \phi_0)] \\ & + \sum_{i=1}^{N_{\text{atoms}}-1} \sum_{j=i+1}^{N_{\text{atoms}}} \left(4\varepsilon_{\text{LJ},ij} \left[\left(\frac{\sigma_{\text{LJ},ij}}{r_{ij}}\right)^{12} - \left(\frac{\sigma_{\text{LJ},ij}}{r_{ij}}\right)^{6} \right] + \frac{1}{4\pi\varepsilon_0} \frac{q_i q_j}{r_{ij}} \right)\end{aligned} \quad (1.9)$$

The Lennard-Jones potential captures both attractive van der Waals type dispersion and hard-sphere repulsion upon atomic contact formation [55]. Parameters for bond and angle terms are typically obtained from spectroscopic and crystallographic data, while the partial charges q_i, the Lennard-Jones coefficients σ_i and ε_i, and the torsion parameters are usually fit to quantum-mechanical calculations. The classical formalism can be extended by including atomic polarizabilities through either dynamically fluctuating charges [56, 57], inducible dipoles and higher-order multipoles [58–60], or Drude oscillators [61, 62].

A major practical issue with the application of $V(r)$ is the calculation of the pairwise non-bonded terms. The Lennard-Jones term contributes significantly up to pairwise distances of about 8 Å [63]. The electrostatic term remains non-negligible up to at least 18 Å in a polar medium with high dielectric screening [64, 65] and much further, up to 30 Å, in non-polar environments [66], which becomes computationally intractable for large systems. In the case of an infinitely periodic system, the total Coulomb interaction can be calculated without a cutoff at moderate cost using the Ewald summation technique [67, 68]. A drawback of this scheme is that a sufficiently large simulation system is required to avoid artifacts when repeated periodically [69]. Nevertheless, simulations with periodic boundaries that use Ewald summation and in particular the efficient particle-mesh Ewald implementation [70, 71] have become the *de facto* standard for explicit solvent simulations.

1.3.2
Water Models

Realistic water models are crucial for the success of explicit aqueous solvent simulations and have been the subject of intensive research over many decades [72, 73]. The most straightforward and most widely used model of water consists of three sites, one at the center of the oxygen atom, and one each at the centers of the hydrogen atoms. Such a three-site model has only eight parameters: partial charges q_O and q_H, Lennard-Jones parameters ε_O, σ_O, ε_H, and σ_H, and the O–H bond distance and H–O–H angle. With such a model, hydrogen bonds are formed as a result of a balance between electrostatic attraction and electronic repulsion through the r^{12} part of the Lennard-Jones potential. The two most popular models, TIP3P [74] and SPC/E [75], are shown in Figure 1.2.

Water models are parameterized and tested by their ability to reproduce experimentally measured water structure (in form of radial distribution functions), density, enthalpy of vaporization, self-diffusion rate, and dielectric constant in bulk solvent. Three-site models can reproduce many of those quantities reasonably well at 298 K and 1 bar pressure, but different models trade good reproduction of one property with poor reproduction of another. For example, TIP3P matches the enthalpy of vaporization and dielectric constant well, but slightly underestimates the density and significantly overestimates the self-diffusion rate [72]. In contrast, SPC/E overestimates the enthalpy of vaporization and underestimates the dielectric constant, but accurately reproduces the density and self-diffusion rates [72]. Three-site models generally have difficulties in correctly describing the temperature dependence of liquid water properties. In particular, the density maximum at 282 K is often poorly reproduced, if present at all [72, 76]. These problems can be partially addressed by including additional off-nucleus electrostatic interaction sites to better represent the electronic structure of water and in particular to model the effect of the oxygen lone pairs. Some variants of these models, such as TIP4P

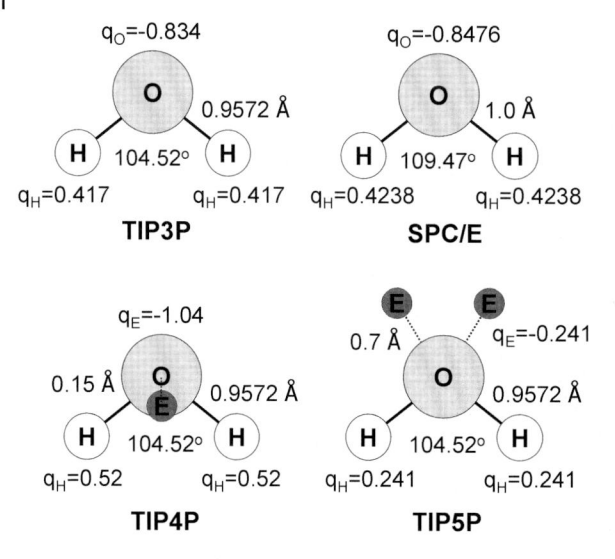

Figure 1.2 Common fixed-charge water models with three (TIP3P [74], SPC/E [75]), four (TIP4P [73]), or five (TIP5P [77]) interaction sites with bonding geometries and partial charges.

[73] and TIP5P [77] (see Figure 1.2), have resulted in significantly improved, although still not perfect, agreement with the experimental properties of bulk water [76]. However, these models are often not used in practice because of the increased computational expense due to the additional interaction sites.

The electron distribution of water can significantly redistribute in response to external electric fields. The polarizability of water is reflected in a change of the dipole moment from 1.85 D in the gas phase to about 3 D in the bulk solvent. Fixed-charge models that do not allow the dipole moment to change and are parameterized to reproduce bulk properties are therefore problematic when used in combination with solutes that may alter the polarization of water, such as ions or hydrophobic solutes. This has motivated the development of polarizable water models, with the most successful models relying on the fluctuating charge approach, such as the TIP4P/FQ model [78]. Polarizable water models are even more expensive than multi-site water models, but may be necessary to correctly reflect water interactions in cases where the environment of water molecules differs significantly from the bulk solvent [79].

1.3.3
Solvent Structure and Dynamics from Simulations

The main advantage of computer simulations is that, given sufficient sampling, solvent structure and dynamics can be extracted directly from the resulting trajectories. Solvent structure is characterized in the form of either radial distribution functions (see Figure 1.1) or volumetric density distributions (see Figure 1.3).

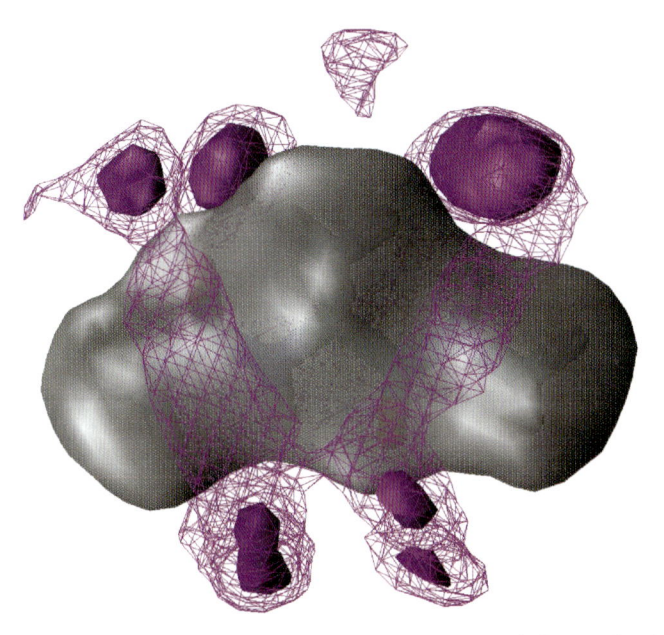

Figure 1.3 Water oxygen density around alanine dipeptide from molecular dynamics simulation. The gray surface encloses the solvent-excluded volume. High-density regions are shown in purple, with the highest densities represented by solid surfaces.

Radial distribution functions are commonly calculated by collecting pairwise distances in a histogram and normalizing the resulting distribution for the bin $[r, r + \Delta r]$ by the number of solvent molecules in a spherical shell of width Δr: $4\pi r^2 \Delta r \cdot \rho_{\text{solvent}} \cdot N_A/M_{\text{solvent}}$, where ρ_{solvent} is the solvent density, M_{solvent} is the molar mass of the solvent, and N_A is Avogadro's number. When radial distribution functions are calculated with respect to solute sites, they reflect the structure of the solvation layer(s).

Solvent density distributions are easily counted with a three-dimensional histogram, but they require either that the solute remains fairly rigid or that the solute is reoriented appropriately for each configuration before the solvent density is accumulated. The resulting volumetric map shows preferential solvent interaction sites as high-density regions and identifies the solvent-excluded volume. An example of the three-dimensional solvent distribution around alanine dipeptide is shown in Figure 1.3. The high-density regions often correlate with crystallographic water sites, but the apparent order does not necessarily imply altered dynamic properties, since a site with a single long-lived solvent molecule is not distinguished from a site where solvent molecules rapidly exchange with the environment.

Dynamic properties of the solvent can also be readily calculated from simulations, typically through time correlation functions. Because solvent dynamics might be substantially altered in the vicinity of a solute, such analyses often consider conditional correlation functions that take solute proximity into account. A property that is commonly calculated in this manner is the distribution of solvent residence times in a specific region according to the correlation function [80–82]

$$c_\alpha(t) = \frac{1}{N_{solv}} \sum_{i=1}^{N_{solv}} \frac{1}{N_{\alpha,i}} \sum_{t'=0}^{t_{max}} p_{\alpha,i}(t', t'+t) \quad (1.10)$$

where $p_{\alpha,i}(t', t' + t)$ is a binary function with a value of 1 if solvent molecule i remains within the confined area α from time t' to $t' + t$ and a value 0 otherwise. $N_{\alpha,i}$ is the number of times that a solvent molecule is present in α and the outer sum is over all solvent molecules i. Other correlation functions can be calculated to determine rotational correlation times, the lifetime of specific interactions such as hydrogen bonds, and so on. In all cases, the resulting correlation functions can be fit to a single, double, or stretched exponential function [82, 83] to extract characteristic time scales that can be compared with experimental measurements. The comparison with experiments, however, relies on a definition of the region α in Equation 1.10 that matches the experimental measurements best, which is not always straightforward.

The self-diffusion coefficient measures the rate of mean square displacement in the long-time limit. It is calculated according to the Einstein relation [84] from

$$D = \frac{1}{6} \lim_{t \to \infty} \frac{d}{dt} \left\langle |\mathbf{r}_i(t_0+t) - \mathbf{r}_i(t_0)|^2 \right\rangle \quad (1.11)$$

where the average is taken over all solvent molecules i and all time origins t_0. It is possible to calculate a local diffusion coefficient from a short-time (1–2 ps) finite-difference approximation to Equation 1.11 and to accumulate the resulting values of D on a grid according to the location of a given solvent molecule to obtain spatially resolved local diffusion coefficients [85, 86]. Furthermore, the three-dimensional diffusion coefficient can be decomposed into components that are parallel and perpendicular to a given biomolecular surface in order to fully capture the variation of solvent dynamics near biomolecular solutes [85].

1.3.4
Free Energy Simulations

Free-energy simulations [87] using thermodynamic integration, free-energy perturbation, or more advanced methods [88] can calculate solvation free energies [89–92] or its components, for example, the solute–solvent component or the charging free energy that results from turning on solute charges in an already solvated system [93]. Compared to the calculation of solvation free energies from integral equation theory, the use of computer simulations is very expensive since extensive sampling is required to obtain converged results [94]. However, in the simulations, no further assumptions are made beyond the classical interaction model, and all multi-body effects are included implicitly. As a result, solvation free energies calculated from free-energy simulations can provide results in close agreement with experimental measurements for small molecules [91]. While experimental solvation free energies of biomolecules are generally not available, calculated free energies of solvation take their place as important reference quantities in the development of solvent models [95].

1.4
Experimental Methods in the Study of Solvation

A wide array of experimental methods have been applied in the study of solvation around biomolecules. The main techniques are described in this section.

1.4.1
X-Ray/Neutron Diffraction and Scattering

The most direct evidence for solvent molecules interacting with biological macromolecules comes from X-ray crystallography. X-ray crystallography averages electron densities over time and over a large number of aligned molecules. Solvent molecules are visible if they are ordered with respect to the crystal cell. Most protein and nucleic acid structures contain water molecules at various sites on the biomolecular surface, and much has been learned about the interaction of solvent with biomolecules from X-ray crystallography [96–98].

Crystal water sites are generally understood to reflect the presence of ordered water molecules, but a static interpretation as a single water molecule occupying a specific site according to the crystallographic coordinates can be misleading. Because of the dynamic nature of water, the solvent-associated electron densities are generally less well defined than for the solute. Therefore, a single water molecule oscillating in place can often not be distinguished from dynamically moving and exchanging solvent atoms that frequently occupy a certain site [99]. In other cases, water sites have been reported to lie at average positions between two actual water sites, with the reported water sites actually rarely being visited according to simulations [100]. Furthermore, in lower-resolution crystal structures, water molecules may be added at seemingly random positions simply to reduce the crystallographic R-factor during the refinement stage [97, 101]. In contrast, the very highest-resolution crystal structures often reveal water sites very clearly and may provide additional information about fractional occupancy and alternate hydration patterns [102, 103].

Most of today's crystal structures are obtained at cryogenic temperatures. While the lowered thermal fluctuations of the solute improve the resolution, the effect on the solvent is less clear. It has been suggested that not just solvent dynamics but also solvent structure are altered significantly [104]. One apparent effect seems to be the formation of ordered, presumably enthalpically stabilized, hexagonal and pentagonal water networks [102, 103] that may otherwise not be present at room temperature. Additional solvent ordering is also likely due to packing forces in the imposed crystal lattice environment.

Neutron diffraction is similar to X-ray crystallography, but scattering occurs at the nuclei rather than the electrons. In contrast to X-ray crystallography, neutron diffraction is sensitive to hydrogen atoms and it is therefore well suited for the study of water structure. Neutron diffraction has been especially useful for the experimental determination of pairwise radial distribution functions [105].

Small-angle X-ray scattering (SAXS) and small-angle neutron scattering (SANS) provide low-resolution structural information, in particular about the distribution of intramolecular distances, from which the size of a given biomolecule can be estimated. The presence of a hydration layer with increased solvent density results in an apparent increase in size in SAXS measurements but a decrease in size in SANS measurements owing to different relative scattering lengths between water and biomolecules with X-rays and neutron radiation. If the structure of a given molecule is known, for example, from X-ray crystallography, SAXS and SANS can therefore provide information about the extent of the hydration layer [106].

Quasi-elastic and other inelastic neutron scattering methods measure the spatially resolved loss of kinetic energy upon scattering in the form of a so-called dynamic structure factor [107, 108]. The dynamic structure factor contains information about solvent dynamics that can be extracted either directly [109] or by interpreting the experimental data with the help of molecular dynamics simulations [110, 111]. Quasi-elastic neutron scattering typically results in estimates of the translational and rotational diffusion rates of interfacial solvent molecules interacting with biomolecules.

1.4.2
Nuclear Magnetic Relaxation

Nuclear magnetic relaxation (NMR) techniques have been used for a long time to study the properties of biomolecular solvation [112–114]. They can provide both structural and dynamic information about water near biological solutes. There are essentially two main NMR-related techniques.

The first relies on the nuclear Overhauser effect (NOE) between solvent and solute atoms based on the relaxation-induced transfer of magnetization. This transfer process is sensitive to short distance separations and therefore allows a localized analysis of solvent molecules interacting directly with a given biomolecular surface. However, NOE measurements are limited in temporal resolution and arguments have been made that NOE measurements cannot fully distinguish between interfacial and bulk solvent, so that the resulting solvent residence times may be significantly overestimated [115].

The other NMR technique, magnetic relaxation dispersion (MRD), measures the longitudinal magnetic relaxation rate of water deuteron or quadrupolar ^{17}O nuclei as a function of resonance frequency. This technique has a shorter time resolution but lacks spatial resolution [116, 117]. Nevertheless, the application of MRD has led to detailed insight into rotational and translational relaxation of water in the vicinity of biomolecules [117].

1.4.3
Optical Spectroscopy

Femtosecond-resolution fluorescence spectroscopy is the third major technique for studying the dynamics of biomolecular solvation [108, 118–120]. It involves the time-resolved analysis of the fluorescence of a laser-excited fluorophore probe located at a biomolecular surface. The location of the fluorescence peak is sensitive to the probe dipole, which in turn depends on the solvent polarization. As the solvent relaxes in response to the excited probe, the fluorescence peak exhibits a (Stokes) shift to longer wavelengths. The dynamics of the fluorescence shift is therefore assumed to be a direct reflection of the solvent relaxation dynamics [119]. Dynamic fluorescence spectroscopy is in principle a very powerful technique, since it combines very high temporal resolution with spatial resolution based on the known location of the fluorescence probe. However, in practice, the results require further interpretation with the help of molecular dynamics simulations [121, 122], since the solvent dynamics extracted from the fluorescence shift measurements depends also on the degree of surface exposure of the probe [122].

Infrared spectroscopy has been used to probe the change in vibrational spectra of both water and solute atoms as a result of solvation. Vibrational spectroscopy is particularly sensitive to the formation of hydrogen bonding and allows the study of specific solvent–solute interactions through site-specific labeling [123, 124]. Such experiments provide mostly qualitative information about the degree of solvation of different parts of a given solute and have been able to suggest two-dimensional networks of hydrogen-bonded water molecules on the surface of solvated proteins [125]. However, more quantitative information is often difficult to extract from such studies [1].

1.4.4
Dielectric Dispersion

Dielectric dispersion measurements are one of the oldest techniques for studying the dynamics of solvated biomolecules [108]. They provide information about characteristic time scales from the frequency dependence of the dielectric constant. However, this method does not provide spatial resolution, the time resolution is limited, and the features observed in the dielectric spectra are often difficult to interpret in terms of specific aspects of solvent dynamics. Therefore, this technique is not widely applied today.

1.5
Hydration of Proteins

Numerous theoretical and experimental studies of protein–water interactions have resulted in a complex picture of protein and peptide hydration that transcends a range of spatial and temporal scales. In the following, the effect on protein folding and peptide conformational sampling is described first before the properties of water near protein surfaces are reviewed.

1.5.1
Protein Folding and Peptide Conformations in Aqueous Solvent

The overall effect of aqueous solvent on proteins in terms of the hydrophobic effect and dielectric screening of electrostatic interactions between polar and charged groups is well understood. For longer peptides and proteins, solvent promotes the formation of regular secondary structure elements and folding into unique native structures [126]. In short peptides, where hydrophobic residues cannot be fully sequestered from the aqueous solvent environment, the electrostatic screening effect is dominant and results in competition for intramolecular salt bridges and hydrogen bonds [127, 128]. In fact, the formation of a single backbone hydrogen bond versus individual solvation of the carbonyl and amide groups in a short peptide is energetically nearly net neutral, with only a slight enthalpic advantage of about 1 kcal mol^{-1} per residue, which is countered by an entropic cost [129]. This gives rise to a delicate energetic balance that allows short peptides to adopt a variety of different conformations with similar relative free energies depending on the amino acid sequence [1, 130]. Recent experimental and computational research suggests a general preference of short peptides for polyproline II (PPII) conformations with backbone dihedral torsion angles in the vicinity of $\phi = -85°$ and $\psi = 140°$ that is apparent in short alanine-based peptides [131–133] but is also found for longer peptides in the unfolded state [134]. It is clear that solvent plays a significant role in stabilizing the PPII conformation, especially in polyalanine peptides, but the fact that the exact mechanism remains a subject of debate [135–137] is testament to the complexity of peptide–water interactions even for relatively simple biomolecular solutes. Longer peptides often show a preference for the formation of α-helices [5, 138, 139] or extended (β) structures [140–143]. α-Helices are frequently observed for alanine-rich peptides [144] and are believed to be stabilized in part as a result of helical dipole formation due to the alignment of multiple backbone hydrogen bonds [145, 146] as well as secondary effects due to side chain–solvent interactions [147]. Extended structures are favored entropically, but the more extensive backbone hydrogen bonding interactions with the solvent incur a penalty in solvent entropy due to ordering of solvent molecules [133]. Therefore, especially the formation of β-hairpins often requires additional stabilizing interactions between amino acid side chains [141, 148–150].

1.5.2
Molecular Properties of Water Near Protein Surfaces

On a molecular level, the interactions between proteins and the surrounding solvent are equally complex. Water forms a distinct hydration layer around proteins that consists of one or more solvent shells containing water molecules with physical characteristics that may differ from those of the bulk solvent. The first hydration layer extends about 3 Å from the protein surface and has an average density that is increased by about 10–15% over the value for bulk solvent [106, 151]. Furthermore, the properties of water molecules within the first hydration layer may vary significantly depending on the type of interactions with the protein. A comparatively small fraction of water molecules in the first hydration shell, on the order of 5–30% depending on the property that is being measured [117, 119, 152], exhibit significantly reduced rotational and translational diffusion times typically on the order of tens of picoseconds [108, 117] and may occupy localized interaction sites relative to the solute [101]. The remaining water molecules retain nearly bulk-like properties, with diffusion times reduced by no more than a factor of 2–3 over bulk solvent despite close interaction with the solute [117, 152, 153]. A related measure is the so-called thermal volume of solvent surrounding proteins, which captures the fraction of water molecules that are thermodynamically behaving as if they were part of the protein. The interpretation of experiments in the context of scaled-particle theory has resulted in a thickness of about 0.6 Å for the thermal volume [154], which again translates into a fraction of 20% of the water molecules in the first hydration layer interacting strongly with a given protein. Finally, some studies have determined the minimum level of hydration necessary to achieve native-like protein dynamics and function as about 0.4 g of water per 1 g of protein, which translates into less than a single layer of hydration [155]. The combination of those observations suggests the following somewhat simplified view of protein hydration. A relatively small, but crucial, number of water molecules strongly associate with part of the protein. Those water molecules exhibit reduced dynamics and occupy well-defined hydration sites that may coincide with crystallographic water sites. The resulting protein–water complex interacts with a near bulk-like hydration layer with only slightly increased density and reduced dynamics, which in turn is surrounded by bulk solvent. The modern view of hydration does not assume an ice-like water layer but allows for fluid-like individual water molecules that visit certain preferred interaction sites where they exhibit temporarily retarded dynamics while continuing to exchange with the surrounding bulk on sub-nanosecond time scales. Furthermore, the notion of reduced solvent dynamics near the protein surface needs to be differentiated further based on observations that water molecules in the first hydration shell largely retain bulk-like rotational dynamics [156] and lateral diffusional dynamics [85] while the diffusional component perpendicular to the protein surface is slowed down most dramatically [85].

In addition to the strongly associated but still exchanging water molecules, there is also a small number of water molecules that do in fact have much longer residence times. These water molecules may be completely buried [157] or bound in deep cavities [158] and often play specific roles that may directly complement protein function – for example, in the form of water molecules that are essential partners in an enzyme active site. Especially the latter roles of water as an essential functional part of proteins have resulted in water being considered the "21st amino acid" [1].

1.5.3
Water Molecules at Protein–Ligand and Protein–Protein Interfaces

The presence of strongly associated water molecules on the protein surface has implications for ligand binding and protein complex formation. Often, water molecules are retained at ligand–protein [15, 158, 159] and protein–protein interfaces [11, 160–162]. A pre-organization of water molecules in the absence of a binding partner reduces the entropic penalty during binding, but the presence of ordered water molecules also essentially changes the molecular shape during molecular recognition [9, 163].

1.6
Hydration of Nucleic acids

Nucleic acids, DNA and RNA, differ significantly from proteins in two major ways. (1) Each nucleotide carries a negative charge associated with the backbone phosphate group. As a result, even a short piece of nucleic acid carries a high net charge compared to proteins and peptides, which are often nearly net neutral. (2) DNA molecules do not form globular structures but rather have flexible rod-like shapes. RNA molecules may fold to more compact shapes, but the RNA folding mechanism and resulting role of solvent are very different from protein folding. RNA folding is not driven by hydrophobic collapse – because of the absence of significant hydrophobic groups – but instead relies on counterion condensation and promiscuous base pairing and stacking interactions.

As a result of the high charge of nucleic acids, interactions with solvent molecules (both ions and water) are stronger than for proteins. While proteins essentially have a single hydration layer where water molecules are perturbed to a significant extent from bulk solvent, solvent around nucleic acids may be perturbed in terms of density and dynamics over two or even three layers [85, 99]. Furthermore, ions play a critical role in nucleic acid solvation as they may interact either directly with the nucleic structure [164] or through a shell of coordinated water molecules [80].

Water molecules in the first hydration layer often occupy very well-defined interaction sites [99, 165, 166]. The most famous hallmark of DNA hydration is the so-called "spine of hydration" in the minor groove that is readily visible in crystal structures [7]. Well-defined hydration sites are also present in the major groove, most prominently in the form of pentagonal water clusters [167], and around the phosphate oxygens in the form of so-called "cones of hydration" [168]. In both grooves, the hydration site patterns vary as a function of nucleotide sequence [99].

The interaction with solvent is also closely related to the equilibrium of the polymorphic A and B right-handed nucleic acid helices. While RNA is essentially locked into A-form helices as a result of steric and electrostatic interactions with the extra 2'-hydroxyl at its ribose unit, DNA is found in both A- and B-forms depending on the environment. It has long been known that A-form DNA is prevalent at low water activities while B-form DNA is preferred at high water activities [6, 169], where "low water activity" means either partial dehydration [6] or the presence of co-solvents such as alcohols and certain ions [170–172]. Consistent with this effect would be a smaller number of strongly interacting water molecules per nucleotide in A-DNA compared to B-DNA according to experimental studies [173]. A different number of water molecules per nucleotide could be rationalized based on the more compact form of A-DNA and increased solvent exposure of the partially hydrophobic ribose ring [174], which would become relatively more favorable under conditions of partial dehydration and/or in the presence of co-solvents. However, simulation studies of A- and B-DNA in aqueous solvent do not necessarily find significantly different numbers of water molecules interacting with A- and B-DNA through hydrogen bonding [175, 176], and A-RNA appears to be hydrated more than B-DNA, presumably due to the presence of the 2'-hydroxyl group [177]. Secondary effects such as increased water bridge formation and increased counterion condensation might therefore be the main driving force for stabilizing A-DNA [176]. The stabilization of A-DNA in the presence of ions would result from additional screening of the repulsive electrostatic interaction between neighboring phosphate groups, since the phosphate–phosphate distance in A-DNA is reduced to 5.9 Å from about 7 Å in B-DNA.

There is also a difference in hydration of A:T versus C:G base pairs [99] that can be correlated to differential propensities to undergo a B- to A-DNA transition. Compared to A:T base pairs, C:G base pairs have an extra hydrogen bond donor in the minor groove from the exocyclic amino group in guanine. Because the total number of water molecules around a base pair is roughly the same, this means that there are more water molecules around A:T base pairs with less favorable, non-hydrogen-bond-forming interactions than around C:G base pairs. This may explain why C:G base pairs more readily assume the A-form upon dehydration and/or addition of co-solvents where polar waters are lost at the expense of hydrophobic waters interacting with the more exposed ribose ring [99].

As in protein hydration, some fraction of the water molecules interacting with nucleic acids exhibit reduced diffusion rates, and, as in the case of protein hydration, reduced mobility and occupancy of well-defined hydration sites are not necessarily correlated [178, 179]. Nevertheless, it has become clear that water molecules located in the grooves have relatively long residence times between tens of picoseconds up to nanoseconds [180–182]. Such relaxation times are about one to two orders of magnitude longer than for water molecules interacting with proteins. The combination of more extensively ordered water molecules and longer residence times around nucleic acids has prompted the notion that "water is an integral part of nucleic acid structures" [183] but also bears implications for the interaction of nucleic acids with proteins and other ligands. As a consequence, proteins scanning nucleic acids, such as transcription factors, may be guided by interactions with the hydration layer exhibiting sequence-specific patterns rather than direct interactions with the nucleic acid.

1.7
Non-Aqueous Solvation

The interaction of biomolecules with non-aqueous solvent is relevant in the biological context (for example, protective solvents that allow cells to survive under freezing conditions or membrane environments) or in biotechnology applications (for example, denaturants such as urea and guanidium hydrochloride) [19]. Many studies have investigated solvation in non-aqueous solvent or co-solvent, but the detailed mechanisms of how certain non-aqueous solvents interact with biomolecules is not always fully understood. One aspect that is often investigated for solvent mixtures is the preferential solvation at the biomolecule–solvent interface. Depending on the type of co-solvent, water may be displaced partially or completely, or water may continue to interact directly with the biomolecule but exhibit altered properties due to secondary interactions with the co-solvent. The specific effects of different types of non-aqueous solvents are reviewed briefly in the following:

1.7.1
Alcohols

Alcohol (methanol, ethanol, 1,1,1,3,3,3-hexafluoropropan-2-ol (HFIP), and 2,2,2-trifluoroethanol (TFE)) mixtures lower the dielectric constant of the solvent and thereby provide reduced screening of electrostatic interactions, which leads to an enhanced propensity to form intramolecular hydrogen bonds and salt bridges within proteins but also reduces the penalty for solvent exposure of hydrophobic side chains. As a result, α-helix-forming peptides display enhanced α-helix propensities in mixtures containing methanol or ethanol [184–188] and secondary structure formation is generally enhanced in the presence of TFE and HFIP [189]. TFE and HFIP overall interact preferentially with peptides [190, 191] while ethanol preferentially interacts with peptide backbone atoms [192], thereby enhancing their effect in creating an environment with reduced polarity.

1.7.2
Urea

Urea is commonly used as a denaturant for proteins. Urea has an amphiphilic character that allows it to interact with both polar and hydrophobic groups. Its effect on protein solvation is actually fairly subtle. Based on simulation studies, it appears that urea affects the stability of proteins both directly and indirectly. The direct effect may involve preferential interaction with hydrophobic residues, thereby destabilizing the hydrophobic core, or interaction with polar constituents and in particular the peptide backbone. The indirect effect alters water structure in such a way that water becomes free to compete more successfully for backbone hydrogen bonds and other intra-protein interactions [193–197].

1.7.3
Glycerol

Glycerol, along with other similar compounds such as trehalose, is a natural compound that has a stabilizing effect on protein structures to allow cells to withstand freezing temperatures or dehydration. Glycerol increases the viscosity of the solvent and thereby dampens the dynamic fluctuations of a given protein [198, 199]. As a result, the protein may be trapped in a native-like conformational state in a glass-like environment that can be maintained at low temperatures. Glycerol also substitutes surface-bound water molecules and thereby reduces the consequences of freezing. A secondary effect of the presence of glycerol appears to be entrapment of water molecules in the first solvation shell due to the increased viscosity that would result in prolonged water residence times [197, 199].

1.8
Summary

This chapter has surveyed the current knowledge about the interplay between solvent and biomolecular structure and dynamics. As an introduction to the remainder of this book, it introduces the theoretical framework, simulation methodologies, and experimental techniques that are commonly applied in the study of biomolecular solvation. With the development of solvent models in mind, the structural and dynamic characteristics of solvent around proteins and nucleic acids are sketched out. In both cases, there are well-defined hydration sites that are preferentially occupied by solvent. Water molecules at those sites may have significantly reduced diffusion rates, but, contrary to the seemingly static pictures portrayed by crystallographic water sites, actual water molecules often continue to exchange with the surrounding solvent and display only moderately reduced rotational diffusion rates. The observation that many water molecules in the first hydration shell or proteins retain near-bulk-like dynamic properties justifies the development of mean-field solvent models that are based on a separation into explicit solute and surrounding bulk-like implicit solvent. However, such an approach essentially ignores the relatively small but significant fraction of water molecules in the first hydration shell that possess significantly perturbed properties and hence do not behave like bulk solvent. In the case of nucleic acids, water molecules are even more ordered and a larger fraction of water molecules is dynamically perturbed. It appears that the most realistic solvent models short of a fully explicit solvent representation would need to take into account the differentiated character of water molecules in the first hydration shell.

References

1 Prabhu, N. and Sharp, K. (2006) Protein–solvent interactions. *Chem. Rev.*, **106** (5), 1616–1623.
2 Levy, Y. and Onuchic, J.N. (2004) Water and proteins: a love–hate relationship. *Proc. Natl. Acad. Sci. U.S.A.*, **101**, 3325–3326.
3 Guo, Z.Y., Thirumalai, D., and Honeycutt, J.D. (1992) Folding kinetics of proteins–a model study. *J. Chem. Phys.*, **97** (1), 525–535.
4 Dill, K.A., Fiebig, K.M., and Chan, H.S. (1993) Cooperativity in protein-folding kinetics. *Proc. Natl. Acad. Sci. U.S.A.*, **90** (5), 1942–1946.
5 Avbelj, F. and Baldwin, R.L. (2003) Role of backbone solvation and electrostatics in generating preferred peptide backbone conformations: distributions of phi. *Proc. Natl. Acad. Sci. U.S.A.*, **100** (10), 5742–5747.
6 Leslie, A.G.W., Arnott, S., Chandrasekaran, R., and Ratliff, R.L. (1980) Polymorphism of DNA double helices. *J. Mol. Biol.*, **143**, 49–72.
7 Drew, H.R. and Dickerson, R.E. (1981) Structure of a B-DNA dodecamer. III. Geometry of hydration. *J. Mol. Biol.*, **151**, 535–556.
8 Cukierman, S. (2006) Et tu, Grotthuss! and other unfinished stories. *Biochim. Biophys. Acta Bioenerg.*, **1757** (8), 876–885.
9 Papoian, G.A., Ulander, J., and Wolynes, P.G. (2003) Role of water mediated interactions in protein–protein recognition landscapes. *J. Am. Chem. Soc.*, **125** (30), 9170–9178.
10 Bhat, T.N., Bentley, G.A., Boulot, G., Greene, M.I., Tello, D., Dallacqua, W., Souchon, H., Schwarz, F.P., Mariuzza, R.A., and Poljak, R.J. (1994) Bound water-molecules and conformational stabilization help mediate an antigen–antibody association. *Proc. Natl. Acad. Sci. U.S.A.*, **91** (3), 1089–1093.
11 Covell, D.G. and Wallqvist, A. (1997) Analysis of protein–protein interactions and the effects of amino acid mutations on their energetics. The importance of water molecules in the binding epitope. *J. Mol. Biol.*, **269** (2), 281–297.
12 Ben-Naim, A. (2002) Molecular recognition – viewed through the eyes of the solvent. *Biophys. Chem.*, **101**, 309–319.
13 Zhao, X., Huang, X.R., and Sun, C.C. (2006) Molecular dynamics analysis of the engrailed homeodomain-DNA recognition. *J. Struct. Biol.*, **155** (3), 426–437.
14 Nguyen, B., Hamelberg, D., Bailly, C., Colson, P., Stanek, J., Brun, R., Neidle, S., and Wilson, W.D. (2004) Characterization of a novel DNA minor-groove complex. *Biophys. J.*, **86** (2), 1028–1041.
15 Tame, J.R.H., Sleigh, S.H., Wilkinson, A.J., and Ladbury, J.E. (1996) The role of water in sequence-independent ligand binding by an oligopeptide transporter protein. *Nat. Struct. Biol.*, **3** (12), 998–1001.
16 Fuxreiter, M., Mezei, M., Simon, I., and Osman, R. (2005) Interfacial water as a "hydration fingerprint" in the noncognate complex of BamHI. *Biophys. J.*, **89** (2), 903–911.
17 Chaplin, M. (2006) Opinion – do we underestimate the importance of water in cell biology? *Nat. Rev. Mol. Cell Biol.*, **7** (11), 861–866.
18 Levy, Y. and Onuchic, J.N. (2006) Water mediation in protein folding and molecular recognition. *Annu. Rev. Biophys. Biomol. Struct.*, **35**, 389–415.
19 Roccatano, D. (2008) Computer simulations study of biomolecules in non-aqueous or cosolvent/water mixture solutions. *Curr. Protein Pept. Sci.*, **9** (4), 407–426.
20 Klibanov, A.M. (2001) Improving enzymes by using them in organic solvents. *Nature*, **409** (6817), 241–246.
21 Carrea, G. and Riva, S. (2000) Properties and synthetic applications of enzymes in organic solvents. *Angew. Chem. Int. Edn*, **39** (13), 2226–2254.
22 Iyer, P.V. and Ananthanarayan, L. (2008) Enzyme stability and stabilization – aqueous and non-aqueous environment. *Process Biochem.*, **43** (10), 1019–1032.

23 Leckband, D. and Israelachvili, J. (2001) Intermolecular forces in biology. *Q. Rev. Biophys.*, **34** (02), 105–267.
24 Chandler, D. and Andersen, H.C. (1972) Optimized cluster expansions for classical fluids. II. Theory of molecular liquids. *J. Chem. Phys.*, **57** (5), 1930–1937.
25 Hirata, F. and Rossky, P.J. (1981) An extended RISM equation for molecular polar fluids. *Chem. Phys. Lett.*, **83** (2), 329–334.
26 Pettitt, B.M. and Rossky, P.J. (1982) Integral-equation predictions of liquid-state structure for waterlike intermolecular potentials. *J. Chem. Phys.*, **77** (3), 1451–1457.
27 Ishizuka, R., Chong, S.H., and Hirata, F. (2008) An integral equation theory for inhomogeneous molecular fluids: the reference interaction site model approach. *J. Chem. Phys.*, **128** (3), 034504.
28 Yu, H.-A., Roux, B., and Karplus, M. (1990) Solvation thermodynamics: an approach from analytic temperature derivatives. *J. Chem. Phys.*, **92**, 5020–5033.
29 Pettitt, B.M., Karplus, M., and Rossky, P.J. (1986) Integral-equation model for aqueous solvation of polyatomic solutes – application to the determination of the free-energy surface for the internal motion of biomolecules. *J. Phys. Chem.*, **90** (23), 6335–6345.
30 Cui, Q.Z. and Smith, V.H. (2003) Solvation structure, thermodynamics, and conformational dependence of alanine dipeptide in aqueous solution analyzed with reference interaction site model theory. *J. Chem. Phys.*, **118** (1), 279–290.
31 Prabhu, N.V., Perkyns, J.S., Pettitt, B.M., and Hruby, V.J. (1999) Structure and dynamics of alpha-MSH using DRISM integral equation theory and stochastic dynamics. *Biopolymers*, **50** (3), 255–272.
32 Imai, T., Kovalenko, A., and Hirata, F. (2006) Hydration structure, thermodynamics, and functions of protein studied by the 3D-RISM theory. *Mol. Simul.*, **32** (10–11), 817–824.
33 Yamazaki, T., Imai, T., Hirata, F., and Kovalenko, A. (2007) Theoretical study of the cosolvent effect on the partial molar volume change of staphylococcal nuclease associated with pressure denaturation. *J. Phys. Chem. B*, **111** (5), 1206–1212.
34 Imai, T., Hiraoka, R., Seto, T., Kovalenko, A., and Hirata, F. (2007) Three-dimensional distribution function theory for the prediction of protein–ligand binding sites and affinities: application to the binding of noble gases to hen egg-white lysozyme in aqueous solution. *J. Phys. Chem. B*, **111** (39), 11585–11591.
35 Kirkwood, J.G. and Buff, F.P. (1951) The statistical mechanical theory of solutions. 1. *J. Chem. Phys.*, **19** (6), 774–777.
36 Newman, K.E. (1994) Kirkwood–Buff solution theory – derivation and applications. *Chem. Soc. Rev.*, **23** (1), 31–40.
37 Bennaim, A. (1977) Inversion of Kirkwood–Buff theory of solutions – application to water–ethanol system. *J. Chem. Phys.*, **67** (11), 4884–4890.
38 Shimizu, S. and Matubayasi, N. (2006) Preferential hydration of proteins: a Kirkwood–Buff approach. *Chem. Phys. Lett.*, **420** (4–6), 518–522.
39 Chitra, R. and Smith, P.E. (2001) Preferential interactions of cosolvents with hydrophobic solutes. *J. Phys. Chem. B*, **105** (46), 11513–11522.
40 Auton, M., Bolen, D.W., and Rosgen, J. (2008) Structural thermodynamics of protein preferential solvation: osmolyte solvation of proteins, aminoacids, and peptides. *Proteins*, **73** (4), 802–813.
41 Abui, M. and Smith, P.E. (2004) A combined simulation and Kirkwood–Buff approach to quantify cosolvent effects on the conformational preferences of peptides in solution. *J. Phys. Chem. B*, **108** (22), 7382–7388.
42 Pierce, V., Kang, M., Aburi, M., Weerasinghe, S., and Smith, P.E. (2008) Recent applications of Kirkwood–Buff theory to biological systems. *Cell Biochem. Biophys.*, **50** (1), 1–22.
43 Kang, M. and Smith, P.E. (2006) A Kirkwood–Buff derived force field for amides. *J. Comput. Chem.*, **27** (13), 1477–1485.
44 Weerasinghe, S. and Smith, P.E. (2005) A Kirkwood–Buff derived force field for

methanol and aqueous methanol solutions. *J. Phys. Chem. B*, **109** (31), 15080–15086.

45 Loftfield, R.B., Eigner, E.A., Pastuszyn, A., Lovgren, T.N.E., and Jakubowski, H. (1980) Conformational-changes during enzyme catalysis – role of water in the transition-state. *Proc. Natl. Acad. Sci. U.S.A.*, **77** (6), 3374–3378.

46 Torelli, A.T., Krucinska, J., and Wedekind, J.E. (2007) A comparison of vanadate to a 2′–5′ linkage at the active site of a small ribozyme suggests a role for water in transition-state stabilization. *RNA-Publ. RNA Soc.*, **13** (7), 1052–1070.

47 Lewendon, A. and Shaw, W.V. (1993) Transition-state stabilization by chloramphenicol acetyltransferase – role of a water molecule bound to threonine-174. *J. Biol. Chem.*, **268** (28), 20997–21001.

48 Zagrovic, B. and Pande, V. (2003) Solvent viscosity dependence of the folding rate of a small protein: distributed computing study. *J. Comput. Chem.*, **24**, 1432–1436.

49 Feig, M. (2007) Kinetics from implicit solvent simulations of biomolecules as a function of viscosity. *J. Chem. Theory Comput.*, **3**, 1734–1748.

50 Vanderzwan, G. and Hynes, J.T. (1983) Non-equilibrium solvation dynamics in solution reactions. *J. Chem. Phys.*, **78** (6), 4174–4185.

51 Garrett, B.C. and Schenter, G.K. (1994) Variational transition-state theory for activated chemical reactions in solution. *Int. Rev. Phys. Chem.*, **13** (2), 263–289.

52 Truhlar, D.G. and Garrett, B.C. (2000) Multidimensional transition state theory and the validity of Grote–Hynes theory. *J. Phys. Chem. B*, **104** (5), 1069–1072.

53 Ponder, J.W. and Case, D.A. (2003) Force fields for protein simulations. *Adv. Prot. Chem.*, **66**, 27–85.

54 Nicholas, M., Arianna, W.R., Marshall, N.R., Augusta, H.T., and Edward, T. (1953) Equation of state calculations by fast computing machines. *J. Chem. Phys.*, **21** (6), 1087–1092.

55 Lennard-Jones, J.E. (1931) Cohesion. *Proc. Phys. Soc.*, **43**, 461–482.

56 Patel, S. and Brooks, C.L. (2004) CHARMM fluctuating charge force field for proteins: I. Parameterization and application to bulk organic liquid simulations. *J. Comput. Chem.*, **25** (1), 1–15.

57 Zhu, S.B., Singh, S., and Robinson, G.W. (1991) A new flexible polarizable water model. *J. Chem. Phys.*, **95** (4), 2791–2799.

58 Dang, L.X., Rice, J.E., Caldwell, J., and Kollman, P.A. (1991) Ion solvation in polarizable water – molecular-dynamics simulations. *J. Am. Chem. Soc.*, **113** (7), 2481–2486.

59 Ren, P.Y. and Ponder, J.W. (2003) Polarizable atomic multipole water model for molecular mechanics simulation. *J. Phys. Chem. B*, **107** (24), 5933–5947.

60 Vanbelle, D., Froeyen, M., Lippens, G., and Wodak, S.J. (1992) Molecular-dynamics simulation of polarizable water by an extended Lagrangian method. *Mol. Phys.*, **77** (2), 239–255.

61 Sprik, M. and Klein, M.L. (1988) A polarizable model for water using distributed charge sites. *J. Chem. Phys.*, **89** (12), 7556–7560.

62 Lamoureux, G., Harder, E., Vorobyov, I.V., Roux, B., and MacKerell, A.D. (2006) A polarizable model of water for molecular dynamics simulations of biomolecules. *Chem. Phys. Lett.*, **418** (1–3), 245–249.

63 Frenkel, D. and Smit, B. (2002) *Understanding Molecular Simulation: From Algorithms to Applications*, 2nd edn, Academic Press, San Diego, CA.

64 Lounnas, V. (1999) Molecular dynamics simulation of proteins in aqueous environment, in *Hydration Process in Biology: Theoretical and Experimental Approaches* (ed. M.-C. Bellissent-Funel), IOS Press, Amsterdam, pp. 261–292.

65 Yonetani, Y. (2005) A severe artifact in simulation of liquid water using a long cut-off length: appearance of a strange layer structure. *Chem. Phys. Lett.*, **406** (1–3), 49–53.

66 Anezo, C., de Vries, A.H., Holtje, H.-D., Tieleman, D.P., and Marrink, S.-J. (2003) Methodological issues in lipid bilayer

simulations. *J. Phys. Chem. B*, **107** (35), 9424–9433.

67 Ewald, P.P. (1921) Die Berechnung optischer und elektrostatischer Gitterpotentiale. *Annalen Physik*, **64**, 253–287.

68 Smith, P.E. and Pettitt, B.M. (1991) Peptides in ionic solutions: a comparison of the Ewald and switching function techniques. *J. Chem. Phys.*, **95**, 8430.

69 Weber, W., Hunenberger, P.H., and McCammon, J.A. (2000) Molecular dynamics simulations of a polyalanine octapeptide under Ewald boundary conditions: influence of artificial periodicity on peptide conformation. *J. Phys. Chem. B*, **104** (15), 3668–3675.

70 Cheatham, T.E., Miller, J.L., Fox, T., Darden, T.A., and Kollman, P.A. (1995) Molecular dynamics simulations on solvated biomolecular systems: the particle mesh Ewald method leads to stable trajectories of DNA, RNA, and proteins. *J. Am. Chem. Soc.*, **117**, 4193–4194.

71 Darden, T.A., York, D., and Pedersen, L.G. (1993) Particle mesh Ewald: an $N\log(N)$ method for Ewald sums in large systems. *J. Chem. Phys.*, **98**, 10089–10092.

72 Guillot, B. (2002) A reappraisal of what we have learnt during three decades of computer simulations on water. *J. Mol. Liq.*, **101**, 219–260.

73 Jorgensen, W.L., Chandrasekhar, J., Madura, J.D., Impey, R.W., and Klein, M.L. (1983) Comparison of simple potential functions for simulating liquid water. *J. Chem. Phys.*, **79** (2), 926–935.

74 Jorgensen, W.L. (1981) Quantum and statistical mechanical studies of liquids. 10. Transferable intermolecular potential functions for water, alcohols, and ethers. Application to liquid water. *J. Am. Chem. Soc.*, **103**, 335–340.

75 Berendsen, H.J.C., Grigera, J.R., and Straatsma, T.P. (1987) The missing term in effective pair potentials. *J. Phys. Chem.*, **91** (24), 6269–6271.

76 Horn, H.W., Swope, W.C., Pitera, J.W., Madura, J.D., Dick, T.J., Hura, G.L., and Head-Gordon, T. (2004) Development of an improved four-site water model for biomolecular simulations: TIP4P-Ew. *J. Chem. Phys.*, **120** (20), 9665–9678.

77 Mahoney, M.W. and Jorgensen, W.L. (2000) A five-site model for liquid water and the reproduction of the density anomaly by rigid, nonpolarizable potential functions. *J. Chem. Phys.*, **112** (20), 8910–8922.

78 Rick, S.W., Stuart, S.J., and Berne, B.J. (1994) Dynamical fluctuating charge force-fields–application to liquid water. *J. Chem. Phys.*, **101** (7), 6141–6156.

79 Patel, S. and Brooks, C.L. (2006) Fluctuating charge force fields: recent developments and applications from small molecules to macromolecular biological systems. *Mol. Simul.*, **32** (3–4), 231–249.

80 Feig, M. and Pettitt, B.M. (1999) Sodium and chlorine ions as part of the DNA solvation shell. *Biophys. J.*, **77**, 1769–1781.

81 Garcia, A.E. and Stiller, L. (1993) Computation of the mean residence time of water in the hydration shells of biomolecules. *J. Comput. Chem.*, **14** (11), 1396–1406.

82 Bizzarri, A.R. and Cannistraro, S. (2002) Molecular dynamics of water at the protein–solvent interface. *J. Phys. Chem. B*, **106** (26), 6617–6633.

83 Rocchi, C., Bizzarri, A.R., and Cannistraro, S. (1998) Water dynamical anomalies evidenced by molecular-dynamics simulations at the solvent–protein interface. *Phys. Rev. E*, **57** (3), 3315–3325.

84 Allen, M.P. and Tildesley, D.J. (1987) *Computer Simulation of Liquids*, 1st edn, Oxford University Press, New York.

85 Makarov, V.A., Feig, M., and Pettitt, B.M. (1998) Diffusion of solvent around biomolecular solutes. A molecular dynamics simulations study. *Biophys. J.*, **75**, 150–158.

86 Lounnas, V., Pettitt, B.M., and Phillips, G.N. (1994) A global model of the protein–solvent interface. *Biophys. J.*, **66**, 601–614.

87 Kollman, P.A. (1993) Free energy calculations: applications to chemical and biochemical phenomena. *Chem. Rev.*, **93**, 2395–2417.

88 Kong, X. and Brooks, C.L., III (1996) λ-Dynamics: a new approach to free energy calculations. *J. Chem. Phys.*, **105**, 2414–2423.

89 Hummer, G., Pratt, L.R., and Garcia, A.E. (1995) Hydration free energy of water. *J. Phys. Chem.*, **99**, 14188–14194.

90 Hummer, G., Pratt, L.R., Garcia, A.E., Berne, B.J., and Rick, S.W. (1997) Electrostatic potentials and free energies of solvation of polar and charged molecules. *J. Phys. Chem.*, **101**, 3017–3020.

91 Villa, A. and Mark, A.E. (2002) Calculation of the free energy of solvation for neutral analogs of amino acid side chains. *J. Comput. Chem.*, **23**, 548–553.

92 Wescott, J.T., Fisher, L.R., and Hanna, S. (2002) Use of thermodynamic integration to calculate the hydration free energies of n-alkanes. *J. Chem. Phys.*, **116**, 2361–2369.

93 Nina, M., Beglov, D., and Roux, B. (1997) Atomic radii for continuum electrostatics calculations based on molecular dynamics free energy simulations. *J. Phys. Chem.*, **101**, 5239–5248.

94 Shirts, M.R., Pitera, J.W., Swope, W.C., and Pande, V.S. (2003) Extremely precise free energy calculations of amino acid side chain analogs: comparison of common molecular mechanics force fields for proteins. *J. Chem. Phys.*, **119**, 5740–5761.

95 Lee, M.S. and Olson, M.A. (2005) Evaluation of Poisson solvation models using a hybrid explicit/implicit solvent method. *J. Phys. Chem. B*, **109** (11), 5223–5236.

96 Berman, H.M. (1994) Hydration of DNA: take 2. *Curr. Opin. Struct. Biol.*, **4**, 345–350.

97 Savage, H. and Wlodawer, A. (1986) Determination of water structure around biomolecules using X-ray and neutron diffraction methods. *Methods Enzymol.*, **127**, 162–183.

98 Gerstein, M. and Chothia, C. (1996) Packing at the protein–water interface. *Proc. Natl. Acad. Sci. U.S.A.*, **93** (19), 10167–10172.

99 Feig, M. and Pettitt, B.M. (1999) Modeling high-resolution hydration patterns in correlation with DNA sequence and conformation. *J. Mol. Biol.*, **286**, 1075–1095.

100 Feig, M. and Pettitt, B.M. (1998) Crystallographic water sites from a theoretical perspective. *Structure*, **6**, 1351–1354.

101 Makarov, V., Pettitt, B.M., and Feig, M. (2002) Solvation and hydration of proteins and nucleic acids: a theoretical view of simulation and experiment. *Acc. Chem. Res.*, **35**, 376–384.

102 Teeter, M.M., Yamano, A., Stec, B., and Mohanty, U. (2001) On the nature of a glassy state of matter in a hydrated protein: relation to protein function. *Proc. Natl. Acad. Sci. U.S.A.*, **98** (20), 11242–11247.

103 Fuhrmann, C.N., Kelch, B.A., Ota, N., and Agard, D.A. (2004) The 0.83 angstrom resolution crystal structure of alpha-lytic protease reveals the detailed structure of the active site and identifies a source of conformational strain. *J. Mol. Biol.*, **338** (5), 999–1013.

104 Halle, B. (2004) Biomolecular cryocrystallography: structural changes during flash-cooling. *Proc. Natl. Acad. Sci. U.S.A.*, **101** (14), 4793–4798.

105 Soper, A.K. (2000) The radial distribution functions of water and ice from 220 to 673 K and at pressures up to 400 MPa. *Chem. Phys.*, **258** (2–3), 121–137.

106 Svergun, D.I., Richard, S., Koch, M.H.J., Sayers, Z., Kuprin, S., and Zaccai, G. (1998) Protein hydration in solution: experimental observation by x-ray and neutron scattering. *Proc. Natl. Acad. Sci. U.S.A.*, **95** (5), 2267–2272.

107 Bee, M. (1988) *Quasi-Elastic Neutron Scattering*. Adam Hilger, Philadelphia, PA.

108 Bagchi, B. (2005) Water dynamics in the hydration layer around proteins and micelles. *Chem. Rev.*, **105** (9), 3197–3219.

109 Paciaroni, A., Orecchini, A., Cinelli, S., Onori, G., Lechner, R.E., and Pieper, J. (2003) Protein dynamics on the picosecond timescale as affected by the environment: a quasielastic neutron

scattering study. *Chem. Phys.*, **292** (2–3), 397–404.
110 Russo, D., Hura, G., and Head-Gordon, T. (2004) Hydration dynamics near a model protein surface. *Biophys. J.*, **86** (3), 1852–1862.
111 Tarek, M., Neumann, D.A., and Tobias, D.J. (2003) Characterization of sub-nanosecond dynamics of the molten globule state of alpha-lactalbumin using quasielastic neutron scattering and molecular dynamics simulations. *Chem. Phys.*, **292** (2–3), 435–443.
112 Halle, B., Andersson, T., Forsen, S., and Lindman, B. (1981) Protein hydration from water O-17 magnetic relaxation. *J. Am. Chem. Soc.*, **103** (3), 500–508.
113 Otting, G., Liepinsh, E., and Wuthrich, K. (1991) Protein hydration in aqueous solution. *Science*, **254** (5034), 974–980.
114 Polnaszek, C.F., and Bryant, R.G. (1984) Nitroxide radical induced solvent proton relaxation – measurement of localized translational diffusion. *J. Chem. Phys.*, **81** (9), 4038–4045.
115 Halle, B. (2003) Cross-relaxation between macromolecular and solvent spins: the role of long-range dipole couplings. *J. Chem. Phys.*, **119** (23), 12372–12385.
116 Modig, K., Liepinsh, E., Otting, G., and Halle, B. (2004) Dynamics of protein and peptide hydration. *J. Am. Chem. Soc.*, **126** (1), 102–114.
117 Halle, B. and Denisov, V.P. (2001) Magnetic relaxation dispersion studies of biomolecular solutions. *Methods Enzymol.*, **338**, 178–201.
118 Pal, S.K. and Zewail, A.H. (2004) Dynamics of water in biological recognition. *Chem. Rev.*, **104** (4), 2099–2123.
119 Bhattacharyya, K. (2008) Nature of biological water: a femtosecond study. *Chem. Commun.*, **25**, 2848–2857.
120 Zhang, L.Y., Wang, L.J., Kao, Y.T., Qiu, W.H., Yang, Y., Okobiah, O., and Zhong, D.P. (2007) Mapping hydration dynamics around a protein surface. *Proc. Natl. Acad. Sci. U.S.A.*, **104** (47), 18461–18466.
121 Sen, S., Andreatta, D., Ponomarev, S.Y., Beveridge, D.L., and Berg, M.A. (2009) Dynamics of water and ions near DNA: comparison of simulation to time-resolved Stokes-shift experiments. *J. Am. Chem. Soc.*, **131**, 1724–1735.
122 Bandyopadhyay, S., Chakraborty, S., Balasubramanian, S., and Bagchi, B. (2005) Sensitivity of polar solvation dynamics to the secondary structures of aqueous proteins and the role of surface exposure of the probe. *J. Am. Chem. Soc.*, **127** (11), 4071–4075.
123 Walsh, S.T.R., Cheng, R.P., Wright, W.W., Alonso, D.O.V., Daggett, V., Vanderkooi, J.M., and DeGrado, W.F. (2003) The hydration of amides in helices; a comprehensive picture from molecular dynamics, IR, and NMR. *Protein Sci.*, **12** (3), 520–531.
124 Starzyk, A., Barber-Armstrong, W., Sridharan, M., and Decatur, S.M. (2005) Spectroscopic evidence for backbone desolvation of helical peptides by 2,2,2-trifluoroethanol: an isotope-edited FTIR study. *Biochemistry*, **44** (1), 369–376.
125 Khoshtariya, D.E., Hansen, E., Leecharoen, R., and Walker, G.C. (2003) Probing protein hydration by the difference O–H (O–D) vibrational spectroscopy: interfacial percolation network involving highly polarizable water–water hydrogen bonds. *J. Mol. Liq.*, **105** (1), 13–36.
126 Onuchic, J.N. and Wolynes, P.G. (2004) Theory of protein folding. *Curr. Opin. Struct. Biol.*, **14** (1), 70–75.
127 Avbelj, F. and Moult, J. (1995) Role of electrostatic screening in determining protein main-chain conformational preferences. *Biochemistry*, **34** (3), 755–764.
128 Sorin, E.J., Rhee, Y.M., Shirts, M.R., and Pande, V.S. (2006) The solvation interface is a determining factor in peptide conformational preferences. *J. Mol. Biol.*, **356** (1), 248–256.
129 Sneddon, S.F., Tobias, D.J., and Brooks, C.L. (1989) Thermodynamics of amide hydrogen-bond formation in polar and apolar solvents. *J. Mol. Biol.*, **209** (4), 817–820.

130 Groth, M., Malicka, J., Rodziewicz-Motowidlo, S., Czaplewski, C., Klaudel, L., Wiczk, W., and Liwo, A. (2001) Determination of conformational equilibrium of peptides in solution by NMR spectroscopy and theoretical conformational analysis: application to the calibration of mean-field solvation models. *Biopolymers*, **60** (2), 79–95.

131 Ramakrishnan, V., Ranbhor, R., and Durani, S. (2004) Existence of specific "folds" in polyproline II ensembles of an "unfolded" alanine peptide detected by molecular dynamics. *J. Am. Chem. Soc.*, **126** (50), 16332–16333.

132 Roe, D.R., Okur, A., Wickstrom, L., Hornak, V., and Simmerling, C. (2007) Secondary structure bias in generalized Born solvent models: comparison of conformational ensembles and free energy of solvent polarization from explicit and implicit solvation. *J. Phys. Chem. B*, **111** (7), 1846–1857.

133 Mezei, M., Fleming, P.J., Srinivasan, R., and Rose, G.D. (2004) Polyproline II helix is the preferred conformation for unfolded polyalanine in water. *Proteins*, **55** (3), 502–507.

134 Ferreon, J.C. and Hilser, V.J. (2003) The effect of the polyproline II (PPII) conformation on the denatured state entropy. *Protein Sci.*, **12** (3), 447–457.

135 Eker, F., Griebenow, K., and Schweitzer-Stenner, R. (2003) Stable conformations of tripeptides in aqueous solution studied by UV circular dichroism spectroscopy. *J. Am. Chem. Soc.*, **125** (27), 8178–8185.

136 Drozdov, A.N., Grossfield, A., and Pappu, R.V. (2004) Role of solvent in determining conformational preferences of alanine dipeptide in water. *J. Am. Chem. Soc.*, **126** (8), 2574–2581.

137 Fleming, P.J., Fitzkee, N.C., Mezei, M., Srinivasan, R., and Rose, G.D. (2005) A novel method reveals that solvent water favors polyproline II over beta-strand conformation in peptides and unfolded proteins: conditional hydrophobic accessible surface area (CHASA). *Protein Sci.*, **14** (1), 111–118.

138 Avbelj, F. (2000) Amino acid conformational preferences and solvation of polar backbone atoms in peptides and proteins. *J. Mol. Biol.*, **300** (5), 1335–1359.

139 Shental-Bechor, D.S., Kirca, S., Ben-Tal, N., and Haliloglu, T. (2005) Monte Carlo studies of folding, dynamics, and stability in alpha-helices. *Biophys. J.*, **88** (4), 2391–2402.

140 Levy, Y., Jortner, J., and Becker, O.M. (2001) Solvent effects on the energy landscapes and folding kinetics of polyalanine. *Proc. Natl. Acad. Sci. U.S.A.*, **98** (5), 2188–2193.

141 Blanco, F.J., Rivas, G., and Serrano, L. (1994) A short linear peptide that folds into a native stable β-hairpin in aqueous solution. *Nat. Struct. Biol.*, **1**, 584–590.

142 Searle, M.S., Zerella, R., Williams, D.H., and Packman, L.C. (1996) Native-like beta-hairpin structure in an isolated fragment from ferredoxin: NMR and CD studies of solvent effects on the N-terminal 20 residues. *Protein Eng.*, **9**, 559–565.

143 Munoz, V., Henry, E.R., Hofrichter, J., and Eaton, W.A. (1998) A statistical mechanical model for β-hairpin kinetics. *Proc. Natl. Acad. Sci. U.S.A.*, **95**, 5872–5879.

144 Marqusee, S., Robbins, V.H., and Baldwin, R.L. (1989) Unusually stable helix formation in short alanine-based peptides. *Proc. Natl. Acad. Sci. U.S.A.*, **86**, 5286–5290.

145 Hol, W.G.J. (1985) The role of the alpha-helix dipole in protein function and structure. *Progr. Biophys. Mol. Biol.*, **45** (3), 149–195.

146 Park, C. and Goddard, W.A. (2000) Stabilization of alpha-helices by dipole–dipole interactions within alpha-helices. *J. Phys. Chem. B*, **104** (32), 7784–7789.

147 Vila, J.A., Ripoll, D.R., and Scheraga, H.A. (2000) Physical reasons for the unusual alpha-helix stabilization afforded by charged or neutral polar residues in alanine-rich peptides. *Proc. Natl. Acad. Sci. U.S.A.*, **97** (24), 13075–13079.

148 Dinner, A.R., Lazaridis, T., and Karplus, M. (1999) Understanding β-hairpin formation. *Proc. Natl. Acad. Sci. U.S.A.*, **96**, 9068–9073.

149 Kobayashi, N., Honda, S., Yoshii, H., and Munekata, E. (2000) Role of side-chains in the cooperative β-hairpin folding of the short C-terminal fragment derived from streptococcal protein G. *Biochemistry*, **39**, 6564–6571.

150 Wu, X.W. and Brooks, B.R. (2004) Beta-hairpin folding mechanism of a nine-residue peptide revealed from molecular dynamics simulations in explicit water. *Biophys. J.*, **86** (4), 1946–1958.

151 Merzel, F. and Smith, J.C. (2002) Is the first hydration shell of lysozyme of higher density than bulk water? *Proc. Natl. Acad. Sci. U.S.A.*, **99** (8), 5378–5383.

152 Pal, S.K., Peon, J., and Zewail, A.H. (2002) Biological water at the protein surface: dynamical solvation probed directly with femtosecond resolution. *Proc. Natl. Acad. Sci. U.S.A.*, **99**, 1763–1768.

153 Nilsson, L. and Halle, B. (2005) Molecular origin of time-dependent fluorescence shifts in proteins. *Proc. Natl. Acad. Sci. U.S.A.*, **102** (39), 13867–13872.

154 Bano, M. and Marek, J. (2006) How thick is the layer of thermal volume surrounding the protein? *Biophys. Chem.*, **120** (1), 44–54.

155 Rupley, J.A. and Careri, G. (1991) Protein hydration and function. *Adv. Protein Chem.*, **41**, 37–172.

156 Persson, E. and Halle, B. (2008) Cell water dynamics on multiple time scales. *Proc. Natl. Acad. Sci. U.S.A.*, **105** (17), 6266–6271.

157 Halle, B. (2004) Protein hydration dynamics in solution: a critical survey. *Philos. Trans. R. Soc. Lond. Ser. B Biol. Sci.*, **359** (1448), 1207–1223.

158 Poornima, C.S. and Dean, P.M. (1995) Hydration in drug design. 3. Conserved water molecules at the ligand-binding sites of homologous proteins. *J. Comput. Aided Mol. Des.*, **9** (6), 521–531.

159 Poornima, C.S. and Dean, P.M. (1995) Hydration in drug design. 1. Multiple hydrogen-bonding features of water molecules in mediating protein–ligand interactions. *J. Comput. Aided Mol. Des.*, **9** (6), 500–512.

160 Li, Z. and Lazaridis, T. (2007) Water at biomolecular binding interfaces. *Phys. Chem. Chem. Phys.*, **9** (5), 573–581.

161 Okimoto, N., Nakamura, T., Suenaga, A., Futatsugi, N., Hirano, Y., Yamaguchi, I., and Ebisuzaki, T. (2004) Cooperative motions of protein and hydration water molecules: molecular dynamics study of scytalone dehydratase. *J. Am. Chem. Soc.*, **126** (40), 13132–13139.

162 Ikura, T., Urakubo, Y., and Ito, N. (2004) Water-mediated interaction at a protein–protein interface. *Chem. Phys.*, **307** (2–3), 111–119.

163 Lemieux, R.U. (1996) How water provides the impetus for molecular recognition in aqueous solution. *Acc. Chem. Res.*, **29** (8), 373–380.

164 Young, M.A., Jayaram, B., and Beveridge, D.L. (1997) Intrusion of counterions into the spine of hydration in the minor groove of B-DNA: fractional occupancy of electronegative pockets. *J. Am. Chem. Soc.*, **119**, 59–69.

165 Schneider, B. and Berman, H.M. (1995) Hydration of the DNA bases is local. *Biophys. J.*, **69**, 2661–2669.

166 Schneider, B., Cohen, D., and Berman, H.M. (1992) Hydration of DNA bases: analysis of crystallographic data. *Biopolymers*, **32**, 725–750.

167 Kennard, O., Cruse, W.B.T., Nachman, J., Prange, T., Shakked, Z., and Rabinovich, D. (1986) Ordered water structure in an A-DNA octamer at 1.7 Å resolution. *J. Biomol. Struct. Dyn.*, **3**, 623–647.

168 Saenger, W. (1987) Structure and dynamics of water surrounding biomolecules. *Annu. Rev. Biophys. Biophys. Chem.*, **16**, 93–114.

169 Pohl, F.M. (1976) Polymorphism of a synthetic DNA in solution. *Nature*, **260**, 365–366.

170 Cheatham, T.E., Crowley, M.F., Fox, T., and Kollman, P.A. (1997) A molecular level picture of the stabilization of A-DNA in mixed ethanol–water solutions. *Proc. Natl. Acad. Sci. U.S.A.*, **94**, 9626–9630.

171 Cheatham, T.E. and Kollman, P.A. (1997) Insight into the stabilization of

A-DNA by specific ion association: spontaneous B-DNA to A-DNA transitions observed in molecular dynamics simulations of d(ACCCGCGGGT)$_2$ in the presence of hexaamminecobalt(III). *Structure*, **15**, 1297–1311.

172 Cruse, W.B.T., Saldudjian, P., Leroux, Y., Leger, G., El Manouni, D., and Prange, T. (1996) A continuous transition from A-DNA to B-DNA in the 1:1 complex between nogalamycin and the hexamer dCCCGGG. *J. Biol. Chem.*, **271**, 15558–15567.

173 Harmouchi, M., Albiser, G., and Premilat, S. (1990) Changes of hydration during conformational transitions of DNA. *Eur. Biophys. J.*, **19**, 87–92.

174 Seba, H.B., Thureau, P., Ancian, B., and Thevand, A. (2006) Combined use of Overhauser spectroscopy and NMR diffusion experiments for mapping the hydration structure of nucleosides: structure and dynamics of uridine in water. *Magn. Reson. Chem.*, **44** (12), 1109–1117.

175 Feig, M. and Pettitt, B.M. (1998) A molecular simulation picture of DNA hydration around A- and B-DNA. *Biopolymers*, **48**, 199–209.

176 Pastor, N. (2005) The B- to A-DNA transition and the reorganization of solvent at the DNA surface. *Biophys. J.*, **88**, 3262–3275.

177 Rozners, E. and Moulder, J. (2004) Hydration of short DNA, RNA and 2′-OMe oligonucleotides determined by osmotic stressing. *Nucleic Acids Res.*, **32** (1), 248–254.

178 Lounnas, V. and Pettitt, B.M. (1994) Distribution function implied dynamics versus residence times and correlations. *Proteins*, **18**, 148–160.

179 Dolenc, J., Baron, R., Missimer, J.H., Steinmetz, M.O., and van Gunsteren, W.F. (2008) Exploring the conserved water site and hydration of a coiled-coil trimerisation motif: a MD simulation study. *Chembiochem*, **9** (11), 1749–1756.

180 Denisov, V.P., Carlstrom, G., and Halle, B. (1997) Kinetics of DNA hydration. *J. Mol. Biol.*, **268**, 118–136.

181 Phan, A.T., Leroy, J.-L., and Gueron, M. (1999) Determination of the residence time of water molecules hydration B′-DNA and B-DNA, by one-dimensional zero-enhancement nuclear Overhauser effect spectroscopy. *J. Mol. Biol.*, **286**, 505–519.

182 Pal, S., Maiti, P.K., Bagchi, B., and Hynes, J.T. (2006) Multiple time scales in solvation dynamics of DNA in aqueous solution: the role of water, counterions, and cross-correlations. *J. Phys. Chem. B*, **110** (51), 26396–26402.

183 Westhof, E. (1988) Water: an integral part of nucleic acid structure. *Annu. Rev. Biophys. Biophys. Chem.*, **17**, 125–144.

184 Gibbs, N., Sessions, R.B., Williams, P.B., and Dempsey, C.E. (1997) Helix bending in alamethicin: molecular dynamics simulations and amide hydrogen exchange in methanol. *Biophys. J.*, **72**, 2490–2495.

185 Kovacs, H., Mark, A.E., Johansson, J., and van Gunsteren, W.F. (1995) The effect of environment on the stability of an integral membrane helix: molecular dynamics simulations of surfactant protein C in chloroform, methanol and water. *J. Mol. Biol.*, **247**, 808–822.

186 Bazzo, R., Tappin, M.J., Pastore, A., Harvey, T.S., Carver, J.A., and Campbell, I.D. (1988) The structure of melittin. A ^{1}H-NMR study in methanol. *Eur. J. Biochem.*, **173**, 139–146.

187 Glättli, A., Chandrasekhar, I., and van Gunsteren, W.F. (2006) A molecular dynamics study of the bee venom melittin in aqueous solution, in methanol, and inserted in a phospholipid bilayer. *Eur. Biophys. J.*, **35**, 255–267.

188 Hirota, N., Mizuno, K., and Goto, Y. (1998) Group additive contributions to the alcohol-induced alpha-helix formation of melittin: implication for the mechanism of the alcohol effects on proteins. *J. Mol. Biol.*, **275**, 365–378.

189 Buck, M. (1998) Trifluoroethanol and colleagues: cosolvents come of age. Recent studies with peptides and proteins. *Q. Rev. Biophys.*, **31** (3), 297–355.

190 Fioroni, M., Diaz, M.D., Burger, K., and Berger, S. (2002) Solvation phenomena of a tetrapeptide in water/trifluoroethanol and water/ethanol

mixtures: a diffusion NMR, intermolecular NOE, and molecular dynamics study. *J. Am. Chem. Soc.*, **124** (26), 7737–7744.

191 Gerig, J.T. (2004) Structure and solvation of melittin in 1,1,1,3,3,3-hexafluoro-2-propanol/water. *Biophys. J.*, **86** (5), 3166–3175.

192 Neuman, R.C. and Gerig, J.T. (2008) Solvent interactions with the Trp-cage peptide in 35% ethanol–water. *Biopolymers*, **89** (10), 862–872.

193 Bennion, B.J. and Daggett, V. (2003) The molecular basis for the chemical denaturation of proteins by urea. *Proc. Natl. Acad. Sci. U.S.A.*, **100** (9), 5142–5147.

194 Lee, M.E. and van der Vegt, N.F.A. (2006) Does urea denature hydrophobic interactions? *J. Am. Chem. Soc.*, **128** (15), 4948–4949.

195 Smith, L.J., Jones, R.M., and van Gunsteren, W.F. (2005) Characterization of the denaturation of human alpha-lactalbumin in urea by molecular dynamics simulations. *Proteins*, **58** (2), 439–449.

196 Caballero-Herrera, A., Nordstrand, K., Berndt, K.D., and Nilsson, L. (2005) Effect of urea on peptide conformation in water: molecular dynamics and experimental characterization. *Biophys. J.*, **89** (2), 842–857.

197 Smolin, N. and Winter, R. (2008) Effect of temperature, pressure, and cosolvents on structural and dynamic properties of the hydration shell of SNase: a molecular dynamics computer simulation study. *J. Phys. Chem. B*, **112** (3), 997–1006.

198 Dirama, T.E., Carri, G.A., and Sokolov, A.P. (2005) Coupling between lysozyme and glycerol dynamics: microscopic insights from molecular-dynamics simulations. *J. Chem. Phys.*, **122** (24), 244910.

199 Scharnagl, C., Reif, M., and Friedrich, J. (2005) Stability of proteins: temperature, pressure and the role of the solvent. *Biochim. Biophys. Acta*, **1749** (2), 187–213.

2
Model-Free "Solvent Modeling" in Chemistry and Biochemistry Based on the Statistical Mechanics of Liquids

Norio Yoshida, Yasuomi Kiyota, Yasuhiro Ikuta, Takashi Imai, and Fumio Hirata

2.1
Introduction

It is a common understanding that solvents play important roles in many phenomena in chemistry, physics, and biology [1]. Especially, in phenomena related to life, the solvent is not just "important" but "essential," as can be understood from the fact that 70 wt% of our human body consists of water. In the body, water is not just a "theater" where biomolecules such as proteins and DNA play out their roles, but also an actor itself, performing a leading role in life's events, for example, enzymatic hydrolysis.

It is no wonder that chemists and biochemists have been concerned with solvents from the earliest stages in their development of theories in chemistry and the life sciences. In the early stage, their models of solvents were entirely phenomenological, based on thermodynamics, electromagnetism, hydrodynamics, and so on [2]. The Stokes–Einstein law is a typical example of a phenomenological theory to describe the effect of solvent on the dynamics of a particle in solution, in which the solvent is modeled with a macroscopic viscosity and boundary conditions such as the Stokes radius. The Debye model for dielectric relaxation, Onsager's reaction field, the Born model for ion hydration, and the Kirkwood model for the hydration free energy of biomolecules are other important contributions to chemistry and biochemistry based on the phenomenological solvation model [3–7]. Those models and theories have been applied to many aspects of chemical processes in solution, with great success. The most typical example is the Marcus theory of electron transfer reactions. Based on the Born model of ion hydration, Marcus could have formulated the non-equilibrium free-energy surface governing the reaction rate by employing an order parameter referred to as the solvent coordinate to realize the fluctuation from an equilibrium state of solvation [8].

The success of the phenomenological or continuum models of solvents lies in their simplicity for describing the physical nature of solvent effects using just a few macroscopic parameters such as the dielectric constant and viscosity. In fact, the model enjoys the status as a standard theory for describing the effects of solvent on chemical reactions and the stability of biomolecules [9]. However, its success is limited just to the physical aspects of life events, as a "background" or an "environment" for the leading characters. Even as a background to the main event, the model is too unrealistic.

The statistical mechanics of liquids, which began in the last century, was expected to replace the continuum models of solvent by including molecular-level descriptions [10–12]. A lot of effort was placed on taking the "granularity" of the liquid or solvent into consideration. However, success was limited largely to the *physical* aspects of liquids, or to liquids consisting of spherical molecules. It is the internal degrees of freedom of molecules that characterize the *chemistry* of liquids: atomic species, bond lengths, bond angles, electronic structures, and so on. So the problem to be solved in order to develop the liquid-state theory in chemistry was essentially how to evaluate the partition function or the configuration integral of intra- and intermolecular degrees of freedom, which are coupled to each other: the former features covalent bonds between atoms, while the latter involves multifold integrals over infinitely large space.

The first breakthrough toward the liquid-state theory in chemistry was made by Chandler and Andersen with their proposition of the reference interaction-site model (RISM) theory [13]. Their idea was to project the intramolecular degrees of freedom onto that of the intermolecular distance between interaction sites or atoms, as an *order parameter*, by taking the average over all molecular orientations. The procedure introduces the intramolecular distribution function, which defines the geometry of molecules consisting of liquids. Thus, the orientational correlation between a pair of molecules is mapped onto the positional correlations, which just depend on the scalar distances between pairs of interaction sites. This theory significantly advanced liquid-state theory in chemistry. However, there was another important feature of molecules to be taken into account, namely the electronic structure or the charge distribution in a molecule. This problem was solved by Hirata and Rossky by applying the renormalization technique developed in the theory of electrolyte solutions [14]. The new theory, referred to as the extended reference interaction-site model (XRISM) theory, has proved its ability to describe the physics and chemistry of liquids in numerous applications: electronic structure of molecules in solution, chemical reactivity, dynamics of water and ions in solution, solvation of biomolecules, and so forth [15]. The theory, however, exposed a weakness inherited from the original RISM theory in some applications: unphysical results were obtained when the theory was applied in a straightforward manner to calculations of the partial molar volume of proteins. The weakness was related to the solute–solvent direct correlation functions, which are not coupled well with the intramolecular correlation of solute or protein.

The weakness was cured by the introduction of the three-dimensional reference interaction-site model (3D-RISM) theory, in which the orientational average is taken only for solvent molecules [15–18]. The power of the 3D-RISM theory has been demonstrated fully in descriptions of the solvation structure and thermodynamics of proteins. Calculated partial molar volumes of proteins in aqueous solutions have exhibited quantitative agreement with corresponding experimental results [19]. This turns out to be the first quantitative results obtained for the thermodynamics of proteins based entirely on statistical mechanics theory. It was a great accomplishment by itself in the sense that it gave confidence in the 3D-RISM for exploring the stability of proteins in solutions. However, it was only a prelude to a discovery that would have an even larger impact. When we were analyzing the 3D distribution of water around hen egg-white lysozyme, we found conspicuous peaks within small cavities of the protein, which no doubt revealed water molecules trapped inside the macromolecule [20]. In fact, the number of water molecules and the positions inside the cavity coincided with those found by X-ray crystallography. This implies that the 3D-RISM has the capability of "detecting" molecules "recognized" by the protein, or the host molecule. In other words, the 3D-RISM theory can be applied to molecular recognition, which is involved in almost all processes in biosystems to maintain life: information transmission, enzymatic reactions, drug binding, and so on.

In this chapter, we present a brief introduction to the RISM and 3D-RISM theories, which is followed by a few sections describing the application of the theories to biomolecular solvation.

2.2
Outline of the RISM and 3D-RISM theories

Let us begin this section by asking the following two questions: "What is the structure of a liquid?" "How can the structure of a liquid be characterized?" These questions are non-trivial, because, unlike molecules or crystals, the liquid state does not form a structure of definite shape. One can readily define the structure of a molecule by giving the bond lengths, bond angles, and dihedral angles, even for the most complex molecules such as proteins. The crystalline structure of a solid can also be defined unambiguously by giving the lattice constants. However, molecules in liquids are in continuous diffusive motion, and thereby no definite geometry among the molecules can be defined. In such a case, we can only use a statistical or probabilistic language.

The probabilistic language for characterizing the structure of a liquid comprises the distribution functions, which are nothing but the moments of the density field $v(\mathbf{r}) = \Sigma_i \delta(\mathbf{r}-\mathbf{r}_i)$ with respect to the Boltzmann weight. If there is no field applied to the system, the first moment or the average density is just constant everywhere in the system, namely, $\rho(\mathbf{r}) \equiv \langle v(\mathbf{r}) \rangle = \rho = N/V$, where V and N are the volume of the container and the number of molecules in the system, respectively, and $\langle ... \rangle$ indicates the thermal average. So, the average density does not convey any information with respect to the liquid. However, the second moment $\rho(\mathbf{r},\mathbf{r}') = \langle v(\mathbf{r})v(\mathbf{r}') \rangle$ carries structural information about the liquid. This quantity is referred to as the density pair distribution function, which has essentially the same physical meaning as the radial distribution function (RDF) obtained from X-ray diffraction measurement. The density pair distribution function $\rho(\mathbf{r},\mathbf{r}')$ is proportional to the probability density of finding two molecules at the two positions $\mathbf{r}$ and $\mathbf{r}'$ at the same time.

When the distance between the two positions becomes so large that there is no "correlation" between the densities of the two positions, it becomes just the product of the average densities:

$$\lim_{|r-r'|\to\infty} \rho(r,r') \to \rho(r)\rho(r') \quad (=\rho^2 \text{ in uniform liquids}) \tag{2.1}$$

The quantity $g(r,r') = \rho(r,r')/\rho^2$ represents a "correlation" of the density at the two positions r and r'. It is referred to as the pair correlation function (PCF), or the radial distribution function when the liquid density is uniform and translational invariance is implied. We further define a function called the total correlation function by $h(r,r') = g(r,r') - 1$, which represents the correlation of the density "fluctuations" at the two positions r and r',

$$h(r,r') = \langle \delta v(r) \delta v(r') \rangle / \rho^2 \tag{2.2}$$

where $\delta v(r)$ ($= v(r) - \rho$) denotes the density fluctuation. The main task of liquid-state theory is to find an equation that governs the function $g(r,r')$ or $h(r,r')$ based on statistical mechanics, and to solve the resulting equation.

An "exact" equation, referred to as the Ornstein–Zernike equation, which relates $h(r,r')$ with another correlation function called the direct correlation function $c(r,r')$, can be "derived" from the grand canonical partition function by means of functional derivatives. Our theory for describing molecular recognition starts from the Ornstein–Zernike equation generalized to a solution of polyatomic molecules, or the molecular Ornstein–Zernike (MOZ) equation [12],

$$h(1,2) = c(1,2) + \int c(1,3)\rho h(3,2) d(3) \tag{2.3}$$

where $h(1,2)$ and $c(1,2)$ are the total and direct correlation functions, respectively, and the numbers in parentheses represent the coordinates of molecules in the liquid system, including both the position R and the orientation Ω. The boldface letters of the correlation functions indicate that they are matrices consisting of the elements labeled by the species in the solution. In the simple case of a binary mixture, the equation can be written down as follows, labeling the solute by "u" and the solvent by "v" (it is straightforward to generalize the equations to multi-component mixtures):

$$h_{vv}(1,2) = c_{vv}(1,2) + \int c_{vv}(1,3)\rho_v h_{vv}(3,2)d(3) + \int c_{vu}(1,3)\rho_u h_{uv}(3,2)d(3) \tag{2.4}$$

$$h_{uv}(1,2) = c_{uv}(1,2) + \int c_{uv}(1,3)\rho_v h_{vv}(3,2)d(3) + \int c_{uu}(1,3)\rho_u h_{uv}(3,2)d(3) \tag{2.5}$$

$$h_{uu}(1,2) = c_{uu}(1,2) + \int c_{uv}(1,3)\rho_v h_{vu}(3,2)d(3) + \int c_{uu}(1,3)\rho_u h_{uu}(3,2)d(3) \tag{2.6}$$

By taking the limit of infinite dilution ($\rho_u \to 0$), one gets

$$h_{vv}(1,2) = c_{vv}(1,2) + \int c_{vv}(1,3)\rho_v h_{vv}(3,2)d(3) \tag{2.7}$$

$$h_{uv}(1,2) = c_{uv}(1,2) + \int c_{uv}(1,3)\rho_v h_{vv}(3,2)d(3) \tag{2.8}$$

The equations essentially depend on six coordinates in Cartesian space, and include a six-fold integral. This integral prevents the theory being used in

2.2 Outline of the RISM and 3D-RISM theories

applications to polyatomic molecules. It was the interaction-site model and the RISM approximation proposed by Chandler and Andersen that enabled solutions to these equations [13]. The idea behind the model is to project the functions onto the one-dimensional space along the distance between the interaction sites, usually placed at the center of atoms, by taking the statistical average over the angular coordinates of molecules with fixed separation between a pair of interaction sites:

$$f_{\alpha\gamma}(r) = \int \delta(\mathbf{R}_1 + \mathbf{l}_1^\alpha) \delta(\mathbf{R}_2 + \mathbf{l}_2^\gamma - \mathbf{r}) f(1,2) d(1) d(2) \quad (2.9)$$

where $\mathbf{l}_i^\alpha$ is the vector displacement of site α in molecule i from the molecular center $\mathbf{R}_i$. It follows that $\mathbf{R}_i + \mathbf{l}_i^\alpha = \mathbf{r}_i^\alpha$ denotes the position of site α in molecule i. The angular average of the second terms in Equations 2.7 and 2.8 is formidable, but the approximation

$$c(1,2) \approx \sum_{\alpha\gamma} c_{\alpha\gamma}(|\mathbf{r}_1^\alpha - \mathbf{r}_2^\gamma|) \quad (2.10)$$

allows one to perform the angular average to lead to the RISM equation

$$\rho h \rho = \omega * c * \omega + \omega * c * \rho h \rho \quad (2.11)$$

where the asterisk denotes the convolution integrals, that is,

$$f * g = \int f(\mathbf{r}_1, \mathbf{r}_3) g(\mathbf{r}_3, \mathbf{r}_2) d\mathbf{r}_3 \quad (2.12)$$

Hereafter, the solvent density is denoted by ρ instead of ρ_v. A new function ω appears in Equation 2.11, which is called the "intramolecular" correlation function and is defined for a pair of atoms α and γ in a molecule by

$$\omega_{\alpha\gamma}(r) = \rho \delta_{\alpha\gamma} \delta(r) + (1 - \delta_{\alpha\gamma}) \delta(r - l_{\alpha\gamma}) \quad (2.13)$$

in which $\delta_{\alpha\gamma}$ and $\delta(r)$ are the Kronecker and Dirac delta functions, respectively. By means of the Dirac delta function, the term $\delta(r - l_{\alpha\gamma})$ imposes a distance constraint $l_{\alpha\gamma}$ between the pair of atoms. So, giving the distance constraints for all pairs of atoms in a molecule defines the molecular structure or geometry in terms of trigonometry. This is the way in which the molecular structure is incorporated into the RISM theory.

The 3D-RISM equation for the solute–solvent system at infinite dilution can be derived from Equation 2.8 by taking the statistical average over the angular coordinates of the "solvent," not of the solute [16, 17, 21]. The equation reads

$$h_\gamma(\mathbf{r}) = \sum_{\gamma'} \int c_{\gamma'}(\mathbf{r'}) [\omega_{\gamma'\gamma}^{vv}(|\mathbf{r'} - \mathbf{r}|) + \rho h_{\gamma'\gamma}^{vv}(|\mathbf{r'} - \mathbf{r}|)] d\mathbf{r'} \quad (2.14)$$

where $h_\gamma(\mathbf{r})$ and $c_{\gamma'}(\mathbf{r'})$ are the total and direct correlation functions of sites γ and γ', respectively, of solvent molecules at two positions $\mathbf{r}$ and $\mathbf{r'}$ in the Cartesian coordinate system with its origin placed at an arbitrary position, generally inside the protein. The functions $\omega_{\gamma'\gamma}^{vv}(r)$ and $h_{\gamma'\gamma}^{vv}(r)$ are the correlation functions for solvent molecules, which appear in Equation 2.11. It is these equations that can be applied to the molecular recognition process. If one views the solute molecule as a source of external force exerted on the solvent molecules, then $\rho g_\gamma(\mathbf{r})$

($= \rho h_\gamma(r) + \rho$) is identified as the density distribution of solvent molecules under the influence of the external force. This realization, called the "Percus trick," is the key concept in describing the molecular recognition process by means of statistical mechanics.

The equations described above contain two unknown functions, $h(r)$ and $c(r)$. Therefore, they are not closed without another equation that relates the two functions. Several approximations have been proposed for the closure relations: hypernetted chain (HNC), Percus–Yervick (PY), mean spherical approximation (MSA), and so on [12].

The HNC closure can be obtained from the diagrammatic expansion of the pair correlation functions with respect to the density and discarding a set of diagrams called the "bridge diagrams," which have multi-fold integrals. It should be noted that the terms kept in the HNC closure relation still include terms up to infinite orders of density. Alternatively, the relation has been derived from the linear response of a free-energy functional to the density fluctuation created by a molecule fixed in space within the Percus trick. The HNC closure relation reads

$$h_\gamma(r) = \exp[-u_\gamma(r)/k_B T + h_\gamma(r) - c_\gamma(r)] + 1 \qquad (2.15)$$

where k_B and T are the Boltzmann constant and temperature, respectively, and $u(r)$ is the interaction potential between pairs of atoms in the system. Equation 2.15 is the relation that incorporates the physical and chemical characteristics of the system into the theory through $u(r)$. The PY approximation can be obtained from the HNC relation simply by linearizing the factor $\exp[h(r) - c(r)]$. The HNC closure has been quite successful for describing the structure and thermodynamics of liquids and solutions, including water. However, the approximation is notorious in the low-density regime. Its limitations may become fatal when one tries to apply the theory to associating liquid mixtures or solutions, especially of dilute concentration, because a solution of dilute concentration is equivalent to a low-density liquid for the minor component. In order to avoid the problem, Kovalenko and Hirata proposed the following approximation, or the KH closure [15, 21]:

$$g_\gamma(r) = \begin{cases} \exp(d_\gamma(r)) & \text{for } d_\gamma(r) \leq 0 \\ 1 + d_\gamma(r) & \text{for } d_\gamma(r) > 0 \end{cases} \qquad (2.16)$$

where $d_\gamma(r) = -u_\gamma(r)/k_B T + h_\gamma(r) - c_\gamma(r)$. The approximation turns out to be quite successful even for mixtures of complex liquids.

The procedure for solving the equations consists of two steps. We first solve the RISM equation (Equation 2.11) for $h_{\gamma\gamma}^{vv}(r)$ of the solvent or a mixture of solvents in the case of solutions. Then, we solve the 3D-RISM equation (Equation 2.14) for $h_\gamma(r)$ of a protein–solvent (solution) system, inserting $h_{\gamma\gamma}^{vv}(r)$ for the solvent into Equation 2.14, which was calculated in the first step. Considering the definition $g(r) = h(r) + 1$, the $g(r)$ thus obtained is the three-dimensional distribution of solvent molecules around a protein in terms of the interaction-site representation of the solvent or a mixture of solvents in the case of solutions. The so-called solvation free energy can be obtained from the distribution function through the following equations [15, 22] that correspond to the two closure relations given in Equations 2.15 and 2.16, respectively:

$$\Delta\mu_{\text{HNC}} = \rho^v k_B T \sum_\gamma \int d\mathbf{r} \left[\frac{1}{2} h_\gamma^{uv}(\mathbf{r})^2 - c_\gamma^{uv}(\mathbf{r}) - \frac{1}{2} h_\gamma^{uv}(\mathbf{r}) c_\gamma^{uv}(\mathbf{r}) \right] \quad (2.17)$$

$$\Delta\mu_{\text{KH}} = \rho^v k_B T \sum_\gamma \int d\mathbf{r} \left[\frac{1}{2} h_\gamma^{uv}(\mathbf{r})^2 \Theta(-h_\gamma^{uv}(\mathbf{r})) - c_\gamma^{uv}(\mathbf{r}) - \frac{1}{2} h_\gamma^{uv}(\mathbf{r}) c_\gamma^{uv}(\mathbf{r}) \right] \quad (2.18)$$

where Θ denotes the Heaviside step function. The other thermodynamic quantities concerning solvation can be readily obtained from the standard thermodynamic derivative of the free energy, except for the partial molar volume.

The partial molar volume (PMV), which is a very important quantity when probing the response of the free energy (or stability) of a protein to pressure, including so-called pressure denaturation, is not a "canonical" thermodynamic quantity for the (V, T) ensemble, since the volume is an independent thermodynamic variable of the ensemble. The partial molar volume of a protein at infinite dilution can be calculated from the Kirkwood–Buff equation [23] generalized to the site–site representation of liquid and solutions [24, 25],

$$\bar{V} = k_B T \chi_T \left[1 - \rho \sum_\gamma \int c_\gamma(\mathbf{r}) d\mathbf{r} \right] \quad (2.19)$$

where χ_T is the isothermal compressibility of pure solvent or solution, which is obtained from the site–site correlation functions of solutions.

Some applications of the 3D-RISM theory require the derivatives of the free energy with respect to the coordinates of solute molecules; for example, the force acting on the protein atoms. The first derivative can be obtained analytically from Equation 2.17 or Equation 2.18. The analytical expression of the free-energy gradient for the RISM theory was presented by Sato *et al.* [26]. The expression can be easily extended to the 3D-RISM formalism [27, 28]. The total Helmholtz free energy A of the system is given as the sum of solute energy E_{solute} and the solvation free energy or the excess chemical potential coming from solute–solvent interaction,

$$A = E_{\text{solute}} + \Delta\mu \quad (2.20)$$

The free energy A can be regarded as a functional of the correlation functions, c, h, and t, as well as the coordinate of solute atoms, where the indirect correlation function t is defined as $t = h - c$. The first derivative of the free energy with respect to the coordinate of the solute atoms $\mathbf{R}_a$ is written as

$$\begin{aligned}\frac{\partial A}{\partial \mathbf{R}_a} =\ & \frac{\partial E_{\text{solute}}}{\partial \mathbf{R}_a} - \frac{\rho}{\beta} \sum_\alpha \int d\mathbf{r} \Big\{ [e^{-\beta u_\alpha(\mathbf{r}) + t_\alpha(\mathbf{r})} - 1 - h_\alpha(\mathbf{r})] \frac{\partial t_\alpha(\mathbf{r})}{\partial \mathbf{R}_a} \\ & + [-t_\alpha(\mathbf{r}) + h_\alpha(\mathbf{r}) - c_\alpha(\mathbf{r})] \frac{\partial h_\alpha(\mathbf{r})}{\partial \mathbf{R}_a} \Big\} \\ & - \frac{1}{(2\pi)^3 \beta} \sum_\alpha \int d\mathbf{k} \Big\{ -\rho \tilde{h}_\alpha(\mathbf{k}) + \sum_\gamma \tilde{c}_\gamma(\mathbf{k}) \tilde{X}_{\gamma\alpha}(\mathbf{k}) \Big\} \frac{\partial \tilde{c}_\alpha(\mathbf{k})}{\partial \mathbf{R}_a} \\ & + \rho \sum_\alpha \int d\mathbf{r} g_\alpha(\mathbf{r}) \frac{\partial u_\alpha(\mathbf{r})}{\partial \mathbf{R}_a}\end{aligned} \quad (2.21)$$

The second, third, and fourth terms on the right-hand side should be zero, because the terms inside the integrals are just a definition of HNC closure, indirect correlation function, and k-space 3D-RISM equation, respectively. Finally, we obtain the simple expression

$$\frac{\partial A}{\partial R_a} = \frac{\partial E_{\text{solute}}}{\partial R_a} + \rho \sum_{\alpha \in \text{solvent}} \int g_\alpha(r) \frac{\partial u_\alpha(r)}{\partial R_a} dr \qquad (2.22)$$

where "a" denotes the atom of a solute molecule. When the KH closure is employed, one obtains an expression identical to Equation 2.22. This analytical formula for the free-energy gradient with respect to the coordinates of solute atoms allows the efficient geometry optimization of solvated molecules. The determination of the structure of a solvated molecule is one of the most important tasks of molecular modeling.

In the following sections, we show applications of the theory described above in order to demonstrate the robustness of the theory.

2.3
Partial Molar Volume of Proteins

The first application is the partial molar volume of a given protein, which can be calculated using Equation 2.19 from $h(r)$, or equivalently from $c(r)$ obtained from the 3D-RISM equation. The partial molar volume of several proteins in water, which appear frequently in the protein research literature, is plotted against the molecular weight in Figure 2.1 [19]. By comparing the results with the experimental ones plotted in the same figure, one can readily see that the theory is capable of reproducing the experimental results at a quantitative level. At first glance, the experimental results seem to be reproduced by just a simple consideration of protein geometry using commercial software to calculate the exclusion volume of the protein. However, this is not the case. The reason is that the partial molar volume is a thermodynamic quantity, not a "geometrical volume." The partial molar volume reflects all the solvent–solvent and solute–solvent interactions as well as all the configurations of water molecules in the system, while the geometrical volume accounts just for the simplified (hard-core-type) repulsive interaction between the solute and solvent. Other factors, such as the attractive interactions between the solute and solvent and the solvent reorganization, are entirely neglected in the geometrical volume. The contributions from solvent reorganization are of particular importance in the partial molar volume of a protein, because this is concerned with the volume associated with cavities inside the protein. As is well appreciated, proteins often have many internal cavities where water molecules may be accommodated.

Let us carry out a simple "thought experiment" with respect to the partial molar volume of protein. Consider the dissolution of a protein in water. Upon dissolution, some of cavities in the protein may be filled by water molecules, but others

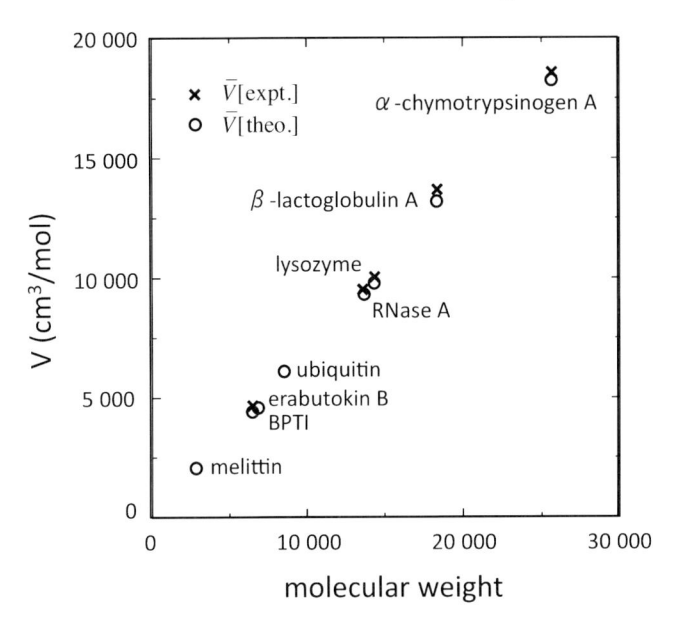

Figure 2.1 Partial molar volume of proteins plotted against the molecular weight. The theoretical results (circles) show quantitative agreement with the experimental ones (crosses).

may not. If a cavity stays empty, then the empty space will contribute to an increased partial molar volume of the protein. On the other hand, if the space is filled by water molecules owing to reorganization of the solvent, this will reduce the entire volume of solution, and compensate for an increase due to the cavity volume. This compensation is non-trivial. If a cavity can accommodate one water molecule, it reduces the volume by $18\,\text{cm}^3\,\text{mol}^{-1}$. In this regard, unless a theory is able to describe the reorganization of water molecules induced by a protein, the theory is useless for accurately predicting the partial molar volume. The excellent quantitative agreement between theory and experiments shown in Figure 2.1 demonstrates that the theory is properly accounting for all the solute–solvent and solvent–solvent interactions as well as solvent reorganization induced by protein, including the accommodation of some water molecules within internal cavities.

2.4
Detecting Water Molecules Trapped Inside Protein

It may not be necessary here to emphasize how important water is for living systems to maintain life. It is no wonder that many scientists in the field of X-ray and neutron diffraction measurement have been trying to determine the position and orientation of water molecules around and inside biomolecules, proteins, and DNA. However, even with modern experimental technology to locate the positions of water molecules, this is not an easy task, partly because of the limited resolution of the diffraction measurements in space as well as in time (see Chapter 1). This is because water molecules at the surface of proteins are not necessarily bound firmly to particular sites of biomolecules, but exchange their positions quite frequently. In fact, this flexibility and fluctuation of water molecules is essential for living systems to control their life. Diffraction measurements can only identify some water molecules that have a long residence time at some particular position of the biomolecules.

In our studies [20, 29], we have carried out a 3D-RISM calculation for a hen egg-white lysozyme immersed in water and obtained the 3D distribution function of the oxygen and hydrogen atoms of water molecules around and inside the protein. The native 3D structure of the protein is taken from the Protein Data Bank (PDB). The protein is known to have a cavity composed of the residues from Y53 to I58 and from A82 to S91, in which four water molecules have been determined by means of X-ray diffraction measurements. In our calculation, those water molecules are not included explicitly.

In Figure 2.2a depicted by green surfaces or spots are the $g(r)$ of the oxygen atoms of water molecules using an isosurface representation, which is very similar to the electron density map obtained from X-ray crystallography. We have drawn $g(r)$ that is greater than a threshold value. The left, center, and right parts correspond, respectively, to $g(r) > 2.0$, $g(r) > 4.0$, and $g(r) > 8.0$. Since $g(r)$ is unity in the bulk, the left part indicates that the probability of finding those water molecules at the surface is more than twice as large compared to bulk water. Furthermore, the water molecules depicted in the right part have eight times higher probability compared to bulk. These water molecules are bound firmly to particular atoms of the protein, presumably due to hydrogen bonds, and they are quite rare, as one can see from the figure. In this sense, the threshold values play the role of the "temperature" in X-ray diffraction measurements: at lower temperatures, one can observe more water molecules that have weaker interactions with the protein. The results suggest that the X-ray and neutron diffraction communities have acquired a powerful theoretical tool to analyze their data for locating the positions and orientations of water molecules, since our theory also provides the distribution of the hydrogen atoms of water molecules.

The results depicted in Figure 2.2a are what we expected before we actually carried out the calculation, although they were novel by themselves in the history of statistical mechanics. Entirely unexpected was that we observed some peaks of water distribution in a cavity inside the protein, which is surrounded by the residues from Y53 to I58 and from A82 to S91. The results are shown in Figure 2.2b. The left panel in Figure 2.2b shows the isosurfaces of $g(r) > 8$ for the oxygen (green) and hydrogen (pink) atoms of water in the cavity. Only the surrounding residues are displayed, except for A82 and L83, which are located in the front. There are four distinct peaks of water oxygen and seven distinct peaks of water hydrogen in the cavity. From the isosurface plot we have reconstructed the most probable model of the hydration structure. It is shown in the center panel of Figure 2.2b, where the four water molecules are numbered in order from left to right. Water 1 is hydrogen-bonded to the main-chain oxygen of Y53 and to the main-

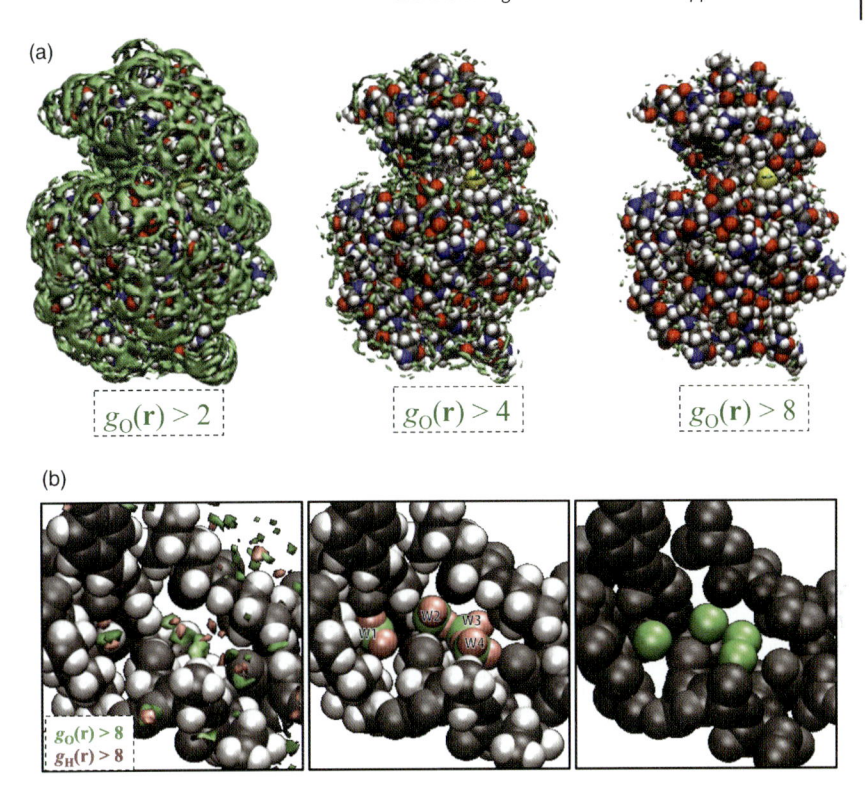

Figure 2.2 (a) Isosurface representation of the 3D distribution function $g(r)$ of water oxygen around lysozyme calculated by the 3D-RISM theory. The green surfaces or spots show the area where the distribution function is larger than 2 (left), 4 (center) and 8 (right). (b) Water molecules in the cavity surrounded by the residues from Y53 to I58 and from A82 to S91. Residues A82 and L83 located in the front side are not displayed. The isosurfaces of water oxygen (green) and hydrogen (pink) for the three-dimensional distributions larger than 8 (left), the most probable model of the hydration structure reconstructed from the isosurface plots (center), and the crystallographic water sites (right).

chain nitrogen of L56. Water 2 forms hydrogen bonds to the main-chain nitrogen of I56 and to the main-chain oxygen of L83, which is not drawn in the figure. Waters 3 and 4 also form hydrogen bonds with protein sites, the former to the main-chain oxygen of S85 and the latter to the main-chain oxygens of A82 (not displayed) and of D87. There is also a hydrogen bond network among waters 2, 3, and 4. The peak of the hydrogen between waters 3 and 4 does not appear in the figure because it is slightly less than 8, which means that the hydrogen bond is weaker or looser than the other hydrogen-bonding interactions. Although the hydroxyl group of S91 is located in the center of the four water molecules, it interacts only weakly with them.

It is interesting to compare the hydration structure obtained by the 3D-RISM theory with crystallographic water sites of X-ray structure [30]. The crystallographic water molecules in the cavity are depicted in the right panel of Figure 2.2b, showing four water sites in the cavity, much as the 3D-RISM theory has predicted. Moreover, the water distributions obtained from theory and experiment are quite similar to each other. Thus the 3D-RISM theory can predict the water binding sites with great success.

It should be noted that one peak of the 3D distribution function does not necessarily correspond to one molecule. If a water molecule transfers back and forth between two sites in the equilibrium state, two peaks correspondingly appear in the 3D distribution function. In fact, the number of water molecules within the cavity calculated from the 3D distribution function is 3.6. This is less than the number of water binding sites, and includes decimal fractions. To explain that observation, we carried out a molecular dynamics (MD) simulation using the same parameters and under the same thermodynamic conditions as the 3D-RISM calculation but with the exception that the four crystallographic water molecules in the cavity as well as the other crystallographic water molecules were initially placed at the crystal positions in the MD simulation. The MD simulation also shows a hydration number less than 4, that is, 3.5 [29]. From the MD trajectory, it is found that the two inner water molecules, waters 1 and 2, remain at fixed sites during the entire simulation time and fluctuate only little around those sites. On the other hand, the two outer water molecules, waters 3 and 4, sometimes enter and leave the sites, and by chance exchange with other water molecules from the bulk phase. As a result, the number of water molecules at the outer sites is 1.5 on average. Thus the 3D-RISM theory can provide reasonable hydration numbers including decimal fractions through statistical mechanical relations, even though the theory takes no explicit account of the dynamics of molecules.

2.5
Selective Ion Binding by Protein

Ion binding is essential for a variety of physiological processes. The binding of calcium ions by some protein triggers the process to induce muscle contraction and enzymatic reactions. The initial process of information transmission through the ion channel is ion binding by channel proteins. Ion binding sometimes plays an essential role in the folding process of a protein by inducing certain secondary structures. Such processes are characterized by highly selective ion recognition by proteins. It is of great importance, therefore, for the life sciences to clarify the origin of ion selectivity in molecular detail.

In this section, we present theoretical results for ion binding by human lysozyme obtained through basically the same procedure as that described in the preceding section [31, 32]. We first prepare the correlation functions for the bulk solutions by solving Equation 2.11, and then we apply those functions within the 3D-RISM equation (Equation 2.14) to obtain the 3D distribution of ions along with water molecules. Special attention, however, should be paid to the treatment of the bulk solution as the reference state, because ion–ion interactions in solution

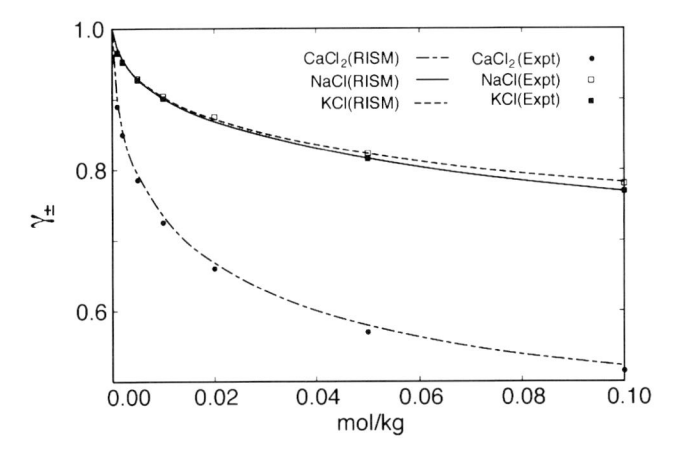

Figure 2.3 Mean activity coefficient of aqueous solutions of NaCl, KCl, and CaCl$_2$.

are Coulomb interactions, and their contribution to the dehydration penalty should not be disregarded even at low concentrations. In order to make sure that the free energy due to ion–ion interaction is reasonably accounted for, we have calculated the excess chemical potential, or the mean activity coefficient, of ions in solutions. The results are shown in Figure 2.3. The theoretical results in general show good agreement with the experimental results. In particular, the theory discriminates the divalent ion from the monovalent ions quite well. On a closer view, however, one can see that the theoretical values for the two monovalent ions are in the inverse order to the experimental ones. This may be due to the potential parameters for the ions. However, the influence of ion recognition by the protein on the results is small, because the process is determined primarily by the free-energy difference of the same ion inside protein and bulk solutions.

The 3D-RISM calculation was carried out for aqueous solutions of three different electrolytes, CaCl$_2$, NaCl, and KCl, and for four different mutants of the protein, wild type, Q86D, A92D, and Q86D/A92D, that have been studied experimentally by Kuroki and Yutani [33]. Figure 2.4a shows the distribution of water molecules and cations inside and around the binding cleft, which consists of the amino acid residues from Q86 to A92. The area where the distribution function, $g(r)$, is greater than 5 is colored differently for each species: oxygen of water, red; Na$^+$ ion, yellow; Ca^{2+} ion, orange; and K$^+$ ion, purple. For the wild type of protein in aqueous solutions of all three electrolytes studied, CaCl$_2$, NaCl and KCl, there are no distributions ($g(r) > 5$) observed for the ions inside the cleft, as is seen in the upper left panel. The Q86D mutant exhibits essentially the same behavior as that of the wild type, but with the water distribution changed slightly (upper right panel). (There is a trace of yellow spot that indicates a slight possibility of finding a Na$^+$ ion in the middle of the binding site, but it would be too small to make a significant contribution to the distribution.) Instead, the distribution corresponding to water oxygen is observed as is shown by the red color in that panel. The distribution covers faithfully the region where the crystallographic water

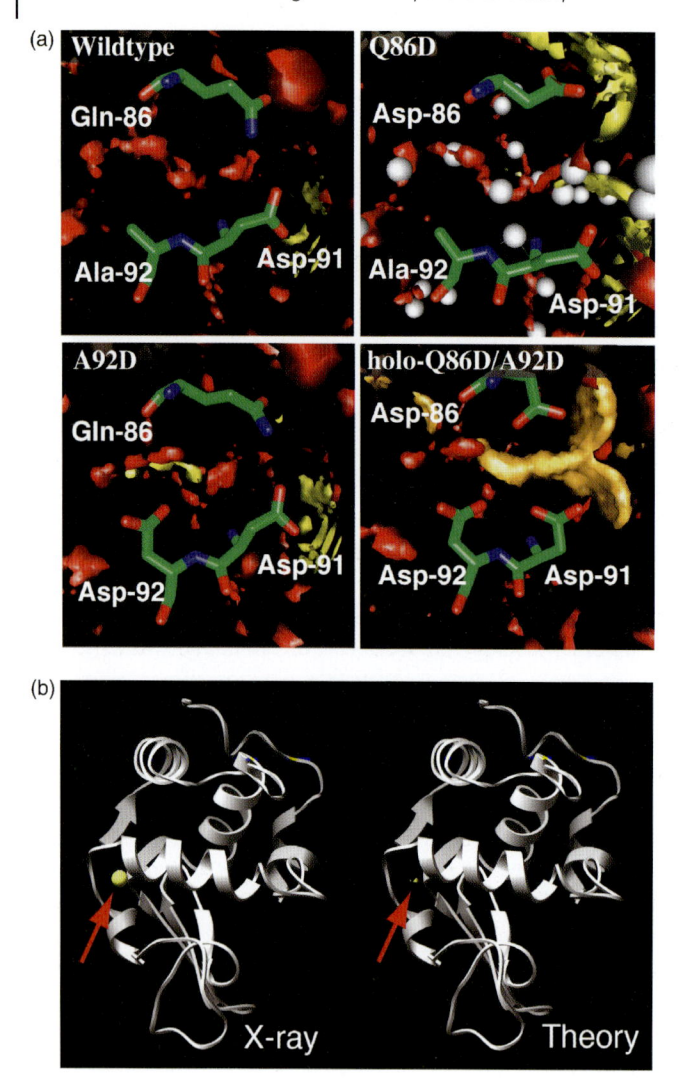

Figure 2.4 (a) Selective ion binding by human lysozyme: upper left, wild type; upper right, Q86D; lower left, A92D; lower right, Q86D/A92D. (b) Comparison of the calcium binding site in the Q86D/A92D mutant detected by X-rays and the 3D-RISM theory.

molecules have been detected, shown as gray spheres. There is a small difference between the theory and the experiment with respect to the crystallographic water bound to the backbone of Asp-91. The theory does not reproduce the water molecule for reasons that are unclear. Except for this difference, the theoretical prediction is consistent with the experimental finding, especially that the protein with the wild-type sequence binds neither Na^+ nor Ca^{2+}.

The A92D mutant in NaCl solution shows a conspicuous distribution of Na^+ ion bound in the recognition site, which is in accord with the experiment (lower left panel of Figure 2.4a). The Na$^+$ ion is apparently bound to the carbonyl oxygen atoms of Asp-92, and is distributed outside of the cleft. There is a water distribution observed in the active site, but the shape of the distribution is entirely changed from that in the wild type. The distribution indicates that the Na$^+$ ion bound in the active site is not bare, but is accompanied by hydrating water molecules. The mutant does not show any indication of binding K$^+$ ion (results not shown). This suggests that the A92D mutant discriminates Na$^+$ binding from K$^+$ binding. This finding demonstrates the capability of the 3D-RISM theory to determine ion selectivity in protein interactions.

The lower right panel in Figure 2.4a shows the distributions of Ca^{2+} ions and of water oxygen in the ion binding site of the holo-Q86D/A92D mutant. The mutant is known experimentally as a calcium binding protein. The protein, in fact, exhibits a strong calcium binding activity, as can be seen in Figure 2.4b. The calcium ion is recognized by the carboxyl groups of the three Asp residues, and is distributed around the oxygen atoms. The water distribution at the center of the triangle made by the three carbonyl oxygen atoms is reduced dramatically, which indicates that the Ca^{2+} ion is coordinated by the oxygen atoms directly, not with water molecules in between. The Ca^{2+} ion, however, also involves interactions with water molecules, as indicated by a persistent water distribution that is observed at least at two positions where original water molecules were located in the wild type of the protein.

2.6
Water Molecules Identified as a Substrate for Enzymatic Hydrolysis of Cellulose

Among the technologies for utilizing solar energy, the decomposition of cellulose due to enzymatic hydrolysis is raising the highest expectations, because the associated biological resource on Earth is essentially inexhaustible, and the reaction proceeds under natural conditions without using precious metals as a catalyst. However, there is a high barrier for such technology to become established as the ultimate substitute for fossil fuel, because of the limited efficiency of the enzyme. In order to improve the enzyme efficiency, one has to clarify the mechanism of the enzymatic hydrolysis reaction. We have investigated the problem based on the 3D-RISM theory, taking a cellulase (Cel44A)–cellohexaose complex as an example [34].

There are two models proposed by experimentalists for the mechanism of the enzymatic hydrolysis reaction of cellulose, namely the inverting and retention processes, which can be distinguished by the distance between the two catalytic residues, and by the position of a water molecule as the substrate for the reaction. In our particular example of the Cel44A–cellohexaose complex, the distance is ~5.5 Å for the retention process, whereas that for the inverting process is ~10.0 Å. The water molecule in the inverting process can make hydrogen bonds with only one of the catalytic residues owing to the large separation between the residues, whereas the water molecule can make hydrogen bonds with both catalytic residues in the retention process (see Figure 2.5).

Figure 2.5 Schematic description of enzymatic hydrolysis of cellulose.

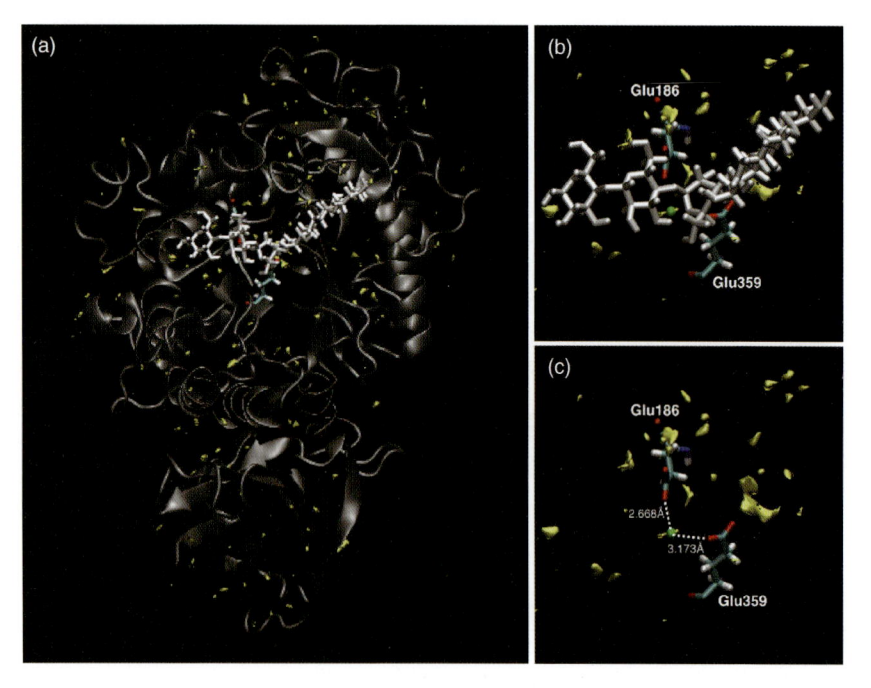

Figure 2.6 (a) The distribution ($g(r) > 8$) of water molecules (yellow spots) inside the Cel44A–cellohexaose complex. (b), (c) Distribution of water around the active site of the complex.

Shown in Figure 2.6a are the 3D-RISM results for the distribution ($g(\mathbf{r}) > 8$) of water molecules (yellow spots) inside the Cel44A–cellohexaose complex. The white "licorice" structure presents the cellohexaose, while the colored "licorice" shows the active sites of the enzyme, Glu186 and Glu359. Many water molecules can be seen inside the complex. Among the distributions, a distinctly high peak ($g(r) > 16$) was observed near the active sites as depicted by green spheres in Figure 2.6b and c. The water molecule is apparently making hydrogen bonds with the two catalytic residues, Glu186 and Glu359. We have identified the water molecule (colored green) as the substrate of the reaction, since the peak is distinctly high among other spots. This is clear support for the retention mechanism as explained above.

2.7
CO Escape Pathway in Myoglobin

Myoglobin (Mb) is a globular protein that has an important biological function, oxygen storage. Owing to its biochemical function, many researchers have made intensive efforts to identify the escape pathway of the ligand, both experimentally and theoretically. Thirty years ago, Perutz *et al.* proposed that the entry and exit pathways of carbon monoxide (CO) into the active site of Mb involves rotation of the distal histidine to form a short and direct channel between the heme pocket and solvent [35]. It has been well recognized that there are several intermediate states separated by activation barriers along the escape pathway, which are referred to as "Xe sites." Four major Xe sites are known, which are called Xe1, Xe2, Xe3, and Xe4. The experimental results indicate that CO spends some time at the Xe sites before escaping to the solvent at room temperature. In mixed Xe and CO solutions, the difference in affinity between Xe and CO for each Xe trapping site makes the CO escape pathway different depending on the Xe concentration. It is believed that the dissociated CO escapes to the solvent through the Xe1 site predominantly under Xe-free conditions. On the other hand, CO escapes through the Xe4 site in a Xe-rich solution. Terazima and coworkers have measured the partial molar volume (PMV) change of the system along the pathway of the CO escape process in Xe solution based on the transient-grating (TG) method [36, 37]. They hypothesized that the intermediate of the pathway is through the Xe4 site, since the experiments were carried out under Xe-rich conditions.

In this section, we apply the 3D-RISM theory to investigate the CO escape pathway of Mb. As mentioned above, the ligand dissociation process of Mb occurs from heme to solvent through some specific cavities. We discuss the CO escape pathway from Mb in terms of the 3D distribution functions of the ligand and the PMV change along the pathway. In this study, the geometry optimization of the Mb structure is also performed by using the analytical formula for the free-energy gradient of the 3D-RISM framework [38].

2.7.1
Effect of Protein Structure on the Distribution of Xe

We first investigate the distribution of Xe in the four Xe sites by means of the 3D-RISM theory. The experimental results from thermodynamic analyses and crystallographic studies for the absorption of Xe by myoglobin suggest that the Xe1 site is 10 times more populated than the Xe4 site in Xe-rich conditions [37].

In order to examine the effects of Mb structure on the Xe distribution, four different structures of Mb are examined [39, 40]:

1) the 1MBC structure from PDB, which was determined under Xe-free conditions (referred to as "Xe-free structure");
2) the 1J52 structure from PDB, which was determined under Xe-rich conditions (referred to as "Xe-rich structure");
3) the structure optimized with 3D-RISM in Xe-rich conditions starting from 1MBC (referred to as "RISM-optimized structure"); and
4) the structure optimized with PCM starting from 1MBC (referred to as "PCM-optimized structure").

The geometry optimization of Mb to obtain the RISM-optimized structure is performed by means of the analytical free-energy gradient with the 3D-RISM theory in aqueous solution of Xe. The probability of finding Xe atoms in the Xe sites is discussed in terms of the 3D distribution function of solvent Xe around and inside Mb. Note that, since 1J52 includes explicit Xe atoms, those atoms were excluded before the 3D-RISM calculation.

The distributions of Xe in the cavities of Mb, evaluated by the 3D-RISM theory, are shown in Figure 2.7 using the vertical sections and the isosurface representation, in which the region with $g > 3$ is indicated. Figure 2.7 focuses on the region around the heme. The 3D-RISM calculations employed four different Mb structures as mentioned above, namely "Xe-free," "Xe-rich," "RISM-optimized," and

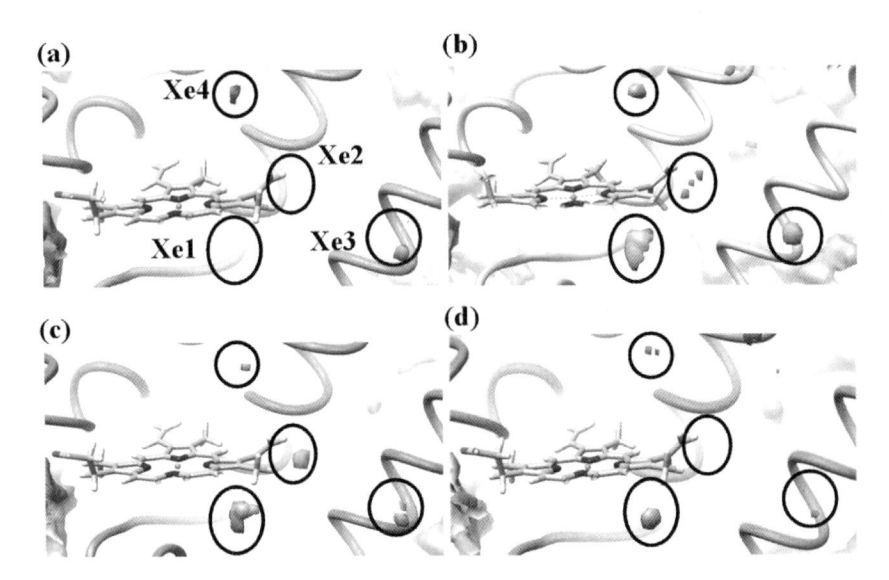

Figure 2.7 Xe distribution around heme in Xe sites using four different X-ray structures: (a) Xe-free structure; (b) Xe-rich structure; (c) RISM-optimized structure; and (d) PCM-optimized structure. The threshold of the 3D distribution functions is 3.0 for Xe.

"PCM-optimized." As is obvious from Figure 2.7, the Xe-free structure does not show any distribution in Xe1 and Xe2 sites, whereas the RISM-optimized structure has the ligand distribution in all Xe sites. The PCM-optimized structure shows smaller distribution of Xe in all cavities. This may be because the effects on the protein structure from solvent molecules inside the protein could not be correctly accounted for by PCM.. The distributions of Xe in the Xe3 and Xe4 sites are not very different.

We analyzed the conformational change including the size of the cavity, which is shown in Table 2.1. The root-mean-square deviations (RMSD) from the Xe-free structure are shown in Table 2.2. These results indicate that the conformational change of side chains is a major factor to determine the size of the cavity, since the change is larger than that of the main chain. Hence, the main difference in the protein structures between the X-ray and the RISM-optimized structures is in the size of the cavities. The Xe1 site is expanded during the 3D-RISM optimization, whereas the Xe4 site is contracted from the Xe-free structure.

2.7.2
Partial Molar Volume Change Through the CO Escape Pathway of Myoglobin

Shown in Figure 2.8 are the coordination numbers (CNs) of Xe and CO in each Xe site of the RISM-optimized structure, which are obtained from the distribution

Table 2.1 Cavity size (Å) evaluated in terms of the inner diameter of the effective sphere that is inscribed on the van der Waals surface.

Site	Xe-free	Xe-rich	RISM-optimized	PCM-optimized
Xe1	2.3	2.9	3.1	2.4
Xe2	2.2	2.5	2.5	2.2
Xe3	2.7	2.6	2.6	2.5
Xe4	2.8	2.6	2.4	2.5

Table 2.2 Root-mean-square deviations between two models of column and row. Hydrogen and heme were ignored in these calculations.

Models		Xe-rich	RISM-optimized	PCM-optimized
Xe-free	Main chain	0.666	0.463	0.499
	Side chain	1.471	0.887	0.978
Xe-rich	Main chain	–	0.658	0.712
	Side chain	–	1.440	1.430
RISM-optimized	Main chain	–	–	0.484
	Side chain	–	–	0.716

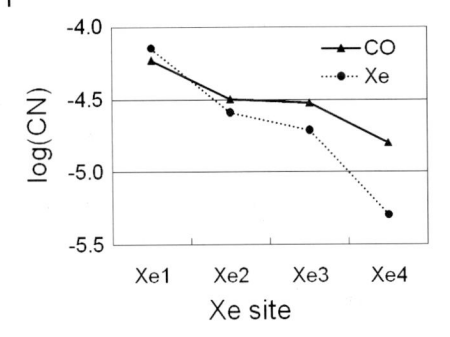

Figure 2.8 Coordination numbers (CN) of CO and Xe molecules in each Xe site, calculated from the radial distribution function at each Xe cavity.

function. Since the RISM-optimized structure was obtained in a Xe–water mixture, these results are regarded as the ligand affinity in a Xe-rich condition. As seen in the figure, the CNs of both ligands show similar behavior: namely the Xe1 site has the highest affinity, while the affinity for the Xe4 site is low. Note that the Xe affinity for the Xe1 site is larger than the CO affinity. Hence the difference of ligand affinities between Xe and CO becomes largest between the Xe1 and Xe4 sites. This result indicates that CO prefers the Xe4 site to Xe1 in a Xe-rich condition. Although the affinity of the CO ligand is largest for the Xe1 site, CO trapping in the Xe1 site would be obstructed by the high Xe affinity. These results are consistent with the experimental observations described above.

In order to evaluate the PMV change through the CO escape pathway, the PMV of the intermediate structure of Mb should be calculated. Here, "intermediate structure" means that one CO molecule is trapped in a specific Xe site of Mb. Therefore, the CO molecule is regarded as part of the solute. The model is referred to as the "explicit ligand model." In the present study, the position of the trapped CO molecule is determined by the peaks of the 3D distribution functions.

We calculated the PMV of the CO associated (MbCO), intermediate (Mb:CO) and CO dissociated (Mb + CO) states. The PMV and its changes are shown in Table 2.3. The PMV of Mb:CO(Xe1) is smaller than that of Mb:CO(Xe4). Because the cavity size of the Xe1 site is large enough, it can accommodate both CO and a water molecule at the same time. On the other hand, in the case of the Xe4 site, there is no space to include a water molecule when the cavity is occupied by a CO molecule. However, a void space is created in the cavity, because the size of the Xe4 site is larger than a CO molecule. This void space contributes to the increase in PMV of Mb:CO(Xe4). The change of PMV through the Xe4 site shows excellent agreement with the experimental data by Terazima *et al.*, which supports Terazima's conjecture regarding the escaping pathway of CO from Mb.

Table 2.3 The PMVs of each model and their changes. Here ΔV_1 represents the PMV change from MbCO to Mb:CO (Xe4); ΔV_2 represents the PMV change from Mb:CO (Xe4) to Mb + CO; and ΔV_{total} represents the PMV change from MbCO to Mb + CO.

Models	PMV (cm^3 mol^{-1})
MbCO	9029.0
Mb:CO (Xe1)	9030.4
Mb:CO (Xe4)	9032.8
Mb + CO	9019.6

	Xe1	Xe4	Expt. [37]
ΔV_1	1.4	3.8	3 ± 1
ΔV_2	−10.8	−13.2	−12.6 ± 1.0
ΔV_{total}	−9.4	−9.4	−10.7 ± 0.5

2.8 Perspective

In this chapter, we have presented model-free "solvent modeling" based on the statistical mechanics of liquids, or the RISM/3D-RISM theory. The theory has exhibited a remarkable ability to predict molecular recognition by biomolecules, which is the most fundamental process in life phenomena. We believe that our results are convincing enough to replace continuum models with a theory based on statistical mechanics.

There are few problems yet to be solved for the complete theory of model-free "solvent modeling." One of those is to take the structural fluctuation of biomolecules into consideration. We have presented one of our efforts to consider protein fluctuations by coupling the 3D-RISM theory simply with Newtonian dynamics [41]. The "Hamiltonian" is constructed from a sum of the conformational energy of the protein and the solvation free energy evaluated from the RISM/3D-RISM theory. The protein dynamics, therefore, takes place on the free-energy surface, in which solvent is always in equilibrium with protein conformation. The dynamics is similar to the Ginzberg–Landau dynamics in spirit, but is distinct from the latter, because the free-energy surface in our dynamics has a microscopic description instead of a phenomenological description from earlier theories. We believe that the theory can be applied successfully to the slow dynamics of macromolecules, such as protein folding, in which detailed solvent dynamics does not play an important role.

On the other hand, there are a number of phenomena in which detailed solvent dynamics and its correlation with the dynamics of the protein *do* play essential roles for the functions of the biomolecule. A typical example are water and ion channels, in which the functions of the protein, and the conductance of ions and water through the channels, are determined by the dynamics of both the biomolecules and the solvent species. In such a case, the dynamics on the free-energy surface described above is insufficient. We have to describe the dynamics of the protein and the solvent on an equal footing. To the best of our knowledge, the generalized Langevin equation is the only theory that meets such a requirement. Efforts to combine the RISM/3D-RISM theory with the generalized Langevin equation in order to realize the correlated dynamics of protein and solvent are in progress in our group [42].

Any of those methods that we have been developing requires the 3D-RISM equations to be solved for many conformations of a protein. Currently, it takes several hours to solve the 3D-RISM equation for *one* conformation of a protein with about 200 residues, using the best available workstation in our group, which has two quad-core 3.0 GHz Xeon CPUs with 64 GB RAM. So, it is not feasible at present to solve such problems as that stated above, even though we might have succeeded in building the theoretical framework. However, there is good news in this respect. A national project to build a next-generation computer is under way in Japan. The new machine to be built and a 3D-RISM program well tuned to the computer will hopefully solve the most important problems in the life sciences.

References

1 Watson, J.D., et al. (eds) (1987) *Molecular Biology of the Gene*, Benjamin/Cummings, Menlo Park, CA.
2 Gurney, R.W. (1953) *Ionic Processes in Solution*, Dover, New York.
3 Prock, A. and McConkey, G. (eds) (1962) *Topics in Chemical Physics: Based on the Harvard Lectures of Peter J.W. Debye*, Elsevier, Amsterdam.
4 Kirkwood, J.G. (1934) Theory of solutions of molecules containing widely separated charges with special application to zwitterions. *J. Chem. Phys.*, **2**, 351.
5 Onsager, L. (1936) Electric moments of molecules in liquids. *J. Am. Chem. Soc.*, **58**, 1486.
6 Frohlich, H. (1958) *Theory of Dielectrics*, Clarendon Press, Oxford.
7 Born, M. (1920) Volumen und Hydratationswärme der Ionen. *Z. Phys.*, **1**, 45.
8 Marcus, R.A. (1964) Chemical and electrochemical electron-transfer theory. *Annu. Rev. Phys. Chem.*, **15**, 155.
9 Tapia, O. and Bertran, J. (eds) (1996) *Solvent Effects and Chemical Reactivity*, Kluwer Academic, Dordrecht.
10 Hill, T. (1987) *Statistical Mechanics*, Dover, New York.
11 McQuarrie, D.A. (1976) *Statistical Mechanics*, Harper and Row, New York.
12 Hansen, J.-P. and McDonald, I.R. (2006) *Theory of Simple Liquids*, 3rd edn, Academic, London.
13 Chandler, D. and Andersen, H.C. (1972) Optimized cluster expansions for classical fluids. II. Theory of molecular liquids. *J. Chem. Phys.*, **57**, 1930.
14 Hirata, F. and Rossky, P.J. (1981) An extended Rism equation for molecular polar fluids. *Chem. Phys. Lett.*, **83**, 329.
15 Hirata, F. (ed.) (2003) *Molecular Theory of Solvation (Understanding Chemical Reactivity)*, Kluwer, Dordrecht.
16 Kovalenko, A. and Hirata, F. (1998) Three-dimensional density profiles of water in contact with a solute of arbitrary shape: a RISM approach. *Chem. Phys. Lett.*, **290**, 237.
17 Beglov, D. and Roux, B. (1997) An integral equation to describe the solvation of polar molecules in liquid water. *J. Phys. Chem. B*, **101**, 7821.
18 Cortis, C.M., Rossky, P.J., and Friesner, R.A. (1997) A three-dimensional reduction of the Ornstein–Zernicke equation for molecular liquids. *J. Chem. Phys.*, **107**, 6400.

19 Imai, T., Kovalenko, A., and Hirata, F. (2005) Partial molar volume of proteins studied by the three-dimensional reference. Interaction site model theory. *J. Phys. Chem. B*, **109**, 6658.

20 Imai, T., Hiraoka, R., Kovalenko, A., and Hirata, F. (2005) Water molecules in a protein cavity detected by a statistical-mechanical theory. *J. Am. Chem. Soc.*, **127**, 15334.

21 Kovalenko, A. and Hirata, F. (1999) Self-consistent description of a metal–water interface by the Kohn–Sham density functional theory and three-dimensional reference interaction site model. *J. Chem. Phys.*, **110**, 10095.

22 Singer, S.J. and Chandler, D. (1985) Free energy functions in the extended RISM approximation. *Mol. Phys.*, **55**, 621.

23 Kirkwood, J.G. and Buff, F.D. (1951) The statistical mechanical theory of solutions. *J. Chem. Phys.*, **19**, 774.

24 Imai, T., Kinoshita, M., and Hirata, F. (2000) Theoretical study for partial molar volume of amino acids in aqueous solution: implication of ideal fluctuation volume. *J. Chem. Phys.*, **112**, 9469.

25 Harano, Y., Imai, T., Kovalenko, A., and Hirata, F. (2001) Theoretical study for partial molar volume of amino acids and polypeptides by the three-dimensional reference interaction site model. *J. Chem. Phys.*, **114**, 9506.

26 Sato, H., Hirata, F., and Kato, S. (1996) Analytical energy gradient for the reference interaction site model multiconfigurational self-consistent-field method: application to 1,2-difluoroethylene in aqueous solution. *J. Chem. Phys.*, **105**, 1546.

27 Gusarov, S., Ziegler, T., and Kovalenko, A. (2006) Self-consistent combination of the three-dimensional RISM theory of molecular solvation with analytical gradients and the Amsterdam density functional package. *J. Phys. Chem. A*, **110**, 6083.

28 Yoshida, N. and Hirata, F. (2006) A new method to determine electrostatic potential around a macromolecule in solution from molecular wave functions. *J. Comput. Chem.*, **27**, 453.

29 Imai, T., Hiraoka, R., Kovalenko, A., and Hirata, F. (2007) Locating missing water molecules in protein cavities by the three-dimensional reference interaction site model theory of molecular solvation. *Proteins Struct. Funct. Bioinform.*, **66**, 804.

30 Wilson, K.P., Malcolm, B.A., and Matthews, B.W. (1992) Structural and thermodynamic analysis of compensating mutations within the core of chicken egg white lysozyme. *J. Biol. Chem.*, **267**, 10842.

31 Yoshida, N., Phongphanphanee, S., Maruyama, Y., Imai, T., and Hirata, F. (2006) Selective ion-binding by protein probed with the 3D-RISM theory. *J. Am. Chem. Soc.*, **128**, 12042.

32 Yoshida, N., Phongphanphanee, S., and Hirata, F. (2007) Selective ion binding by protein probed with the statistical mechanical integral equation theory. *J. Phys. Chem. B*, **111**, 4588.

33 Kuroki, R. and Yutani, K. (1998) Structural and thermodynamic responses of mutations at a Ca^{2+} binding site engineered into human lysozyme. *J. Biol. Chem.*, **273**, 34310.

34 Ikuta, Y., Karita, S., Kitago, Y., Watanabe, N., and Hirata, F. (2008) A water molecule identified as a substrate of enzymatic hydrolysis of cellulose: a statistical-mechanics study. *Chem. Phys. Lett.*, **465**, 279.

35 Perutz, M.F. and Matthews, F.S. (1966) An x-ray study of azide methaemoglobin. *J. Mol. Biol.*, **21**, 199.

36 Sakakura, M., Yamaguchi, S., Hirota, N., and Terazima, M. (2001) Dynamics of structure and energy of horse carboxymyoglobin after photodissociation of the carbon monoxide. *J. Am. Chem. Soc.*, **123**, 4286.

37 Nishihara, Y., Sakakura, M., Kimura, Y., and Terazima, M. (2004) The escape process of carbon monoxide from myoglobin to solution at physiological temperature. *J. Am. Chem. Soc.*, **126**, 11877.

38 Kiyota, Y., Hiraoka, R., Yoshida, N., Maruyama, Y., Imai, T., and Hirata, F. (2009) Theoretical study of CO escaping pathway in myoglobin with the 3D-RISM theory. *J. Am. Chem. Soc.*, **131**, 3852.

39 Tilton, R.F., Kuntz, I.D., and Petsko, G.A. (1984) Cavities in proteins: structure of a

metmyoglobin–xenon complex solved to 1.9 Å. *Biochemistry*, **23**, 2849.

40 Cohen, J., Arkhipov, A., Braun, R., and Schulten, K. (2006) Imaging the migration pathways for O_2, CO, NO, and Xe inside myoglobin. *Biophys. J.*, **91**, 1844.

41 Miyata, T. and Hirata, F. (2008) Combination of molecular dynamics method and 3D-RISM theory for conformational sampling of large flexible molecules in solution. *J. Comput. Chem.*, **29**, 871.

42 Kim, B.-S., Chong, S.H., Ishizuka, R., and Hirata, F. (2008) An attempt toward the generalized Langevin dynamics simulation. *Condens. Matter Phys.*, **11**, 179.

3
Developing Force Fields From the Microscopic Structure of Solutions: The Kirkwood–Buff Approach

Samantha Weerasinghe, Moon Bae Gee, Myungshim Kang, Nikolaos Bentenitis, and Paul E. Smith

3.1
Introduction

What is the structure of a solution? How can we understand and predict the effects of solvation at the atomic level? These are important issues that need to be addressed if we are to develop accurate and useful models of solution mixtures. In principle, computer simulation can provide this information in exquisite detail. However, the results of computer simulations are determined by the quality of the force field (FF), and the extent of sampling achieved during the simulation. The latter is not usually a problem for the majority of solution mixtures using currently available computational resources. On the other hand, the quality of the force fields used to represent solution mixtures is debatable. Current force fields for small molecules reproduce many of the properties of solution quite well. However, recently, we have found that they do a relatively poor job in reproducing other important characteristics of solution mixtures, in particular, the activity of a cosolvent in a solvent [1–4].

Our interest in developing improved force fields for solution mixtures arose from studies concerning the interactions of cosolvents with biomolecules in solution. We have simulated the interactions of small solutes and biomolecules with several simple salts in an effort to understand the underlying features of the Hofmeister series [5–7]. These initial simulations used a mix of biomolecular and small-molecule (salt) force fields. Some interesting interactions between peptides and salt ions were observed during the simulations, including possibly unphysical behaviors as described most recently for the AMBER ion parameters. However, it quickly became clear that one had to trust the underlying force fields, as there was no quantitative way to relate what was observed during the simulations to real experimental data. Experimental structural data are not typically available, as many of the salts form only weak interactions with biomolecules, that is, there are no well-defined binding sites to investigate [8]. Exactly how one could relate the computer simulation data to the corresponding experimental thermodynamic data was also unclear.

Modeling Solvent Environments. Applications to Simulations of Biomolecules. Edited by Michael Feig
Copyright © 2010 WILEY-VCH Verlag GmbH & Co. KGaA, Weinheim
ISBN: 978-3-527-32421-7

It was at this point that we turned to Kirkwood–Buff (KB) theory. KB theory is an exact theory of solutions that provides a link from radial distribution functions (RDFs) between the various species present in the solution to the thermodynamic properties of the solution [9, 10]. In this way one has a direct connection between structure, in terms of the RDFs, and thermodynamics, in terms of changes in chemical potentials. This link has been recognized by others and has been used to develop the concept of preferential solvation [11, 12]. Using KB theory to determine the properties of solution mixtures involving a variety of cosolvents of biomolecular interest, we soon discovered that the majority of the force fields available at that time performed poorly when trying to reproduce cosolvent activities in water [1–3]. Unfortunately, as we shall see later, this is one of the most important solution properties required to accurately reproduce the thermodynamic effects of cosolvents on peptides and proteins. Furthermore, the errors were not trivial and indicated an incorrect balance between cosolvent self-association and solvation of the cosolvent. Therefore, one has no choice but to improve the quality of the force fields in order to study these types of effects.

Our ultimate goal is to provide a force field for peptides and proteins based on the properties of solution mixtures using KB theory as a guide during the parameterization process. Hence, we have labeled this as the Kirkwood–Buff derived force field (KBFF) approach. The advantages and disadvantages of this particular type of approach will be discussed later. Here we outline our progress to date and discuss some of the reasons, in our opinion, for the deficiencies in current force fields. We, and others [13, 14], believe that this type of approach has the potential for generating highly accurate biomolecular force fields and represents an exciting new direction in force-field development. These new force fields should provide a more realistic picture of the microscopic structure of solution mixtures, which can then be used to develop our understanding of the properties of a variety of solutions.

3.2
Biomolecular Force Fields

The current discussion will focus on force fields typically designed for the study of biomolecular systems, as they have a somewhat unique set of problems to overcome. There are many force fields in the literature, which have been designed to study a variety of different systems or properties by computer simulation. While the quality of the force fields has consistently improved, significant advances are still possible and are required to increase the reliability of computer simulation results [15–17]. Biomolecular simulations are now advancing to the stage where quantitative data on folding equilibria is obtainable [18]. Hence, there is an increased need to improve the accuracy of force fields to further facilitate exact comparisons with experiment. Force-field improvements will be essential to model the delicate balance between the large forces that favor folding, and the equally large forces that favor the unfolded state. In addition, it is well known in the drug design field that, while docking studies of host–guest complexes generally result in very reasonable structures, the corresponding binding free energies are less satisfactory and force-field dependent [19]. This is generally attributed to deficiencies in current force fields [20].

The most popular approach for improving current force fields seems to be the inclusion of explicit polarization effects in the simulations [21, 22]. While this is a welcome step forward, the results still have to be carefully validated, and even then this approach does not solve the problem completely. For instance, the majority of folding simulations now involve implicit solvation models to help increase the time scales accessible to simulation [23]. Here, explicit polarization effects are difficult to include. Furthermore, explicit treatment of polarization effects is computationally more demanding. Hence, there is still a desire for simple effective charge force fields that accurately model the interactions present in biomolecular systems, especially in the field of rational drug design.

Protein force fields use simple equations to describe intra- and intermolecular interactions so as to provide an efficient determination of the energy and corresponding atomic forces [24–28]. This is required to ensure a reasonable degree of sampling (many nanoseconds) for these usually large (10 000–200 000 atoms) simulations. Hence, effective solution-phase partial atomic charges are the most common approximation, with the significantly slower explicit treatment of polarization being less prevalent [21, 22]. Obviously, care must therefore be used when studying systems for which large changes in polarization might be expected – ions at liquid surfaces [29], for example. All protein force fields typically use the transferable and additive intermolecular potential approximations (see Equation 1.9). Calculations and simulations on small fragment molecules are performed and used to refine parameters. It is then assumed that the description of a large molecule, such as a protein, can be considered as just the sum of the smaller fragments. The small-fragment calculations usually involve quantum-mechanical determinations of the gas-phase charge distributions. These are then sometimes scaled to mimic polarization effects in the condensed phase [24]. We note that this is an efficient, but very approximate, model for polarization effects in polar solvents such as water. Finally, simulations of the pure liquid are then performed and compared with experiment. In some cases, further tests are performed for a mixture of the fragment with water if possible. However, it is often assumed that the properties of the fragment and water mixtures will be reasonable if the properties of the two pure liquids are in good agreement with experiment.

In attempting to generate an accurate protein force field, it is difficult to determine if one has succeeded. Clearly, if the peptide–peptide group interactions are too strong, then tests on the native states of proteins will probably produce good comparisons with experiment – root-mean-square deviation (RMSD) from the crystal structure, and so on. However, they will not accurately reproduce conformational equilibria, as they will tend to favor the folded state, and will therefore be of questionable value in protein folding studies. Here we present an approach that helps to ensure that these types of interactions are correctly modeled by the force field, therefore helping to avoid the possibility of the peptide or other groups displaying too much self-aggregation or too much solvation.

3.3
Examples of Problems with Current Force Fields

Before describing our approach, we present some evidence that there are significant problems with current force fields (FFs) that need to be addressed. The earliest indications came from studies performed by ourselves and others [1, 2, 30, 31]. There was evidence that cosolvents tend to over-aggregate in solution. More recently, other examples have been observed, and some are displayed in Figure 3.1. Cheatham and coworkers observed the formation of salt aggregates resembling salt crystals during their simulations of NaCl and KCl around nucleic acids [32]. In addition, Cerutti et al. have observed high degrees of aggregation in ammonium sulfate solutions when they attempted to mimic the conditions in real

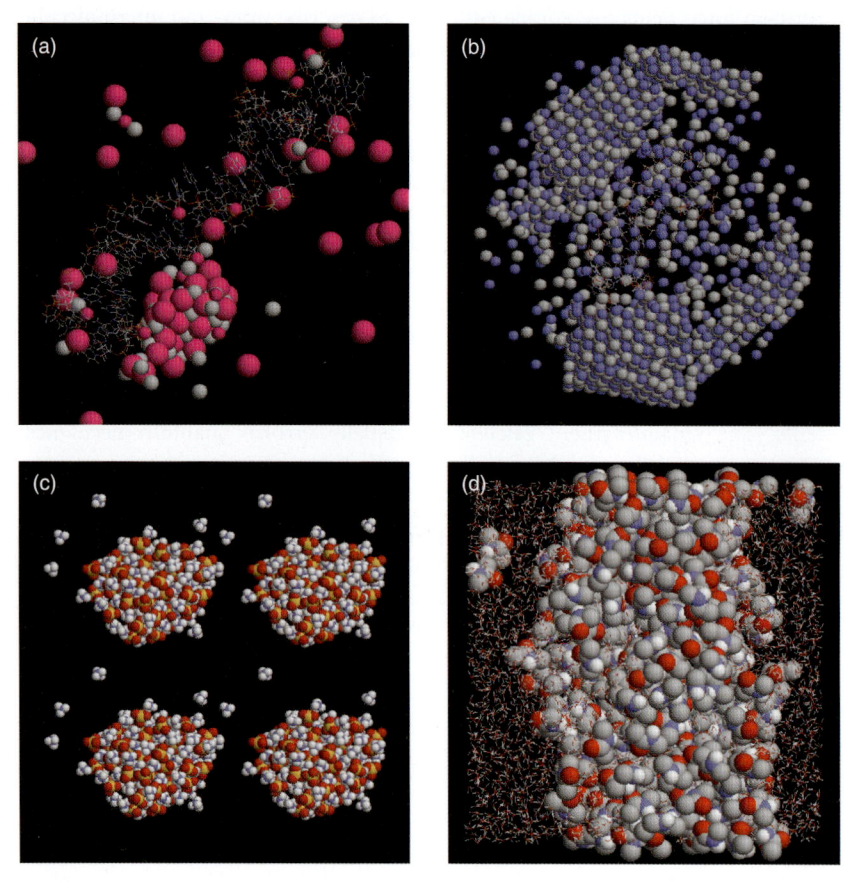

Figure 3.1 Some indications of the problems with current force fields. (a) A simulation of RNA in KCl solution using the AMBER FF. (b) A simulation of DNA in NaCl solution using the AMBER FF. (c) A replicated configuration obtained from a simulation of ammonium sulfate solution. (d) Phase separation of NMA (solid) in water as described by the GROMOS 45a3 force field. All display an excessive aggregation of the cosolvent.

protein crystals [33]. Our recent studies of N-methylacetamide (NMA), a model for the peptide group, and water mixtures provided by different protein FFs have indicated a range of behaviors – the most significant being a phase separation [34].

Some of these examples obviously provide an incorrect picture of the microstructure of common solutions, which is evident on visual analysis. However, in many cases the problems may not be so obvious. Therefore, we need an approach to determine when the solution distributions are reasonable, and when there are problems with the balance of interactions between the different species present in solution. One way of quantifying the relative distribution of different species in solution and comparing with experimental data is through the use of Kirkwood–Buff theory.

3.4 Kirkwood–Buff Theory

Kirkwood–Buff theory was first published in 1951 [9]. The theory lay relatively unused until 1977 when Ben-Naim illustrated how one can use KB theory to analyze experimental data on binary solution mixtures [35]. This has been extended to include three- and higher-component solutions [36, 37]. Subsequently, KB theory has been used extensively in the chemistry and chemical engineering fields to provide information on intermolecular distributions and preferential solvation in solution [12, 38].

The specific advantages of KB theory include the following:

1) It is an exact theory.
2) It can be applied to any stable solution mixture involving any number of components.
3) It can be applied to molecules of all sizes and complexity.
4) It does not assume pairwise additivity of interactions.
5) It is well suited for the analysis of computer simulation data.

Kirkwood–Buff theory provides relationships between particle distribution functions in the grand canonical ensemble (μVT) and derivatives of the chemical potentials of all species involved in either open, semi-open or closed systems [9, 39, 40]. The primary quantity of interest is the Kirkwood–Buff integral (G_{ij}) between species i and j given by [10]

$$G_{ij} = G_{ji} = 4\pi \int_0^\infty [g_{ij}^{\mu VT}(r) - 1] r^2 dr \qquad (3.1)$$

where g_{ij} is the corresponding center-of-mass based radial distribution function (RDF). The above integral relates the distribution of j molecules around a central i molecule. An excess coordination number can be defined as $N_{ij} = \rho_j G_{ij} \neq N_{ji}$, where $\rho_j = n_j/V$ is the number density of species j. The N_{ij} values quantify the excess number of j molecules observed around an i molecule in the open system, above that observed within an equivalent volume of a bulk reference solution at the same chemical potential [41]. It is a measure of how the addition of a single i molecule affects the distribution of i and j molecules around it with reference to the corresponding bulk distribution. For small molecules (certainly not proteins), a positive value of N_{ij} typically indicates an increase in the local density of j around i above that of their bulk solution ratios, and vice versa. This can be viewed as the result of some favorable net interaction or affinity between the two species.

As the KB integrals correspond to distributions in open systems, the expressions for solution properties in open systems are generally rather simple. These expressions can then be converted to other ensembles using a series of thermodynamic transformations. The expressions become more complicated as we move to closed systems and/or increase the number of components in the solution. A general matrix formulation is available for chemical potential derivatives in closed systems [10]. Recently, we have suggested a stepwise transformation process that provides expressions for preferential interaction parameters in semi-open and closed systems in a simple manner [40, 42, 43].

3.5
Applications of Kirkwood–Buff Theory

KB theory can be applied to study a wide variety of solution behaviors. We have recently reviewed the application of KB theory to biological systems [44]. Our desire to develop a protein FF has involved the study of small cosolvents, representative of peptide fragments, in solution mixtures with water. KB theory applied to binary solution mixtures of a cosolvent (2) in a solvent (1) provides the following expressions for derivatives of the chemical potential or activity of the cosolvent (μ_2 or a_2) with respect to molarity (ρ) or mole fraction (x), the partial molar volumes $\overline{V}$, and the isothermal compressibility (κ_T) at a given temperature (T) and pressure (P) [9, 10]:

$$a_{22} = \beta\left(\frac{\partial \mu_2}{\partial \ln \rho_2}\right)_{T,P} = \left(\frac{\partial \ln a_2}{\partial \ln \rho_2}\right)_{T,P} = \frac{1}{1+\rho_2(G_{22}-G_{12})} \tag{3.2}$$

$$\beta\left(\frac{\partial \mu_2}{\partial \ln x_2}\right)_{T,P} = \frac{\rho_1+\rho_2}{\eta} \tag{3.3}$$

$$\overline{V_1} = \frac{1+\rho_2(G_{22}-G_{12})}{\eta} \quad \text{and} \quad \overline{V_2} = \frac{1+\rho_1(G_{11}-G_{12})}{\eta} \tag{3.4}$$

$$RT\kappa_T = \frac{1+\rho_1 G_{11}+\rho_2 G_{22}+\rho_1\rho_2(G_{11}G_{22}-G_{12}^2)}{\eta} \tag{3.5}$$

where

$$\eta = \rho_1 + \rho_2 + \rho_1\rho_2(G_{11}+G_{22}-2G_{12}) \tag{3.6}$$

and $\beta = 1/RT$, with R being the gas constant.

We have mainly used the above equations as a route to changes in the cosolvent chemical potential or activity (Equations 3.2 and 3.3). In principle, one can calculate the cosolvent chemical potential from particle insertion or thermodynamic integration approaches [45]. However, typical changes in the chemical potential are usually small compared to the current precision provided by these approaches. The above expressions relate the thermodynamic properties of solution mixtures to combinations of KB integrals and number densities. Clearly, if a particular FF reproduces the KB integrals and number densities, it will also provide an accurate description of changes in the cosolvent (and solvent) activities. In addition, the relative molecular distributions in solution should be correct.

KB theory can be applied to study a variety of problems in mixed solutions. We will briefly discuss some common applications – as they indicate the importance of correctly modeling the solution mixture if reasonable simulation results are to be obtained. Our initial studies of cosolvent effects on the hydrophobic hydration of small non-polar solutes (S) resulted in the expression [30]

$$\beta\left(\frac{\partial \mu_S^*}{\partial \ln \rho_2}\right)_{T,P}^{\infty} = -\frac{\rho_2(G_{S2} - G_{S1})}{1 + \rho_2(G_{22} - G_{21})} = -\Gamma_{S2} a_{22} \qquad (3.7)$$

which is valid for infinitely dilute solutes and where μ^* is the pseudo chemical potential, as developed by Ben-Naim [10]. This was later developed into a general expression for the effect of a cosolvent on the molar solubility (S_S) of a solute at any concentration of solute, cosolvent or solvent [46]:

$$\left(\frac{\partial \ln S_S}{\partial \rho_2}\right)_{T,P,\mu_S} = \frac{G_{S2} - G_{S1}}{1 + \rho_2(G_{22} - G_{21})} \qquad (3.8)$$

One of our main interests lies in the effects of cosolvents on biomolecular equilibria. KB theory can be used to understand the change in the equilibrium constant ($K = \rho_D/\rho_N$) between native (N) and denatured (D) forms on addition of a cosolvent [47, 48],

$$\left(\frac{\partial \ln K}{\partial \ln \rho_2}\right)_{T,P}^{\infty} = (\Gamma_{D2} - \Gamma_{N2}) a_{22} \qquad (3.9)$$

and is only valid for infinitely dilute biomolecule concentrations. More recently, we have applied KB theory to the analysis of cosolvent adsorption or exclusion from interfaces. The corresponding change in surface tension (γ) can be written [49] as

$$\beta\left(\frac{\partial \gamma}{\partial \rho_2}\right)_{T,P} = -\frac{\Gamma_{2,1}}{\rho_2} a_{22} \qquad (3.10)$$

where $\Gamma_{2,1}$ is the relative surface adsorption per unit area. In all cases, the value of Γ indicates the change in the distribution of cosolvent and solvent molecules around a solute or interface from that expected according to the bulk solution distribution.

It is quite remarkable that all the above effects can be related through similar exact expressions provided by KB theory. In particular, they all involve the combination of two terms (Γ and a_{22}). The first (Γ) measures the degree of affinity of the cosolvent (and solvent) for the solute(s) or an interfacial region and determines the direction of the effect. The second (a_{22}), which must be positive, is a regulating property of the bulk solution. All of the above expressions are exact. Therefore, it is clear that one cannot correctly represent the thermodynamics of the process (left-hand side), and the affinity of the cosolvent for the solute or surface (Γ), unless the bulk solution properties (a_{22}) are accurately reproduced. This is one of the major aims of the KBFF approach.

Before leaving this section, we wish to further emphasize the relationship between the KB integrals and the microscopic structure of binary solution mixtures. It is instructive to compare with the distributions observed for ideal solutions. First, we note that the G_{ij} values are not zero for ideal solutions [39]. Second, ideality depends on the choice of concentration scale. For ideal behavior to be observed on the molar concentration scale, that is, when the molar activity coefficient is unity for all compositions, one must have $G_{22} = G_{12}$. For ideal behavior to be observed on the mole fraction scale, one must have $G_{11} + G_{22} - 2G_{12} = 0$. The latter is referred to as the symmetric ideal (SI) case [39]. A general expression for the KB integrals in SI solutions with any number of components has been presented [37]. The result for binary solutions is

$$G_{ij}^{SI} = RT\kappa_T - \overline{V}_i - \overline{V}_j + \rho_1 \overline{V}_1^2 + \rho_2 \overline{V}_2^2 \tag{3.11}$$

where the partial molar volumes refer to the pure solution components at the appropriate T and P. This can provide useful reference values for comparison, although one should be careful when interpreting deviations of the experimental KB integrals from the SI values [50]. It is clear in both cases (Equations 3.2 and 3.3) that an increase in the cosolvent activity coefficient with cosolvent concentration will be observed when $G_{12} > G_{22}$ (and G_{11}), that is, that the affinity of the cosolvent for solvent molecules is larger than the affinity for other cosolvent molecules. The opposite is also true.

3.6
The General KBFF Approach

The basic idea behind the KBFF approach is to use the experimental KB integrals for solution mixtures as a guide during the parameterization procedure. The KB integrals provide additional data for the fitting procedure. They also provide a route to changes in activity – a key thermodynamic property of solution mixtures. Furthermore, in our experience the KB integrals are sensitive measures of the solution distributions that arise from the intermolecular interactions [2, 3, 51]. For example, it is quite common for different FFs to provide equal accuracy for the densities, diffusion constants, dielectric properties, and so on of solution mixtures or pure solvents. However, they will usually display quite different KB integrals. To date we have published force fields for urea [3], acetone [4], NaCl [52], guanidinium chloride [53], methanol [54], and several amides [55].

The philosophy we have followed during the KBFF approach is as follows:

3.6 The General KBFF Approach

1) The FF should be simple and efficient to allow for large long-time simulations of biomolecules.
2) Bond and angle parameters and atom sizes are reasonably well known.
3) Dihedral potentials can be obtained from gas-phase quantum-mechanical calculations.
4) Existing hydrocarbon parameters should be reasonable, as these molecules are non-polar and their charge distributions play a minor role.
5) The major difficulty is determining the effective charge distributions of polar groups.
6) Adopt the SPC/E (simple point charge/extended) water model for all simulations.

The KBFF approach uses the same basic equations and the non-polarizable approximation as used by many other force fields in an effort to keep the model computationally efficient and consistent with existing protein force fields and water models. The non-bonded interactions are described by a simple Lennard-Jones (LJ) 6–12 plus Coulomb potential using geometric combination rules for both σ and ε parameters. The bonded terms involve a simple harmonic potential for the bond angle and improper dihedral potentials, with the usual Fourier expansion for dihedral angle rotations. The KBFF approach is mainly concerned with determining the non-bonded parameters that best reproduce the relative distribution of molecules in solution. This distribution is insensitive to reasonable changes in bonded interactions. Hence, the required bonded terms have been borrowed from the GROMOS force field of Van Gunsteren and coworkers [25]. Also, electrically neutral united-atom hydrocarbon non-bonded parameters are taken from a recent parameterization of alkanes, which involved extensive simulations of short- and long-chain hydrocarbons [56]. The KBFF approach is specifically designed to reproduce the experimental KB integrals for solution mixtures in addition to other properties. During our initial investigations into the properties of urea solutions, it was observed that the KB parameters were relatively insensitive to reasonable variations in the σ (≈ 0.01 nm) and ε (≈ 0.1 kJ mol^{-1}) parameters [3]. Considering the above observations, the σ and ε parameters for new atom types are typically determined from a simple scaling scheme [3]. Finally, the SPC/E water model has been adopted for all simulations, as it was considered the best three-site water model available when this work was initiated [57].

Using the previous philosophy, the general KBFF approach involves the following steps:

1) Obtain experimental KB integrals from the experimental densities, activities, and compressibilities for the solution mixture of interest.
2) Assign bond and angle parameters from the literature.
3) Assign any hydrocarbon parameters from the literature.
4) Assign LJ 6–12 parameters for polar atoms using a simple scaling scheme.
5) Assign an initial "guessed" charge distribution.
6) Modify any dihedral potentials to fit gas-phase quantum-mechanical calculations.

7) Perform simulations of the cosolvent and water systems at multiple compositions.
8) Determine the simulated density and KB integrals.
9) Compare with experimental KB integrals.
   - If accurate, determine other properties of the mixture.
   - If inaccurate, repeat steps 5 through 9 until reasonable results are obtained.

The simulations are typically performed for 1–20 ns on systems of 5000–20 000 atoms at 300 K and 1 atm. The number of guessed charge distributions can range from as few as five to as many as 25 in some cases. The above process has to be modified for salts with monatomic cations or anions where the charge distribution cannot be modified. In these cases we have chosen to fit the crystal structure data (unit-cell dimensions) in addition to the solution KB integrals [52]. This involves varying the LJ σ and ε parameters, while retaining a simple scaling of σ related to the ionic radii of the ions.

3.7
Technical Aspects of the KBFF Approach

The KB integrals are defined in a system open to all species. Only after suitable thermodynamic transformations do the integrals provide information on closed systems. In this case the KB integrals correspond to an equivalent open system in which the average $\langle N_i \rangle$ equals the fixed N_i in the closed system, and so on. The vast majority of simulations are performed in closed systems. Integration to infinity in closed systems provides $G_{ii} = -1$ and $G_{i \neq j} = 0$ for the KB integrals. This issue can be avoided by truncation of the integral when g_{ij} becomes approximately unity. Hence, we have

$$G_{ij} = 4\pi \int_0^\infty [g_{ij}^{\mu VT}(r) - 1] r^2 dr \approx 4\pi \int_0^R [g_{ij}^{NPT}(r) - 1] r^2 dr \tag{3.12}$$

The distance at which this condition is met varies from system to system and between different components. We generally examine the behavior of the integral as a function of integration distance, R. In most situations one observes a reasonable plateau region in which the value of G_{ij} remains relatively constant. In cases where the integral is still oscillating slightly, we have averaged the integral over a small range of distances (0.3–0.5 nm) at large values of r.

A slight complication arises when applying KB theory to the study of salt solutions. KB theory can be applied to salt solutions as long as one does not treat the individual ions as independent thermodynamic variables [39]. There are several approaches to this problem [12, 38, 58–61]. We have used the indistinguishable-ion approach in the case that the cosolvent is a salt [48]. This involves treating the salt as a collection of ions, without acknowledging the differences between anions and cations. This has particular advantages when analyzing computer simulation data [48]. In our opinion, this is the simplest approach for salts and provides equivalent expressions to other methods [38, 58, 62]. Further details can be found elsewhere [48].

The FFs generated using the KBFF approach should provide accurate descriptions of changes in the cosolvent chemical potentials as a function of cosolvent concentration. We note that this includes changes in the internal partition function for the cosolvent—specifically vibrational and rotational modes. Hence, these factors are *implicitly* included in the KBFF models. In addition, as we have used the SPC/E model for water, which includes a polarization correction term, this type of correction should also be included when determining properties of the pure KBFF cosolvent solutions – in particular, the enthalpy of vaporization [54].

When comparing simulation and experimental data, it is important to note that there are statistical uncertainties in both. The simulated KB integrals display fluctuations over the nanosecond time scale. The experimental KB integrals can be sensitive to the fitting procedure for the cosolvent activity data [63]. In both cases, the uncertainties increase as the concentration of either one of the components decreases. Hence, we prefer to compare the experimental and simulated N_{ij} values and focus on the most statistically relevant data, that is, N_{11} and N_{21} when ρ_2 is small.

Finally, we would like to note that, in principle, many different RDFs could lead to the same value of the corresponding KB integral. However, this has not occurred in our studies to date. In fact, it is generally difficult to find just one charge distribution that reproduces the experimental density and KB integrals. This is probably a reflection of the fact that we typically do not vary the LJ σ and ε parameters during the fitting process.

3.8
Results for Urea and Water Binary Solutions

The KBFF approach has been applied to derive FFs for several systems. To illustrate the kind of results that can be obtained, we will present data for urea and water mixtures as an example [3]. The RDFs obtained from simulations using the final charge distribution for urea in water are displayed in Figure 3.2a. The RDFs can then be integrated to provide the corresponding KB integrals, which are also shown in Figure 3.2b. The simulated and experimental KB integrals are compared in Figure 3.3 as excess coordination numbers N_{ij}, which have the interpretation of the change in the number of j particles in a given volume on the addition of a central i particle from the number of j particles found in an equivalent volume of bulk solution. It can be seen that the agreement is very good within the statistical error. It is also observed that the OPLS (optimized potentials for liquid simulations) urea model [64] does not reproduce the KB integrals when used with the SPC/E model of water (or others). The N_{ii} values are much too positive, while the N_{ij} value is too negative, indicating an excessive self-aggregation or association of urea (and water) molecules. In our opinion, this illustrates convincingly that the balance of solvation and self-association in the solution can be in significant error

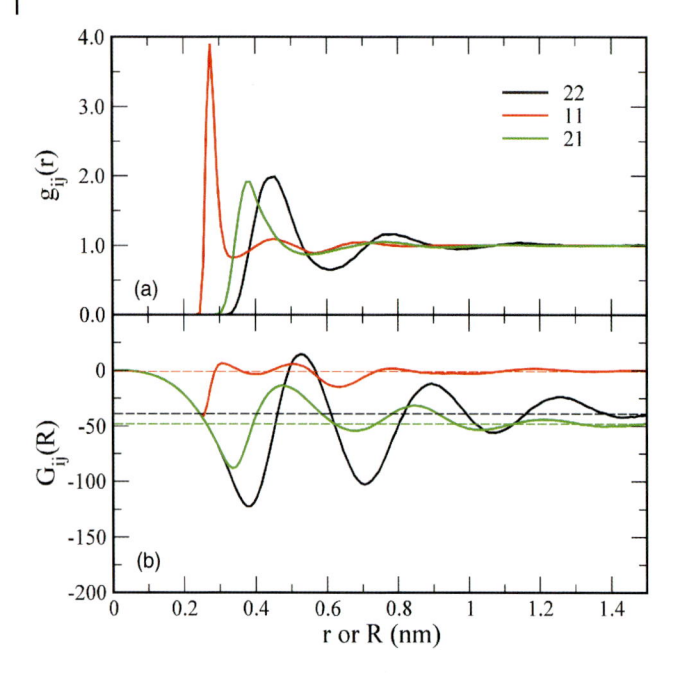

Figure 3.2 (a) Simulated RDFs obtained for an 8 M mixture of urea and water using the KBFF model for urea. (b) The corresponding KB integrals (cm^3 mol^{-1}) as a function of integration distance, R. The dashed lines indicate the final simulated values.

with conventional FFs. This problem can be overcome by adjusting the effective molecular charge distributions of the cosolvent.

The KB integrals and activity derivatives obtained for urea and water mixtures indicate that urea and water mixtures display essentially ideal behavior on the molarity scale. This arises from the fact that $G_{22} \approx G_{12}$, that is, the affinity of urea for other urea molecules is balanced by the affinity of water for urea. This balance is reproduced by the KBFF model (see Figure 3.2). If required, the activity of the model as a function of concentration can be obtained by fitting the simulated activity derivatives (a_{22}) to the corresponding expression obtained from the same fitting function as used to represent the experimental activity data.

The KBFF approach is specifically designed to reproduce the experimental KB integrals for small cosolvents in solution mixtures. Equations 3.7 to 3.10 suggest that this is an important property to model correctly. However, if one has to sacrifice agreement with experiment for other thermodynamic or physical properties, then the approach may not be beneficial for all computational studies. Fortunately, this does not seem to be the case. In Figure 3.4 we provide data for additional properties of urea and water mixtures that were not used during the parameterization procedure [3]. The agreement with experiment is as good as, if not better than, those obtained using conventional FFs. Clearly, one does not have to sacrifice agreement with experiment for other properties in order to reproduce

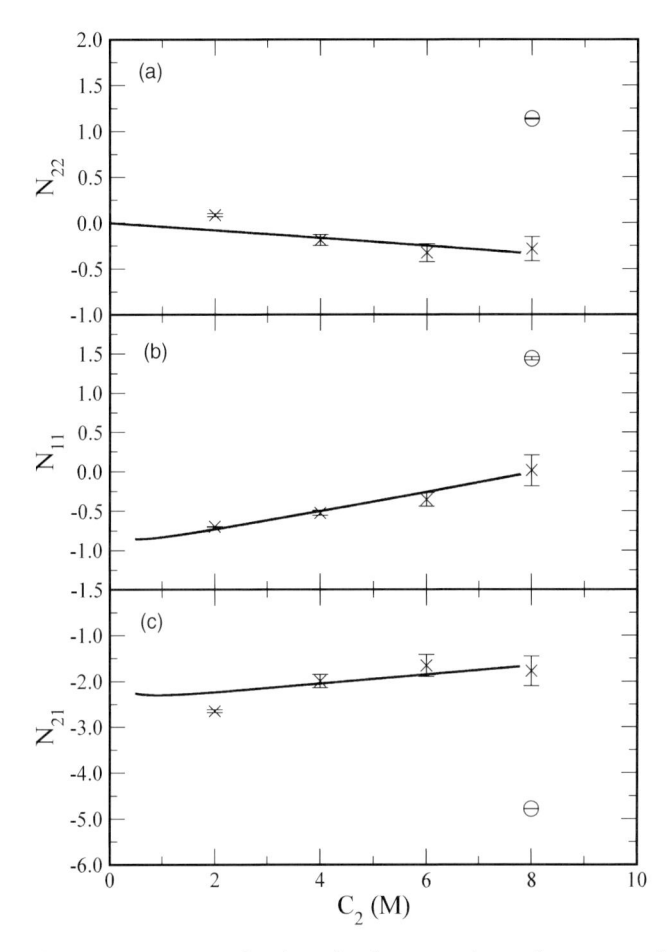

Figure 3.3 Experimental and simulated KB integrals (N_{ij}) for mixtures of urea and water as a function of urea molarity (C_2). The solid lines represent the experimental data, the crosses are the values obtained for the KBFF model, while the circles are for the OPLS model.

the experimental KB integrals. The experimental KB integrals merely provide additional data, which are sensitive to the FF parameters, for use in the parameterization procedure. The quality of the urea KBFF model has been confirmed by others [65, 66].

The KBFF cosolvent models have been developed using the SPC/E water model. For several of these models, we have investigated the effect of changing the water model [3, 4, 52]. In general, the corresponding changes in the KB integrals are relatively small for other comparable three-site models. Hence, reasonable results should still be obtained in these cases. This is probably due to the relatively small differences in solvent polarity between the models. In contrast, cosolvent models can vary significantly in their effective charge distributions.

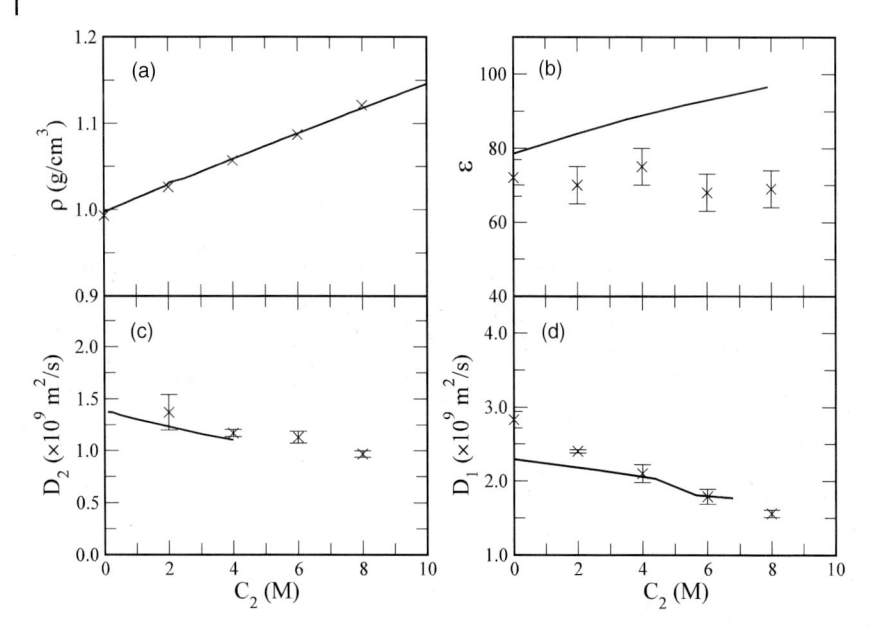

Figure 3.4 Properties of urea and water mixtures obtained with the KBFF model: (a) density, (b) relative permittivity, (c) urea diffusion, and (d) water diffusion, as functions of urea molarity (C_2).

3.9
Preferential Interactions of Urea

In the previous section we illustrated that the charge distribution, and other parameters, can have an effect on the KB integrals observed for solution mixtures. One may ask, however, if this makes a significant difference to simulated results obtained for cosolvent effects on small solutes and biomolecules? The expressions presented in Equations 3.7 to 3.10 clearly suggest that this should be the case. In this section, we present evidence that this is indeed the case using the following complementary numerical studies.

In several of our earlier studies we investigated the change in the excess chemical potential for the insertion of non-polar solutes into solutions containing a variety of cosolvents [30, 51, 67]. One of those studies determined the free energy for cavity formation in a series of urea solutions simulated using different urea models [51]. The urea models possessed the same LJ and bonded parameters, but a range of charge distributions. The corresponding activity derivatives (a_{22}) varied substantially between the models, even though their dipole moments were similar (≈ 4.5 D). The particle insertion results for the urea models are presented in Figure 3.5. All the urea models displayed the same free energy for cavity formation. Hence, the left-hand side of Equation 3.7 is identical in all cases. However, the degree of exclusion of urea from the cavity provided by Γ_{s2} is different for each model. This is a direct consequence of the difference in the magnitude of a_{22}

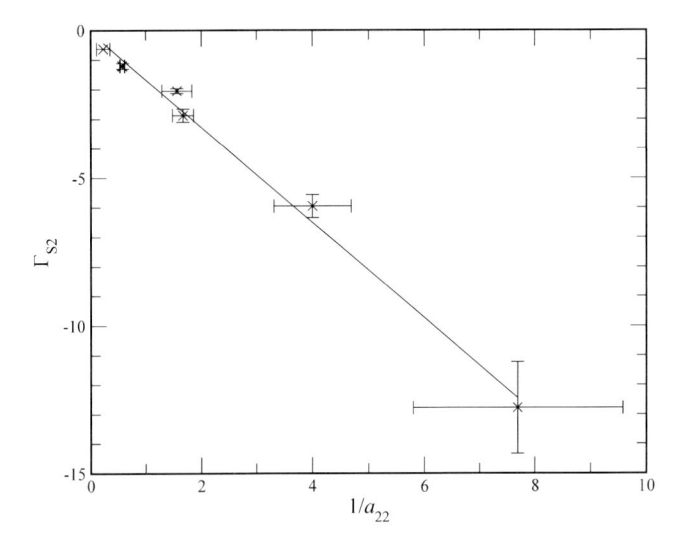

Figure 3.5 The variation in cosolvent exclusion from a solute cavity (Γ_{S2}) as a function of the bulk solution activity derivative (a_{22}) obtained for a variety of urea models.

between the models. Hence, urea models that display a high degree of urea self-affinity (G_{22}) will produce a smaller value of a_{22} and therefore a larger negative value of Γ_{S2}. This clearly illustrates that, even when the free energy changes for cavity formation are the same, if the bulk properties of the solution are incorrect, then the observed cosolvent binding or exclusion from the solute will also be incorrect.

The previous results illustrate that different solution distributions can be observed around solutes, or near interfaces, depending on the quality of the FF used. We have also observed this kind of behavior for the distribution of cosolvent (and solvent) molecules around biomolecules [68]. Again, we use urea as an example. The experimental value of Γ_{N2} for urea around native lysozyme at pH 7 is known [69]. We have performed two simulations of urea around native lysozyme using the GROMOS FF and SPC/E water model. One was performed using the KBFF model, and the other using the OPLS model for urea. The latter has been validated for use with GROMOS [70]. The corresponding RDFs and Γ_{N2} values are displayed in Figure 3.6. From the RDFs it is clear that the OPLS urea model displays an increased distribution of urea around lysozyme compared to the KBFF model. This is accompanied by a reduced distribution of water. Integration of these RDFs produces a very different degree of urea affinity for the protein between the two models. The OPLS model produces Γ_{N2} values that are several times higher than those from the KBFF model. Not only are the values different, but the OPLS model suggests that changes in the cosolvent and solvent distributions extend several solvation shells away from the protein surface, whereas the KBFF model suggests that changes are limited to the first solvation shell. Both models provide values that are larger than experiment ($\Gamma_{N2} = 16$). The results obtained using the

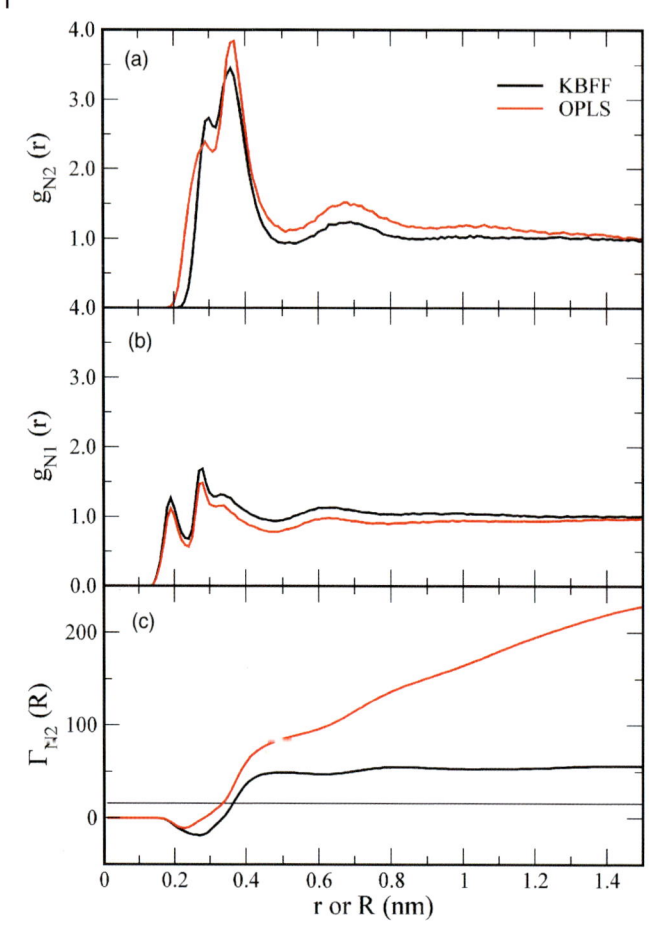

Figure 3.6 (a, b) The RDFs for urea (2) and water (1) around native lysozyme (N) at pH 7 obtained from simulations using the KBFF and OPLS models of urea. (c) The corresponding binding (Γ_{N2}) for the two models, along with the experimental value (thin horizontal line).

KBFF model represent an improvement over the OPLS model. In our opinion, this is due to an improved description of the bulk solution properties (a_{22}). It is hoped that, once we have a complete FF for proteins, then the simulated values of Γ will improve further, and thereby be in closer agreement with experiment.

It is interesting to conclude our discussion by determining the changes in chemical potential of native lysozyme provided by both the simulation and experimental data. If we assume a linear change in protein chemical potential with urea concentration, then one can estimate the total change for the addition of 8 M urea at 300 K using a simple finite-difference approximation to Equation 3.7. The values of a_{22} are 0.93 and 0.16 for the KBFF and OPLS models, compared to 1.16 obtained experimentally. From Figure 3.6 the values of Γ_{N2} are approximately 50 and >250 for the KBFF and OPLS models, respectively, compared to the experimental value of 16. Only an upper limit can be placed on the OPLS model, as the value of Γ_{N2} has not fully converged (see Figure 3.6). Using these numbers, the final estimates for $\Delta\mu_N^*$ are −116 and <−100 kJ mol^{-1} for the KBFF and OPLS models, respectively. These are to be compared with a value of −46 kJ mol^{-1} obtained from the experimental data. Both models are substantially in error. However, it appears that the OPLS actually provides slightly better agreement with experiment. Unfortunately, this is primarily due to incorrect descriptions of both Γ_{N2} and a_{22}, which provide a significant degree of cancellation.

3.10
Conclusions and Future Directions

We have outlined how one can use KB integrals and KB theory to quantify a variety of solution behavior. Subsequently, this has illustrated a series of problems in traditional approaches to developing FF parameters. In our opinion, this is directly related to the effective charge distributions obtained from gas-phase quantum calculations, which do not correctly model the complicated issue of polarization effects in solution. If this is the case, an explicitly polarizable FF should produce reasonable KB integrals for solution mixtures. We are currently comparing our KBFF models with the results from polarizable FFs to establish the validity of this assumption.

The simulated KB integrals appear to be most sensitive to the charge distribution used. Fortunately, by varying the charge distributions for many different cosolvents, one can accurately reproduce the experimental KB integrals over a range of concentrations. Correspondingly, the solution thermodynamics, including the activity of the cosolvent, are well described. This also ensures the correct balance between solvation and self-association of the cosolvent molecules. In addition, other physical properties (density, diffusion constants, compressibility) are also accurately reproduced. These represent the major advantages of the KBFF approach. The disadvantages are primarily related to the need for larger simulations, performed for several nanoseconds, which are required to obtain the KB integrals from simulation, coupled with the rather random search for the ideal charge distribution. However, we consider the current FF improvements essential if one is to study issues such as peptide and protein aggregation in quantitative detail.

Another area where the KBFF models could be useful is helping to understand the role of solvent in biomolecular processes. In particular, implicit models of water are being developed that are efficient enough to fold peptides and small proteins. In most cases, it is assumed that the role of solvent (water) molecules is correctly reproduced in the corresponding explicit solvent FF, and therefore the derived implicit solvent model can be compared with experimental data for validation. In our opinion, this approach can be problematic. There are clearly imbalances in the original explicit solvent models, which, if fixed by the implicit solvent model, will lead to incorrect models of solvation. We are currently examining some of the common implicit solvation models to see if they can reproduce the KB integrals between cosolvent molecules (G_{22}).

It is instructive to review what we have learned about solution structure in mixed solvent systems. First, the expressions presented in Equations 3.1 to 3.10 clearly indicate two things. First, changes in the relative distributions of different species are important and, in principle, include contributions over many solvation shells. Second, the effect of changes in the solvent distribution (G_{12}) plays a role and has to be included. It is impossible in a simulation to reproduce the thermodynamic changes on the left-hand sides of Equations 3.7 to 3.10, to provide an accurate description of the binding/adsorption (Γ), without reasonable values for the bulk solution properties (a_{22}). These issues illustrate the power of KB theory. However, they also make it difficult to understand the properties of solutions. One cannot simply examine the number or strength of the hydrogen bonds between two cosolvent molecules, or between a cosolvent molecule and a solvent molecule, in the first solvation shell. Other more subtle longer-range effects can be equally important, but are much more difficult to understand and explain. Finally, we have observed that the cosolvent polarity (dipole moment) is not typically a good indicator of solvation [51]. One does not always observe an increase in solvation energy on increasing the cosolvent dipole moment, and therefore better solvation in terms of a lower N_{21} value. It is actually the charge *distribution* that appears to be more important. Hence, one cannot just scale gas-phase charge distributions in order to mimic solution-phase distributions, as different regions of the cosolvent can be polarized to varying degrees.

We hope to have a first-generation KBFF available for proteins in the near future. Only then will one be able to fully determine the significance of the improvements we have attempted to make. In our opinion, it is clear that the KB integrals provide, at the very least, a stern test of force fields for the description of liquid mixtures and should therefore be used during the parameterization procedure. Only then can one be assured of a correct balance between intermolecular forces, and thereby the appropriate molecular distributions in solution.

Acknowledgments

P.E.S. would like to thank Tom Cheatham, Dave Cerutti, and Terry Lybrand for providing configurations for Figure 3.1, and also Bob Mazo and Arieh Ben-Naim for many useful discussions. Since the submission of the manuscript for this chapter, we were made aware of a publication that provides additional insight into the issue of salt parameters in AMBER [71]. The project described herein was supported by Grant R01GM079277 (P.E.S.) from the National Institute of General Medical Sciences. The content is solely the responsibility of the authors and does not necessarily represent the official views of the National Institute of General Medical Sciences or the National Institutes of Health.

References

1 Chitra, R. and Smith, P.E. (2000) Molecular dynamics simulations of the properties of cosolvent solutions. *J. Phys. Chem. B*, **104**, 5854–5864.

2 Chitra, R. and Smith, P.E. (2001) Properties of 2,2,2-trifluoroethanol and water mixtures. *J. Chem. Phys.*, **114**, 426–435.

3 Weerasinghe, S. and Smith, P.E. (2003) A Kirkwood–Buff derived force field for mixtures of urea and water. *J. Phys. Chem. B*, **107**, 3891–3898.

4 Weerasinghe, S. and Smith, P.E. (2003) A Kirkwood–Buff derived force field for mixtures of acetone and water. *J. Phys. Chem. B*, **118**, 10663–10670.

5 Smith, P.E. and Pettitt, B.M. (1991) Effects of salt on the structure and dynamics of the bis(penicillamine) enkephalin zwitterion–a simulation study. *J. Am. Chem. Soc.*, **113**, 6029–6037.

6 Smith, P.E. and Pettitt, B.M. (1992) Amino acid side-chain populations in aqueous and saline solution: bis-penicillamine enkephalin. *Biopolymers*, **32**, 1623–1629.

7 Smith, P.E., Marlow, G.E., and Pettitt, B.M. (1993) Peptides in ionic solutions–a simulation study of a bis(penicillamine) enkephalin in sodium acetate solution. *J. Am. Chem. Soc.*, **115**, 7493–7498.

8 Timasheff, S.N. (1998) Control of protein stability and reactions by weakly interacting cosolvents: the simplicity of the complicated. *Adv. Protein Chem.*, **51**, 355–432.

9 Kirkwood, J.G. and Buff, F.P. (1951) The statistical mechanical theory of solutions. I. *J. Chem. Phys.*, **19**, 774–777.

10 Ben-Naim, A. (1992) *Statistical Thermodynamics for Chemists and Biochemists*, Plenum Press, New York.

11 Ben-Naim, A. (1988) Theory of preferential solvation of nonelectrolytes. *Cell Biophys.*, **12**, 255–269.

12 Matteoli, E. and Mansoori, G.A. (1990) *Fluctuation Theory of Mixtures*, Taylor & Francis, New York.

13 Perera, A., Sokolic, F., Almasy, L., Westh, P., and Koga, Y. (2005) On the evaluation of the Kirkwood–Buff integrals of aqueous acetone mixtures. *J. Chem. Phys.*, **123**, 24503.

14 Lee, M.E. and van der Vegt, N.F.A. (2005) A new force field for atomistic simulations of aqueous tertiary butanol solutions. *J. Chem. Phys.*, **122**, 114509.

15 MacKerell, A.D. (2004) Empirical force fields for biological macromolecules: overview and issues. *J. Comput. Chem.*, **25**, 1584–1604.

16 Cheatham, T.E. and Brooks, B.R. (1998) Recent advances in molecular dynamics simulation towards the realistic representation of biomolecules in solution. *Theor. Chem. Acc.*, **99**, 279–288.

17 Wang, W., Donini, O., Reyes, C.M., and Kollman, P.A. (2001) Biomolecular simulations: recent developments in force fields, simulations of enzyme catalysis, protein–ligand, protein–protein, and protein–nucleic acid noncovalent interactions. *Annu. Rev. Biophys. Biomol. Struct.*, **30**, 211–243.

18 Daura, X., Gademann, K., Schafer, H., Jaun, B., Seebach, D., and Van Gunsteren, W.F. (2001) The beta-peptide hairpin in solution: conformational study of a beta-hexapeptide in methanol by NMR spectroscopy and MD simulation. *J. Am. Chem. Soc.*, **123**, 2393–2404.

19 Bonnet, P. and Bryce, R.A. (2004) Molecular dynamics and free energy analysis of neuraminidase–ligand interactions. *Protein Sci.*, **13**, 946–957.

20 Lazaridis, T. (2002) Binding affinity and specificity from computational studies. *Curr. Org. Chem.*, **6**, 1319–1332.

21 Patel, S., Mackerell, A.D., Jr., and Brooks, C.L., III (2004) CHARMM fluctuating charge force field for proteins: II Protein/solvent properties from molecular dynamics simulations using a nonadditive electrostatic model. *J. Comput. Chem.*, **25**, 1504–1514.

22 Kaminski, G.A., Stern, H.A., Berne, B.J., Friesner, R.A., Cao, Y.X.X., Murphy, R.B., Zhou, R.H., and Halgren, T.A. (2002) Development of a polarizable force field for proteins via ab initio quantum

chemistry: first generation model and gas phase tests. *J. Comput. Chem.*, **23**, 1515–1531.

23 Feig, M. and Brooks, C.L. (2004) Recent advances in the development and application of implicit solvent models in biomolecule simulations. *Curr. Opin. Struct. Biol.*, **14**, 217–224.

24 MacKerell, A.D., Bashford, D., Bellott, M., Dunbrack, R.L., Evanseck, J.D., Field, M.J., Fischer, S., Gao, J., Guo, H., Ha, S., Joseph-McCarthy, D., Kuchnir, L., Kuczera, K., Lau, F.T.K., Mattos, C., Michnick, S., Ngo, T., Nguyen, D.T., Prodhom, B., Reiher, W.E., Roux, B., Schlenkrich, M., Smith, J.C., Stote, R., Straub, J., Watanabe, M., Wiorkiewicz-Kuczera, J., Yin, D., and Karplus, M. (1998) All-atom empirical potential for molecular modeling and dynamics studies of proteins. *J. Phys. Chem. B*, **102**, 3586–3616.

25 Scott, W.R.P., Hunenberger, P.H., Tironi, I.G., Mark, A.E., Billeter, S.R., Fennen, J., Torda, A.E., Huber, T., Kruger, P., and Van Gunsteren, W.F. (1999) The GROMOS biomolecular simulation program package. *J. Phys. Chem. A*, **103**, 3596–3607.

26 Cornell, W.D., Cieplak, P., Bayly, C.I., Gould, I.R., Merz, K.M., Ferguson, D.M., Spellmeyer, D.C., Fox, T., Caldwell, J.W., and Kollman, P.A. (1995) A second generation force field for the simulation of proteins, nucleic acids, and organic molecules. *J. Am. Chem. Soc.*, **117**, 5179–5197.

27 Kaminski, G.A., Friesner, R.A., Tirado-Rives, J., and Jorgensen, W.L. (2001) Evaluation and reparametrization of the OPLS-AA force field for proteins via comparison with accurate quantum chemical calculations on peptides. *J. Phys. Chem. B*, **105**, 6474–6487.

28 Ponder, J.W. and Case, D.A. (2003) Force fields for protein simulations. *Adv. Protein Chem.*, **66**, 27–85.

29 Jungwirth, P. and Tobias, D.J. (2002) Chloride anion on aqueous clusters, at the air–water interface, and in liquid water: solvent effects on Cl⁻ polarizability. *J. Phys. Chem. A*, **106**, 379–383.

30 Chitra, R. and Smith, P.E. (2001) Preferential interactions of cosolvents with hydrophobic solutes. *J. Phys. Chem. B*, **105**, 11513–11522.

31 Sokolic, F., Idrissi, A., and Perera, A. (2002) Concentrated aqueous urea solutions: a molecular dynamics study of different models. *J. Chem. Phys.*, **116**, 1636–1646.

32 Auffinger, P., Cheatham, T.E., and Vaiana, A.C. (2007) Spontaneous formation of KCl aggregates in biomolecular simulations: a force field issue? *J. Chem. Theory Comput.*, **3**, 1851–1859.

33 Cerutti, D.S., Trong, I.L., Stenkamp, R.E., and Lybrand, T.P. (2008) Simulations of a protein crystal: explicit treatment of crystallization conditions links theory and experiment in the streptavidin–biotin complex. *Biochemistry*, **47**, 12065–12077.

34 Kang, M., and Smith, P.E. (2009) To be published.

35 Ben-Naim, A. (1977) Inversion of the Kirkwood–Buff theory of solutions: application to the water–ethanol system. *J. Chem. Phys.*, **67**, 4884–4890.

36 Matteoli, E. and Lepori, L. (1995) Kirkwood–Buff integrals and preferential solvation in ternary nonelectrolyte mixtures. *J. Chem. Soc. Faraday Trans.*, **91**, 431–436.

37 Smith, P.E. (2008) On the Kirkwood–Buff inversion procedure. *J. Chem. Phys.*, **129**, 124509.

38 Newman, K.E. (1994) Kirkwood–Buff solution theory – derivation and applications. *Chem. Soc. Rev.*, **23**, 31–40.

39 Ben-Naim, A. (2006) *Molecular Theory of Solutions*, Oxford University Press, New York.

40 Smith, P.E. (2006) Equilibrium dialysis data and the relationships between preferential interaction parameters for biological systems in terms of Kirkwood–Buff integrals. *J. Phys. Chem. B*, **110**, 2862–2868.

41 Hall, D.G. (1971) Kirkwood–Buff theory of solutions – alternative derivation of part of it and some applications. *Trans. Faraday Soc.*, **67**, 2516–2524.

42 Smith, P.E. (2006) Chemical potential derivatives and preferential interaction parameters in biological systems from Kirkwood–Buff theory. *Biophys. J.*, **91**, 849–856.

43 Kang, M. and Smith, P.E. (2008) Kirkwood–Buff theory of four and higher component mixtures. *J. Chem. Phys.*, **128**, 244511.

44 Pierce, V., Kang, M., Aburi, M., Weerasinghe, S., and Smith, P.E. (2008) Recent applications of Kirkwood–Buff theory to biological systems. *Cell Biochem. Biophys.*, **50**, 1–22.

45 Widom, B. (1982) Potential-distribution theory and the statistical-mechanics of fluids. *J. Phys. Chem.*, **86**, 869–872.

46 Smith, P.E. and Mazo, R.A. (2008) On the theory of solute solubility in mixed solvents. *J. Phys. Chem. B*, **112**, 7875–7884.

47 Aburi, M. and Smith, P.E. (2004) A combined simulation and Kirkwood–Buff approach to quantify cosolvent effects on the conformational preferences of leucine enkephalin. *J. Phys. Chem. B*, **108**, 7382–7388.

48 Smith, P.E. (2004) Cosolvent interactions with biomolecules: relating computer simulation data to experimental thermodynamic data. *J. Phys. Chem. B*, **108**, 18716–18724.

49 Chen, F. and Smith, P.E. (2008) Theory and computer simulation of solute effects on the surface tension of liquids. *J. Phys. Chem. B*, **112**, 8975–8984.

50 Ben-Naim, A. (2008) Comment on "The Kirkwood–Buff theory of solutions and the local composition of liquid mixtures". *J. Phys. Chem. B*, **112**, 5874–5875.

51 Weerasinghe, S. and Smith, P.E. (2003) Cavity formation and preferential interactions in urea solutions: dependence on urea aggregation. *J. Chem. Phys.*, **118**, 5901–5910.

52 Weerasinghe, S. and Smith, P.E. (2003) A Kirkwood–Buff derived force field for sodium chloride in water. *J. Chem. Phys.*, **119**, 11342–12349.

53 Weerasinghe, S. and Smith, P.E. (2004) A Kirkwood–Buff derived force field for the simulation of aqueous guanidinium chloride solutions. *J. Chem. Phys.*, **121**, 2180–2186.

54 Weerasinghe, S. and Smith, P.E. (2005) A Kirkwood–Buff derived force field for methanol and aqueous methanol solutions. *J. Phys. Chem. B*, **109**, 15080–15086.

55 Kang, M. and Smith, P.E. (2006) A Kirkwood–Buff derived force field for amides. *J. Comput. Chem.*, **27**, 1477–1485.

56 Schuler, L.D., Daura, X., and Van Gunsteren, W.F. (2001) An improved GROMOS96 force field for aliphatic hydrocarbons in the condensed phase. *J. Comput. Chem.*, **22**, 1205–1218.

57 Berendsen, H.J.C., Grigera, J.R., and Straatsma, T.P. (1987) The missing term in effective pair potentials. *J. Phys. Chem.*, **91**, 6269–6271.

58 Kusalik, P.G. and Patey, G.N. (1987) The thermodynamic properties of electrolyte solutions: some formal results. *J. Chem. Phys.*, **86**, 5110–5116.

59 Perry, R.L., Cabezas, H., and O'Connell, J.P. (1988) Fluctuation thermodynamic properties of strong electrolyte solutions. *Mol. Phys.*, **63**, 189–203.

60 Lee, L.L. (2000) Thermodynamic consistency and reference scale conversion in multisolvent electrolyte solutions. *J. Mol. Liq.*, **87**, 129–147.

61 Perry, R.L. and O'Connell, J.P. (1984) Fluctuation thermodynamic properties of reactive components from species correlation-function integrals. *Mol. Phys.*, **52**, 137–159.

62 Behera, R. (1998) On the calculation of thermodynamic properties of electrolyte solutions from Kirkwood–Buff theory. *J. Chem. Phys.*, **108**, 3373–3374.

63 Matteoli, E. and Lepori, L. (1984) Solute–solute interactions in water. II. An analysis through the Kirkwood–Buff integrals for 14 organic solutes. *J. Chem. Phys.*, **80**, 2856–2863.

64 Duffy, E.M., Severance, D.L., and Jorgensen, W.L. (1993) Urea: potential functions, log P, and free energy of hydration. *Isr. J. Chem.*, **33**, 323–330.

65 Mountain, R.D. and Thirumalai, D. (2004) Importance of excluded volume on the solvation of urea in water. *J. Phys. Chem. B*, **108**, 6826–6831.

66 Kokubo, H., Rosgen, J., Bolen, D.W., and Pettitt, B.M. (2007) Molecular basis of the apparent near ideality of urea solutions. *Biophys. J.*, **93**, 3392–3407.

67 Smith, P.E. (1999) Computer simulation of cosolvent effects on hydrophobic hydration. *J. Phys. Chem. B*, **103**, 525–534.

68 Kang, M. and Smith, P.E. (2007) Preferential interaction parameters in biological systems by Kirkwood–Buff theory and computer simulation. *Fluid Phase Equilibria*, **256**, 14–19.

69 Timasheff, S.N. and Xie, G.F. (2003) Preferential interactions of urea with lysozyme and their linkage to protein denaturation. *Biophys. Chem.*, **105**, 421–448.

70 Smith, L.J., Berendsen, H.J.C., and Van Gunsteren, W.F. (2004) Computer simulation of urea–water mixtures: a test of force field parameters for use in biomolecular simulation. *J. Phys. Chem. B*, **108**, 1065–1071.

71 Chen, A.A. and Pappu, R.V. (2007) Parameters of monovalent ions in the AMBER-99 forcefield: assessment of inaccuracies and proposed improvements. *J. Phys. Chem. B*, **111**, 11884–11887.

4
Osmolyte Influence on Protein Stability: Perspectives of Theory and Experiment

Char Hu, Jörg Rösgen, and B. Montgomery Pettitt

4.1
Introduction

The mechanism governing the transition of proteins from their unfolded state to their native state still remains as one of the unanswered fundamental biophysical questions. As the misfolding of proteins has been associated with a number of diseases [1, 2], the understanding of the folding process has important potential therapeutic implications. Although a significant effort by the computational biology community has been focused on folding proteins based upon sequence, an emerging trend is to use the advantages of *in silico* biophysics to provide unique insight into the reversible process of folding as it may occur in the cell. In order to accurately describe protein folding in the true native environment, the cellular milieu must be modeled as a thermodynamically non-ideal system as opposed to dilute ideal solutions. By recognizing the importance of the influence of cosolvents on protein stability, a more accurate understanding of protein folding may be accomplished. Recent computational and theoretical descriptions of the non-ideal behavior have begun to provide accurate descriptions of experimentally observed behaviors of such systems [3–5].

A specific class of cosolvents known as osmolytes are particularly interesting because of their usage as cellular and protein stabilizers in all kingdoms of life [6]. There are two classes of osmolytes, delineated by their effect on protein stability: protecting osmolytes and denaturing osmolytes. A protecting osmolyte is known to increase protein stability, whereas a denaturing osmolyte shifts the reversible process of folding toward the unfolded state. The exact mechanisms by which osmolytes affect protein stability are still unknown.

Knowledge gained by studying the interactions between osmolytes and proteins have provided an emerging paradigm shift in the understanding of protein folding [7]. Thermodynamic measurements carried out on proteins and osmolyte aqueous solutions have determined that the driving force behind osmolyte activity is the interactions of the osmolyte with the peptide backbone [8, 9]. Osmolyte-induced stability is brought about by unfavorable interactions with the protein backbone (the

Modeling Solvent Environments. Applications to Simulations of Biomolecules. Edited by Michael Feig
Copyright © 2010 WILEY-VCH Verlag GmbH & Co. KGaA, Weinheim
ISBN: 978-3-527-32421-7

osmophobic effect), whereas denaturation occurs by favorable interactions of the osmolyte with the protein backbone [10]. The result is a de-emphasis of the role of the amino side chains, and a refocusing on the importance of the protein backbone for protein folding as envisioned at the onset of protein structural biophysics [11].

This chapter will focus on the solvation properties of protecting and denaturing osmolytes and how these properties allow for an understanding of the mechanism of influence of osmolytes on protein stability. Recent computational and theoretical advances will be highlighted, with significant related experimental studies also examined. Several recent reviews have highlighted important topics concerning the relationship between cosolvents and protein stability. Notable issues covered in these include: the backbone-based model for protein folding [12, 13], analysis of denatured states [14–16], the role of water in biological chemistry [17], simulations of folding and unfolding [18, 19], solution theory of osmolytes [20], and energetics of protein folding [21]. Therefore, we mention the above topics only in passing, not out of neglect, but as they relate to osmolyte solvation and protein stability; we direct the reader to more specific reviews for further information.

4.2
Denaturing Osmolytes

The only molecule belonging to the class of denaturing organic osmolytes is urea. Interestingly, despite the denaturing effects of urea, it is still accumulated in high concentrations in several species [22], for instance, in mammalian kidneys [23, 24], marine elasmobranches [23], and amphibians [23]. The protein destabilizing ability of urea is well established; however, the mechanism by which urea denatures proteins remains unclear. The prevailing theme in the literature has been that there are two competing mechanisms by which urea denatures proteins: the direct and indirect mechanisms. The direct mechanism states that urea directly interacts with the protein, thereby destabilizing the native state. The indirect mechanism involves a weakening of the protein-stabilizing hydrophobic effect caused by a decrease in the structure of water. After considering current theoretical and experimental studies concerned with the mechanism of urea denaturation, we conclude that the growing body of data suggests that the mechanism for urea denaturation is due to its direct interaction with the protein backbone [25, 26], although some authors continue to regard the side chains as equally important [27, 28, 29], although the activity-corrected data do not appear to support that idea [25, 26]. We note in passing that guanidinium appears to act via a π-stacking mechanism rather than by hydrogen bonding.

4.2.1
Does Urea Weaken Water Structure?

Mounting evidence by theoretical, simulation, and experimental studies has demonstrated that urea does not appreciably perturb water structure, thus contradicting the notion that urea denaturation is caused by a weakening of the water structure and thus disturbance of the hydrophobic effect [30]. Some supporting evidence for the water structure breaking properties of urea has come from simulation studies that have shown a decrease in the number and strength of the hydrogen bonding network of water [31, 32], and even suggested that water is the initial denaturant of proteins [31]. However, a recent theoretical framework based upon Kirkwood–Buff theory [33] has proved that this cannot be the mechanism by which denaturing osmolytes unfold proteins [4, 34]. The advantage of utilizing Kirkwood–Buff theory is that it allows for the connection of experimental thermodynamics and simulation results for finite-concentration solutions without further approximation [35]. Under this rigorous framework, equations for protein stability and chemical activity were derived completely devoid of terms describing water self-hydration [34]. This clearly indicated that the fundamental origin of protein stability effects by osmolytes is not dictated by the change in the structure of water [34].

Further analysis of the solvation of urea has provided more compelling evidence that urea, rather than breaking the water structure, actually resides harmoniously within the existing water network. Experimental and simulation evidence has shown that, even though a urea molecule is significantly larger than a water molecule, it still fits well within the water network [36–38]. Similarly, the unfolding of a charged hydrocarbon in solution was not marked by a strong perturbation of the structure of water [39].

Two recent simulation studies were undertaken to accurately describe the solvation properties of urea by using two separate models for urea, OPLS (optimized potentials for liquid simulations) and KBFF (Kirkwood–Buff derived force field), thus reducing the dependence of the results upon force-field parameters [40, 41]. The KBFF model has been specifically parameterized according to experimental activity coefficient data for urea, thereby allowing for highly accurate simulation results [42]. It was shown that the OPLS model of urea self-associated more readily than the KBFF model [41]. Surprisingly, the origin of this behavior was shown to be the differences in the van der Waals parameters of the two models, as opposed to the charges, and suggested a role for the excluded-volume effect in the denaturation process [41]. Prior to this study, the use of an excluded-volume term, in conjunction with the preferential interactions of osmolytes, was advocated by Schellman in order to describe the nature by which cosolvents affect protein stability [43]. Through this theoretical formalism, it was shown that the contributions from the excluded-volume effect stabilize the native state, whereas preferential solvation has a destabilizing effect [43]. However, when the calculations were applied to simulations, it was shown that the excluded volume contributions to the preferential interaction parameter were lower than predicted by the theory, perhaps due to the ability of the simulations to account for the flexibility of the protein in solution [44].

The second such study analyzed urea clustering thoroughly and saw no evidence of long-lived urea clusters in either molecular model, though the OPLS model for urea had higher amounts of medium-sized clusters [40]. Importantly, careful analysis of the solvation properties of urea showed that, for both models of urea, there was no appreciable weakening of the average number of hydrogen bonds, or of the lifetime of these hydrogen bonds, at a variety of concentrations of urea [40, 41]. Similarly, free-energy simulations of concentrated urea solution were carried out, and the activity coefficients were determined in order to validate the Kirkwood–Buff theoretical model mentioned above [4], while providing insight into the nature of ideality of urea solutions [3]. The activity coefficients of urea in solution agreed very well with experimentally determined values. The authors saw a decrease in the favorability of the electrostatic component and an increase in the favorability of the van der Waals component with increasing concentration of urea [3]. Furthermore, the calculated activity coefficients in the molar scale were very similar to an ideal solution, which indicates a lack in disruption of the water network [3] (see below). Activity coefficients are notoriously sensitive solution measurements, and therefore the accuracy of these chemical potential calculations are a testament to the robustness of the urea model and the simulation techniques [3].

It is important to recognize that the meaning of *ideality* depends strongly on the concentration scale chosen. It is only in the molar scale that the activity coefficient is free of extraneous contributions from concentration scale conversion [3]. This is a major reason why the molar concentration scale should be preferred for theoretical considerations over other concentration units [45]. In the molal concentration scale, the urea activity coefficient can be modeled as originating from dimerizing urea molecules [46, 47]. The corresponding steep drop of the molal activity coefficient of urea was originally interpreted as indicating that indeed such association takes place in solution [46]. Conversely, an increasing activity coefficient was taken as a sign that an osmolyte "binds" hydration waters [48].

Osmolytes in general have been found to match such simple interaction models very well [4, 47]. However, the only rigorous conclusion possible from such a finding is that the osmolyte partition function in water is a polynomial of second order (at most) in osmolyte activity [4, 47]. The question then is what such observation means. Invoking Kirkwood–Buff theory [33] reveals that such low-order partition function results from a remarkably constant solvation behavior of osmolytes that depends little on concentration [49]. The remarkable feature of urea solution is that the urea and water self-solvation, as well as their mutual solvation, are extraordinarily similar [50]. So, considerations of solution ideality indicate that urea is far from disturbing the water structure, but rather mimics water in terms of its overall solvation behavior.

4.2.2
Effect of Urea on Hydrophobic Interactions

In light of the large amount of supporting data suggesting that urea does not weaken the water network, reviewed above, it is important to directly study the effect of urea on hydrophobic interactions. Simulation studies have provided unique molecular-level insight into the effects of urea on hydrophobicity. Building upon the initial simulations of urea and hydrophobic solutes [51, 52], recent simulation studies have again demonstrated that urea does not denature proteins by weakening the hydrophobic effect. This is consistent with the experimental finding that hydrophobic amino acid side chains make a minimal (if any) contribution to urea-dependent protein stability [25].

Experimental thermodynamic calculations by Nozaki and Tanford [53] provided the initial evidence suggesting that urea denatures proteins by decreasing the hydrophobic effect. These measurements, known as group transfer free energies (GTFEs), indicated that the solubility of amino acid side chains in urea was increased over that in pure water [53], thus laying the foundation for the indirect mechanism of urea-induced unfolding. However, these calculations were only apparent GTFEs, because the activity coefficients for the peptide backbone were unknown [53]. Recently, Auton and Bolen were able to calculate true GTFE values by using the activity coefficient data available for glycine as a backbone model [25] and showed significant differences from the initial transfer free energies of Nozaki and Tanford [53]. Specifically, that the solubility of amino acid side chains are *not* increased in the presence of urea [25]. Using these corrected GTFEs, as well as data demonstrating direct interaction of urea with proteins reviewed above, it is clear that it is the large favorable interactions of urea with the peptide backbone that causes urea-induced protein denaturation as opposed to urea solubilizing the amino acid side chains. Utilizing an off-lattice side-chain representation for proteins, and experimental side-chain transfer free-energy values, a recent theoretical model for the denatured state [28] was able to calculate concentration-dependent transfer free energies. The calculated m-values demonstrated reasonable agreement with previous experimentally calculated results. However, the authors used GTFE values different from those cited in their text. The side-chain GTFEs used in their calculations had not been corrected with the activity coefficient data of glycine [26]. Usage of the correct values for side-chain GTFEs should allow for improved correlation with experiment and predicted results.

To determine whether urea decreases the interactions of hydrophobic solutes, the thermodynamics of methane self-association has been simulated in aqueous urea solutions [36, 38, 54]. It was shown that urea actually increases the stability of methane–methane association, as opposed to destabilizing these interactions [38, 54]. Even at very high concentrations of urea, no effect was seen on the thermodynamic interactions between methane molecules [36]. The fact that urea is solvated well by water molecules could potentially explain its lack of effect on the hydrophobic interactions [38, 55]. Though the notion that the effect of urea on hydrophobic interactions may be dependent upon the size of the hydrophobic solute has been suggested [54], it has been shown recently that neopentane pairs are also stabilized in the presence of urea [56]. An interesting study was carried out on a model hydrocarbon to directly determine the effect of urea on hydrophobic interactions [39]. Linear hydrocarbon chains that either had their end-groups blocked or had a modest charge were simulated in solution. Significant destabilization of the model hydrocarbon was only shown to occur for those hydrocarbons with modestly charged end-groups. Therefore, urea was not able to "unfold" the hydrocarbon by altering hydrophobic interactions, but rather relied on direct interaction with the linear chain [39].

4.2.3
Direct Interaction of Urea with Proteins

Early simulations of urea, protein, and water systems demonstrated the direct interaction of urea with proteins [57, 58], which corresponds well to recent experiments demonstrating preferential interaction of urea with proteins [59]. Current simulations have shown an increased urea concentration around a blocked valine peptide that was marked by strong electrostatic interactions, manifested as hydrogen bonds, between the peptide backbone and urea [55]. Even simulations in which water structure was suggested to be weakened by urea saw significant protein–urea interactions [31] and increased urea concentration surrounding the peptide [32]. O'Brien et al. demonstrated that the unfolding of a short stable helix was caused by the direct hydrogen bonding of urea molecules to the side chains and backbone of the helix [36].

Recent simulations of urea and all 20 amino acids demonstrated clear preferential interaction, via contacts, of urea with the backbone and side chains of each amino acid [60]. However, upon calculating the strength of the hydrogen bonds, the authors determined that peptide–water hydrogen bonds were the strongest, whereas peptide–urea bonds were no more favorable than peptide–peptide interactions [60]. The inability of the strength of hydrogen bonding to account for the preferential interaction of urea with the peptides led to the conclusion that urea operates by a counterbalance between its ability to solvate hydrophobic groups and yet still to fit within the water hydrogen bonding network [60].

A novel interpretation of the direct interaction of urea with proteins has come from all-atom microsecond molecular dynamics simulations of urea and hen lysozyme [29]. Concurrent with the unfolding of lysozyme, an increase in the concentration in the local domain of the protein of urea was seen [29]. By analyzing the time series of hydrogen bonding between species, an increase in protein–urea hydrogen bonds was seen prior to an increase in protein–water hydrogen bonds, suggesting a two-phase mechanism [29]. This model states that preferential interaction and hydrogen bonding of urea to the protein is followed by a second phase in which water aids in the solvation of the exposed protein groups, thus facilitating denaturation [29]. Though the direct interaction of urea with the protein backbone and side chains is viewed as the driving force [29], implicit in this model is the ability of water to aid in the denaturing of proteins, the validity of which has been addressed above. Additionally, a change in the van der Waals interaction energy distribution between protein and urea in the first solvation shell versus bulk solution, and a corresponding lack of change in the electrostatic potential energy distributions, was cited as the reason that urea is preferentially enhanced surrounding the protein. This means that van der Waals dispersion forces are responsible for the direct interaction between urea and protein [29]. However, it is not surprising that the profile of van der Waals interactions of urea is significantly different in the local region of the protein, as it was demonstrated that there are more urea molecules present surrounding the protein.

4.3 Protecting Osmolytes

Though much of the knowledge regarding protein stability has been established through the study of protein unfolding caused by the denaturing osmolyte, urea, a growing amount of insight into the reversible process of folding has been obtained through the study of protecting osmolytes. In this section, recent data concerning the mechanism of the solvation of protecting osmolytes is summarized and further analyzed in order to gain perspective on the current understanding of osmolyte-conferred stability.

Of the protecting osmolytes, trimethylamine N-oxide (TMAO) is among the strongest at bolstering protein stability, and is also one of the most studied. Several experimental studies have demonstrated the ability of TMAO to fold intrinsically unstructured proteins and peptides [61–63], as well as thermodynamically unfolded proteins [64]. TMAO has even aided in structural determination using nuclear magnetic resonance [65] and X-ray crystallography [66, 67]. The remarkable ability of TMAO to induce protein folding has been extended to the elucidation of protein folding disease mechanisms of prion-associated diseases [68–70] and Alzheimer's disease [71–73], and even hint at possible novel therapeutic methodologies [74]. The sugar trehalose is another protecting osmolyte with the unique function in preserving biomaterials in organisms that undergo a state of suspended animation known as anhydrobiosis [23, 75]. Another common osmoprotectant is glycine betaine, which effectively protects bacteria such as *Escherichia coli* [23, 76].

Thus, the ability of protecting osmolytes to bias the protein folding transition toward the native state is increasingly being chronicled, yet the mechanism by which these osmolytes induce increased protein stability is uncertain. Below, we analyze recent experimental and theoretical experiments and their implications for protein stability. Though most of the examples given below concern TMAO, it is assumed that the implications extend to all protecting osmolytes.

4.3.1 Do Protecting Osmolytes Increase Water Structure?

Most of the contention regarding the mechanism of osmolyte-conferred stability is centered upon the protecting osmolyte's influence on water structure, and whether this is indeed the mechanism of stabilization. Tangential issues such as its effect on hydrophobic interactions are also unresolved. It has been proposed that protecting osmolytes increase the water hydrogen bonding network, thereby increasing protein stability [37, 77–79]. Others believe that the unfavorable interactions between the osmolyte and protein backbone, and thus the exclusion of protecting osmolytes from the local vicinity of the protein, are the main driving force for increased stability [7, 8, 10, 13]. The osmophobic effect has been extensively reviewed [7, 10, 12], so we will focus on studies purely concerned with osmolyte effect on water structure.

Initial insight into the solvation properties of TMAO came from studies concerned with the solvation property differences between TMAO and *tert*-butyl alcohol (TBA) [80–86]. These two molecules are quite similar in their geometry, with both having large hydrophobic moieties comprising three methyl groups. However, TMAO possesses a significant dipole moment that is three times larger than that of TBA [83]. Interestingly, though TBA and TMAO appear to be quite similar, they have profoundly different effects on protein stability: TBA is known to destabilize proteins [83] and TMAO is a protecting osmolyte. Thus, analyzing the solvation differences between TBA and TMAO should provide an ideal metric for understanding the solution properties of TMAO.

It has been shown, both experimentally [80, 81, 83, 84, 86] and using simulations [85–87], that TMAO possesses a stronger and more compact hydration layer than TBA, which explains why TBA is prone to aggregation whereas TMAO is not. This strong hydration layer of TMAO is caused by the much stronger dipole moment as compared to TBA [83, 84, 86]. Correspondingly, the number of water molecules within the first hydration layer of TMAO remains relatively constant with increasing concentration of osmolyte, whereas the number of waters within the first hydration layer of TBA decreases with increasing concentration at a much more rapid rate [81, 82, 85]. Another indication of a stronger hydration of TMAO as compared to TBA comes from the three-dimensional solvent density profile obtained from molecular dynamics simulations [82]. When comparing two protecting osmolytes (TMAO and glycine betaine) to TBA, it is shown that the hydrophobic groups of the protecting osmolytes are hydrated significantly better than those of TBA. This could be due to the nitrogen of TMAO helping to decrease the hydrophobicity of the protecting osmolytes [82]. However, when comparing high concentrations of both TBA and TMAO, simulations demonstrate that TBA coordinates water much tighter than does TMAO [82].

Using a combination of Fourier-transform infrared spectroscopy and molecular dynamics simulations, it was shown that aqueous TMAO solutions possess a differing distribution in the angles of water–water hydrogen bonds as compared to pure water [37]. The authors have suggested that the disturbance of water structure brought about by a solute may best be quantified via the distribution of the angle of water hydrogen bonds within the hydration layer of the solute [37]. Thus, when describing the structure of the water network, the authors state that the water hydrogen bond angle distribution, as opposed to the distance of the water–water hydrogen bond, is a more accurate gage of water structure [37]. Using this distribution, the authors utilized molecular dynamics simulations to better assign the spectra from their experiments and demonstrate that the water hydrogen bond angle distribution is reduced within the hydration layer of TMAO. A reduction in the hydrogen bond angle is indicative of more strongly interacting waters, and thus an increase in the strength of the hydrogen bond network within the first hydration layer of TMAO. The authors did not see a significant disturbance in the hydrogen bond energies with the addition of TMAO [37]. However, the same lab found later that water in the first solvation shell of urea has the same change in angle distribution as in the case of TMAO [88]. The change in the angle distribution therefore does not seem to be a measure of the stabilizing or destabilizing ability of an osmolyte.

Increased water structure was also noted in the first simulations of a peptide-like unit (cyclic dialanine and cyclic diglycine) in protecting osmolyte solution [79]. Upon calculation of water–water radial distribution functions (functions that measure local density of solvent species as compared to the bulk), it was shown that, with the addition of TMAO, the first peak, associated with the first hydration layer of water, was higher than that of pure water [79]. This increase in peak was stated to indicate more coordination and ordering of water molecules upon the addition of TMAO. A similar increase in the water–water radial distribution function upon the addition of TMAO was also seen in simulations carried about by Athawale *et al.* [89]. However, the authors cautioned against the use of the radial distribution function alone for assigning orientation and also cited a potential bias that may occur due to differences in bulk density [89]. An orientational order parameter was also used to complement the information gained from the radial distribution function. It was shown using this orientational order parameter that the addition of TMAO actually reduces the structure of water [89]. Furthermore, the addition of the denaturing osmolyte urea to pure water also increases the water–water radial distribution function [39, 90]. Therefore, the use of this function to describe water structure is limited and must be supplemented with further information.

The simulations of cyclic dipeptides mentioned above also found an increase in the average number of hydrogen bonds per water molecule and in the strength of these interactions, indicated by a shortening of the distance of the hydrogen bond [79]. Additionally, a significant increase in the lifetime of the hydrogen bonds was also cited. Subsequent work was not able to re-create an increase in the strength of water hydrogen bonds [90], number of hydrogen bonds per water molecule [78, 85], or such a large increase in the lifetime of hydrogen bonds [90].

Additional evidence demonstrating a lack of increased structure induced by TMAO comes from experiments using a novel technique to determine the extent of water structure, known as pressure perturbation calorimetry [91]. Utilizing a wide variety of both denaturing and stabilizing cosolvents, the authors deduced that there was no clear relationship between a solute's influence on water structure and its subsequent effect on protein stability [91]. In fact, hydration is more sensitive to the osmolyte size rather than to the class of osmolyte [92]. *Ab initio* and empirical computational calculations on osmolytes and water also demonstrate a lack of relationship between the interactions of the solute with water and their corresponding effect on protein stability [93].

Further evidence that the effect on water structure is not a good indicator of protein stability comes from experimental infrared spectroscopy data [80, 81]. It has been shown that *both* TBA and TMAO demonstrate the spectra characteristic of low-temperature water (more ordered water), although TMAO and TBA have completely differing effects on protein stability [81]. Similarly, two known protecting osmolytes, glycine betaine and TMAO, have differing effects on the structure of water, with TMAO demonstrating a spectrum similar to lower-temperature water and glycine betaine demonstrating a profile similar to that of higher-temperature water [80].

4.3.2
Effect of Protecting Osmolytes on Hydrophobic Interactions

The effect of protecting osmolytes on hydrophobic interactions has been studied in order to determine whether protecting osmolytes increase the hydrophobic effect and thus favor folding. This has been accomplished by studying the effect of protecting osmolytes on the micellization process and hydration and interactions of hydrophobic solutes.

The same hydrophobic properties that govern the micellization process of surfactants have been related to the protein folding process [80, 84], specifically, concerning the burial of non-polar protein groups and sequestration from hydration waters [84]. Using infrared spectroscopy, it was determined that increasing concentrations of TMAO had no effect on the critical micelle concentration (CMC) of sodium dodecyl sulfate, whereas TBA disrupted the micellization process. Subsequent experiments also demonstrated the neutrality of the protecting osmolytes TMAO and glycine betaine toward the association of surfactants [80]. The authors concluded that there was no direct correlation between the cosolvent's effect on water structure and the micellization process with protein stability [80].

Similar neutrality toward hydrophobic interactions was demonstrated using simulations of methane molecules in osmolyte solution [89]. No effect was seen on the thermodynamics of hydration or of the interactions between the hydrophobic solutes in protecting osmolyte solution [89]. Additionally, the folding dynamics on a 25-mer hydrophobic polymer were analyzed and demonstrated that TMAO had a negligible effect on the thermodynamic hydration of the polymer [89]. The advantage of simulation studies is that they allow for the modification of particular molecular parameters (such as atomic size and charge), thus allowing for a precise annotation of factors that give rise to particular results. In this manner, the atomic charges were modified on TMAO and the subsequent effects on the thermodynamics of the hydrophobic polymer were monitored [89]. It was shown that the amphiphilic character of TMAO allows for its neutrality toward hydrophobic interactions [89]. Recently, it was demonstrated that the neutrality of TMAO on the hydration and thermodynamics of hydrophobic solutes is because of a precise balancing of the entropic and enthalpic components [94]. Molecular dynamics simulations of larger hydrophobic solutes demonstrated that TMAO completely destroyed the hydrophobic attractions between neopentane molecules [95]. Taken together, these studies conclusively demonstrate that protecting osmolytes do indeed increase protein stability; however, they do not increase protein stability by increasing the hydrophobic effect [95].

Historically, TMAO has been categorized as a counteracting osmolyte for its influence on protein stability (specifically counteraction of urea denaturation), whereas osmolytes that protect against environmental extremes are termed compatible osmolytes as they have marginal effects on biological activity [96]. However, the neutrality of TMAO on hydrophobic interactions indicates that it shares traits with compatible solutes in that it could possibly allow TMAO to be accumulated at high concentrations without negatively affecting hydrophobic processes, such as membrane formation [80, 89, 94]. Protecting osmolytes are known as compatible solutes in that they can be accumulated at high concentration within the cell, but still negligibly affect other important biomolecular interactions [89]. This neutrality could explain the successful usage of protecting osmolytes by all taxa of life.

4.4
Mixed Osmolytes

Many organisms have developed the usage of multiple osmolytes within their cells to help counterbalance the effect of the other. For instance, cartilaginous fishes that are rich in urea utilize methylammonium solutes to oppose the deleterious effects of urea on protein stability [23]. Similarly, mammalian kidney cells utilize protecting osmolytes to counteract urea denaturation [23, 63]. This natural usage of osmolytes in tandem to maintain protein stability by a multitude of organisms presents a very unique means by which to study the manner in which selective adaptation has affected protein stability. Do the osmolytes interfere to neutralize each other? Or are their interactions independent of the presence of the other?

Numerous experimental measurements using differing techniques have shown that mixtures of protecting and denaturing osmolytes do not interfere with each other. This is true for ternary urea and TMAO solutions [22, 96, 97] and for urea and sarcosine solutions [63]. Thermodynamic measurements on proteins in mixed osmolyte solution have shown that the stability of the protein in mixed osmolyte is roughly equivalent to the addition of the protein in each osmolyte solution separately [22, 96]. Simulations have also demonstrated that the addition of urea to TMAO solution does not change the manner in which TMAO affects neopentane interactions, and therefore it is unlikely that TMAO counteracts urea denaturation by mitigating the hydrophobic effect [95].

However, simulations have shown that TMAO increases the solution structure when added to urea aqueous solutions [77]. This is demonstrated through somewhat stronger urea–water hydrogen bonds and water–water hydrogen bonds [77]. It was stated that this increase in structure by TMAO allows for water to initially attack and denature the protein with urea molecules to follow [77]. Because urea and water interact strongly with TMAO, and TMAO does not self-associate, it has also been suggested that urea and water then prefer to solvate TMAO and not the protein [90].

The demonstration by both computational [77] and experimental evidence [22, 63, 96, 97] that, in mixed osmolyte solutions, protecting and denaturing osmolytes act independently of one another, rules out the two mechanisms presented in the previous paragraph (urea and water preferentially solvating TMAO and increased water structure). These suggest differing mechanisms of action when the osmolytes are combined and when they are alone in solution. However, the low binding constants and multitude of potential interaction sites along the protein allow for a lack in competition for interaction sites. This allows each osmolyte to exert its effect on protein stability in the same manner, regardless of the presence or absence of another osmolyte [63]. Further analysis of the interactions of osmolytes in mixed solutions, and the energetics associated with these interactions, is needed in order to develop a complete picture of mixed osmolyte solutions.

4.5
Conclusions

The interactions between osmolytes and proteins, and the subsequent effects on protein stability, have been reviewed, while the importance for local concentration fluctuations and its relationship to protein folding have also been emphasized. Though the relatively small binding constants associated with the interaction of osmolytes and proteins [15] initially obscured the mechanisms behind osmolytes and protein stability, it is clear that framing the effect of osmolytes on protein folding as purely a deviation in water structure is too simple [40, 98]. This model, whereby osmolytes affect protein stability by altering water structure, does not take into account binding or exclusionary events that are known to occur. Therefore, once these interactions are explicitly considered, the mechanisms for osmolyte-affected protein stability become clear.

Acknowledgments

The authors acknowledge partial support for this work by NIH Research Grants GM049760 (to J.R.) and GM 37675 (to B.M.P.) and a grant from the Robert A. Welch Foundation E-1028 (to B.M.P.). C.H. is grateful for a training grant from NIH through the Houston Area Molecular Biophysics Training Program administered by the Keck Center for Interdisciplinary Bioscience in Houston.

References

1 Herczenik, E. and Gebbink, M.F.B.G. (2008) Molecular and cellular aspects of protein misfolding diseases. *FASEB J.*, **22**, 2115–2133.
2 Chiti, F. and Dobson, C.M. (2006) Protein misfolding, functional amyloid, and human disease. *Annu. Rev. Biochem.*, **75**, 333–366.
3 Kokubo, H., Rosgen, J., Bolen, D., and Pettitt, B. (2007) Molecular basis of the apparent near ideality of urea solutions. *Biophys. J.*, **93**, 3392–3407.
4 Rosgen, J., Pettitt, B., and Bolen, D. (2004) Uncovering the basis for nonideal behavior of biological molecules. *Biochemistry*, **43**, 14472–14484.
5 Smith, P.E. (2004) Cosolvent interactions with biomolecules: relating computer simulation data to experimental thermodynamic data. *J. Phys. Chem. B*, **108**, 18716–18724.
6 Yancey, P., Clark, M., Hand, S., Bowlus, R., and Somero, G. (1982) Living with water stress: evolution of

osmolyte systems. *Science*, **217**, 1214–1222.

7 Rose, G., Fleming, P., Banavar, J., and Maritan, A. (2006) A backbone-based theory of protein folding. *Proc. Natl. Acad. Sci. U.S.A.*, **103**, 16623–16633.

8 Auton, M. and Bolen, D.W. (2004) Additive transfer free energies of the peptide backbone unit that are independent of the model compound and the choice of concentration scale. *Biochemistry*, **43**, 1329–1342.

9 Liu, Y. and Bolen, D. (1995) The peptide backbone plays a dominant role in protein stabilization by naturally occurring osmolytes. *Biochemistry*, **34**, 12884–12891.

10 Bolen, D. and Baskakov, I. (2001) The osmophobic effect: natural selection of a thermodynamic force in protein folding. *J. Mol. Biol.*, **310**, 955–963.

11 Pauling, L., Corey, R.B., and Branson, H.R. (1951) The structure of proteins: two hydrogen-bonded helical configurations of the polypeptide chain. *Proc. Natl. Acad. Sci. U.S.A.*, **47**, 205–211.

12 Auton, M. and Bolen, D.W. (2007) Application of the transfer model to understand how naturally occurring osmolytes affect protein stability. *Methods Enzymol.*, **428**, 397–418.

13 Bolen, D.W. and Rose, G.D. (2008) Structure and energetics of the hydrogen-bonded backbone in protein folding. *Annu. Rev. Biochem.*, **77**, 339–362.

14 Bowler, B. (2007) Thermodynamics of protein denatured states. *Mol. Biosyst.*, **3**, 88.

15 Schellman, J. (2002) Fifty years of solvent denaturation. *Biophys. Chem.*, **96**, 91–101.

16 Shi, Z., Chen, K., Liu, Z., and Kallenbach, N.R. (2006) Conformation of the backbone in folded proteins. *Chem. Rev.*, **106**, 1877–1897.

17 Ball, P. (2008) Water as an active constituent in cell biology. *Chem. Rev.*, **108**, 74–108.

18 Beck, D., Bennion, B., Alonso, D., and Daggett, V. (2007) Simulations of macromolecules in protective and denaturing osmolytes: properties of mixed solvent systems and their effects on water and protein structure and dynamics. *Methods Enzymol.*, **428**, 373–396.

19 Daggett, V. (2006) Protein folding – simulation. *Chem. Rev.*, **106**, 1898–1916.

20 Rosgen, J. (2007) Molecular basis of osmolyte effects on protein and metabolites. *Methods Enzymol.*, **428**, 459–486.

21 Baldwin, R. (2007) Energetics of protein folding. *J. Mol. Biol.*, **371**, 283–301.

22 Lin, T. and Timasheff, S. (1994) Why do some organisms use a urea–methylamine mixture as osmolyte? Thermodynamic compensation of urea and TMAO interactions with protein. *Biochemistry*, **33**, 12695–12701.

23 Hochachka, P.W. and Somero, G.N. (2002) *Biochemical Adaptation. Mechanism and Process in Physiological Evolution*, Oxford University Press, New York.

24 MacMillen, R.E. and Lee, A.K. (1967) Australian desert mice: independence of exogenous water. *Science*, **158**, 383–385.

25 Auton, M., Holthauzen, L., and Bolen, D. (2007) Anatomy of energetic changes accompanying urea-induced protein denaturation. *Proc. Natl. Acad. Sci. U.S.A.*, **104**, 15317–15322.

26 Auton, M. and Bolen, D. (2005) Predicting the energetics of osmolyte-induced protein folding/unfolding. *Proc. Natl. Acad. Sci. U.S.A.*, **102**, 15065–15068.

27 Rossky, P.J. (2008) Protein denaturation by urea: slash and bond. *Proc. Natl. Acad. Sci. U.S.A.*, **105**, 16825–16826.

28 O'Brien, E.P., Ziv, G., Haran, G., Brooks, B.R., and Thirumalai, D. (2008) Effects of denaturants and osmolytes on proteins are accurately predicted by the molecular transfer model. *Proc. Natl. Acad. Sci. U.S.A.*, **105**, 13403–13408.

29 Hua, L., Zhou, R., Thirumalai, D., and Berne, B.J. (2008) Urea denaturation by stronger dispersion interactions with proteins than water implies a 2-stage unfolding. *Proc. Natl. Acad. Sci. U.S.A.*, **105**, 16928–16933.

30 Frank, H. and Franks, F. (1968) Structural approach to the solvent power of water for hydrocarbons; urea as a structure breaker. *J. Chem. Phys.*, **48**, 4746.

31 Bennion, B.J. and Daggett, V. (2003) The molecular basis for the chemical denaturation of proteins by urea. *Proc. Natl. Acad. Sci. U.S.A.*, **100**, 5142–5147.

32. Caballero-Herrera, A., Nordstrand, K., Berndt, K., and Nilsson, L. (2005) Effect of urea on peptide conformation in water: molecular dynamics and experimental characterization. *Biophys. J.*, **89**, 842–857.

33. Kirkwood, J.G. and Buff, F.P. (1951) The statistical mechanical theory of solutions. 1. *J. Chem. Phys.*, **19**, 774–777.

34. Rosgen, J., Pettitt, B.M., and Bolen, D.W. (2005) Protein folding, stability, and solvation structure in osmolyte solutions. *Biophys. J.*, **89**, 2988–2997.

35. Smith, P.E. (2004) Local chemical potential equalization model for cosolvent effects on biomolecular equilibria. *J. Phys. Chem. B*, **108**, 16271–16278.

36. O'Brien, E., Dima, R., Brooks, B., and Thirumalai, D. (2007) Interactions between hydrophobic and ionic solutes in aqueous GdnCl and urea solutions: lessons for protein denaturation mechanism. *J. Am. Chem. Soc.*, **129**, 7346–7353.

37. Sharp, K., Madan, B., Manas, E., and Vanderkooi, J. (2001) Water structure changes induced by hydrophobic and polar solutes revealed by simulations and infrared spectroscopy. *J. Chem. Phys.*, **114**, 1791.

38. Wallqvist, A., Covel, D., and Thirumalai, D. (1998) Hydrophobic interactions in aqueous urea solutions with implications for mechanism of protein denaturation. *J. Am. Chem. Soc.*, **120**, 427–428.

39. Mountain, R.D. and Thirumalai, D. (2003) Molecular dynamics simulations of end-to-end contact formation in hydrocarbon chains in water and aqueous urea solution. *J. Am. Chem. Soc.*, **125**, 1950–1957.

40. Kokubo, H. and Pettitt, B. (2007) Preferential solvation in urea solutions at different concentrations: properties from simulation studies. *J. Phys. Chem. B*, **111**, 5233–5242.

41. Mountain, R. and Thirumalai, D. (2004) Importance of excluded volume on the solvation of urea in water. *J. Phys. Chem. B*, **108**, 6826–6831.

42. Weerasinghe, S. and Smith, P.E. (2003) A Kirkwood–Buff derived force field for mixtures of urea and water. *J. Phys. Chem. B*, **107**, 3891–3898.

43. Schellman, J. (2003) Protein stability in mixed solvents: a balance of contact interaction and excluded volume. *Biophys. J.*, **85**, 108–125.

44. Kang, M. and Smith, P.E. (2007) Preferential interaction parameters in biological systems by K-B theory and computer simulation. *Fluid Phase Equilibria*, **256**, 14–19.

45. Ben-Naim, A. (1978) Standard thermodynamics of transfer. Uses and misuses. *J. Phys. Chem.*, **82**, 792–803.

46. Schellman, J. (1955) The thermodynamics of urea solutions and the heat of formation of the peptide hydrogen bond. *Compt. Rend. Lab. Carlsberg. Ser. Chim.*, **29**, 225–229.

47. Rosgen, J., Pettitt, B., Perkyns, J., and Bolen, D.W. (2004) Statistical thermodynamic approach to the chemical activities in two-component solutions. *J. Phys. Chem. B*, **108**, 2048–2055.

48. Stokes, R.H. and Robinson, R.A. (1966) Interactions in aqueous nonelectrolyte solutions. Solute–solvent equilibria. *J. Phys. Chem.*, **70**, 2126–2131.

49. Rosgen, J., Pettitt, B., and Bolen, D.W. (2005) Protein folding, stability, and solvation structure in osmolyte solutions. *Biophys. J.*, **89**, 2988–2997.

50. Rosgen, J., Pettitt, B., and Bolen, D.W. (2007) An analysis of the molecular origin of osmolyte-dependent protein stability. *Protein Sci.*, **16**, 733–643.

51. Kuharski, R. and Rossky, P. (1984) Solvation of hydrophobic species in aqueous urea solution: a molecular dynamics study. *J. Am. Chem. Soc.*, **106**, 5794–5800.

52. Kuharski, R. and Rossky, P. (1984) Molecular dynamics study of solvation in urea–water solution. *J. Am. Chem. Soc.*, **106**, 5786–5793.

53. Nozaki, Y. and Tanford, C. (1963) The solubility of amino acids and related compounds in aqueous urea solutions. *J. Biol. Chem.*, **238**, 4074–4081.

54. Ikeguchi, M., Nakamura, S., and Shimizu, K. (2001) Molecular dynamics study on hydrophobic effects in aqueous urea solutions. *J. Am. Chem. Soc.*, **123**, 677–682.

55. Tobi, D., Elber, R., and Thirumalai, D. (2003) The dominant interaction between peptide and urea is electrostatic in nature: a MD study. *Biopolymers*, **68**, 359–369.

56 Lee, M.E. and van der Vegt, N.F. (2006) Does urea denature hydrophobic interactions? *J. Am. Chem. Soc.*, **128**, 4948–4949.

57 Caflisch, A. and Karplus, M. (1999) Structural details of urea binding to barnase: a molecular dynamics study. *Structure*, **7**, 477–488.

58 Tirado-Rives, J., Orozco, M., and Jorgensen, W.L. (1997) Molecular dynamics simulations of the unfolding of barnase in water and 8M aqueous urea. *Biochemistry*, **36**, 7313–7329.

59 Hong, J., Capp, M., Anderson, C., and Record, M. (2003) Preferential interactions in aqueous solutions of urea and KCl. *Biophys. Chem.*, **105**, 517–532.

60 Stumpe, M.C. and Grubmüller, H. (2007) Interaction of urea with amino acids: implications for urea-induced protein denaturation. *J. Am. Chem. Soc.*, **129**, 16126–16131.

61 Celinski, S.A. and Scholtz, J.M. (2002) Osmolyte effects on helix formation in peptides and the stability of coiled-coils. *Protein Sci.*, **11**, 2048–2051.

62 Henkels, C., Kurz, J., Fierke, C., and Oas, T. (2001) Linked folding and anion binding of the *Bacillus subtilis* ribonuclease P protein. *Biochemistry*, **40**, 2777–2789.

63 Holthauzen, L. and Bolen, D. (2007) Mixed osmolytes: the degree to which one osmolyte affects the protein stabilizing ability of another. *Protein Sci.*, **16**, 1–6.

64 Baskakov, I. and Bolen, D.W. (1998) Forcing thermodynamically unfolded proteins to fold. *J. Biol. Chem.*, **273**, 4831–4834.

65 Schein, C.H., Oezguen, N., Volk, D.E., Garimella, R., Paul, A., and Braun, W. (2006) NMR structure of the viral peptide linked to the genome (VPg) of poliovirus. *Peptides*, **27**, 1676–1684.

66 Doolittle, R.F. (2003) Some notes on crystallizing fibrinogen and fibrin fragments. *Biophys. Chem.*, **100**, 307–313.

67 Hill, C.M., Bates, I.R., White, G.F., Hallet, F.R., and Harauz, G. (2002) Effects of the osmolyte trimethylamine-N-oxide on conformation, self-association, and two-dimensional crystallization of myelin basic protein. *J. Struct. Biol.*, **139**, 13–26.

68 Granata, V., Palladino, P., Tizzano, B., Negro, A., Berisio, R., and Zagari, A. (2006) The effect of the osmolyte trimethylamine N-oxide on the stability of the prion protein at low pH. *Biopolymers*, **82**, 234–240.

69 Nandi, P., Bera, A., and Sizaret, P. (2006) Osmolyte trimethylamine *N*-oxide converts recombinant α-helical prion protein to its soluble β-structured Form at high temperature. *J. Mol. Biol.*, **362**, 810–820.

70 Tatzelt, J., Prusiner, S.B., and Welch, W.J. (1996) Chemical chaperones interfere with the formation of scrapie prion protein. *EMBO J.*, **15**, 6363–6373.

71 Eidenmuller, J., Fath, T., Hellwig, A., Reed, J., Sontag, E., and Brandt, R. (2000) Structural and functional implications of tau hyperphosphorylation: information for phosphorylation-mimicking mutated tau proteins. *Biochemistry*, **39**, 13166–13175.

72 Scaramozzino, F., Peterson, D., Farmer, P., Gerig, J., Graves, D., and Lew, J. (2006) TMAO promotes fibrillization and microtubule assembly activity in C-terminal repeat region of tau. *Biochemistry*, **45**, 3684–3691.

73 Tseng, H.C. and Graves, D.J. (1998) Natural methylamine osmolytes, trimethylamine-N-oxide and betaine, increase tau-induced polymerization of microtubules. *Biochem. Biophys. Res. Commun.*, **250**, 726–730.

74 Ignatova, Z. and Gierasch, L. (2007) Effects of osmolytes on protein folding and aggregation in cells. *Methods Enzymol.*, **428**, 355–372.

75 Cottone, G., Ciccotti, G., and Cordone, L. (2002) Protein–trehalose–water structures in trehalose coated carboxy-myoglobin. *J. Chem. Phys.*, **117**, 9862–9866.

76 Felitsky, D., Cannon, J., Capp, M., Hong, J., Wynsberghe, A., Anderson, C., and Record, M. (2004) The exclusion of glycine betaine from anionic biopolymer surface: why glycine betaine is an effective osmoprotectant but also a compatible solvent. *Biochemistry*, **43**, 14732–14743.

77 Bennion, B. and Daggett, V. (2004) Counteraction of urea-induced protein denaturation by TMAO: a chemical chaperone at atomic resolution. *Proc. Natl. Acad. Sci. U.S.A.*, **101**, 6433–6438.

78 Bennion, B., Demarco, M., and Daggett, V. (2004) Preventing misfolding of the prion protein by trimethylamine N-oxide. Biochemistry, 43, 12955–12963.

79 Zou, Q., Bennion, B., Daggett, V., and Murphy, K. (2002) The molecular mechanism of stabilization of proteins by TMAO and its ability to counteract the effects of urea. J. Am. Chem. Soc., 124, 1192.

80 Di Michele, A., Freda, M., Onori, G., Paolantoni, M., Santucci, A., and Sassi, P. (2006) Modulation of hydrophobic effect by cosolutes. J. Phys. Chem. B, 110, 21077–21085.

81 Di Michele, A., Freda, M., Onori, G., and Santucci, A. (2004) Hydrogen bonding of water in aqueous solutions of TMAO and TBA: a near-infrared study. J. Phys. Chem. A, 108, 6145–6150.

82 Fornili, A., Civera, M., Sironi, M., and Fornili, S. (2003) Molecular dynamics simulation of aqueous solutions of TMAO and tert-butyl alcohol. Phys. Chem. Chem. Phys., 5, 4905–4910.

83 Freda, M., Onori, G., and Santucci, A. (2001) Infrared study of the hydrophobic hydration and hydrophobic interactions in aqueous solutions of tert-butyl alcohol and TMAO. J. Phys. Chem. B, 105, 12714–12718.

84 Freda, M., Onori, G., and Santucci, A. (2002) Hydrophobic hydration and hydrophobic interaction in aqueous solutions of tert-butyl alcohol and TMAO: a correlation with the effect of these two solutes on the micellization process. Phys. Chem. Chem. Phys., 4, 4979–4984.

85 Paul, S. and Patey, G. (2006) Why tert-butyl alcohol associates in aqueous solution but TMAO does not. J. Phys. Chem. B, 110, 10514–10518.

86 Sinibaldi, R., Casieri, C., Melchionna, S., Onori, G., Segre, A., Viel, S., Mannina, L., and De Luca, F. (2006) the role of water coordination in binary mixtures. A study of two model amphiphilic molecules in aqueous solutions by MD and NMR. J. Phys. Chem. B, 110, 8885–8892.

87 Kocherbitov, V., Veryazov, V., and Soderman, O. (2007) Hydration of trimethylamine-N-oxide and of dimethyldodecylamine-N-oxide: an ab initio study. J. Mol. Struct. (Theochem), 808, 111–118.

88 Gallagher, K.R. and Sharp, K.A. (2003) A new angle on heat capacity changes in hydrophobic solvation. J. Am. Chem. Soc., 125, 9853–9860.

89 Athawale, M.V., Dordick, J., and Garde, S. (2005) Osmolyte TMAO does not affect the strength of hydrophobic interactions: origin of osmolyte compatibility. Biophys. J., 89, 858–866.

90 Paul, S. and Patey, G. (2007) Structure and interaction in aqueous urea–TMAO solutions. J. Am. Chem. Soc., 129, 4476–4482.

91 Batchelor, J.D., Olteanu, A., Tripathy, A., and Pielak, G.J. (2004) Impact of protein denaturants and stabilizers on water structure. J. Am. Chem. Soc., 126, 1958–1961.

92 Rosgen, J., Pettitt, B., and Bolen, D. (2007) An analysis of the molecular origin of osmolyte-dependent protein stability. Protein Sci., 16, 733–643.

93 Maclagan, R., Malardier-Jugroot, C., Whitehead, M., and Lever, M. (2004) Theoretical studies of the interaction of water with compensatory and noncompensatory solutes for proteins. J. Phys. Chem. A, 108, 2514–2519.

94 Athawale, M.V., Sarupria, S., and Garde, S. (2008) Enthalpy–entropy contributions to salt and osmolyte effects on molecular-scale hydrophobic hydration and interactions. J. Phys. Chem. B, 112, 5661–5670.

95 Paul, S. and Patey, G. (2007) The influence of urea and TMAO on hydrophobic interactions. J. Phys. Chem. B, 111, 7932.

96 Wang, A. and Bolen, D. (1997) A naturally occurring protective system in urea-rich cells: mechanism of osmolyte protection of proteins against urea denaturation. Biochemistry, 36, 9101–9108.

97 Mello, C. (2003) Measuring the stability of partly folded proteins using TMAO. Protein Sci., 12, 1522–1529.

98 Chitra, R. and Smith, P.E. (2002) Molecular association in solution: a K-B analysis of sodium chloride, ammonium sulfate, guanidinium chloride, urea, and 2,2,2-trifluoroethanol in water. J. Phys. Chem. B, 106, 1491–1500.

5
Modeling Aqueous Solvent Effects through Local Properties of Water

Sergio A. Hassan and Ernest L. Mehler

5.1
The Role of Water and Cosolutes on Macromolecular Thermodynamics

Whether *in vivo* or *in vitro*, biomolecular interactions occur in aqueous environments that vary broadly in composition. The aqueous medium affects the thermodynamic and kinetic behavior of biomolecules, and controls their chemical reactions [1–3]. Changes in the composition of the medium can have profound effects on biomolecular properties. For example, changing the concentration of salts may affect protein solubility [4], and high concentrations may induce protein aggregation and precipitation. Changing the makeup of the solution can also affect the thermodynamics and structural stability of single proteins [5]. For example, protein denaturation [6] and dissociation of multimeric proteins [7] can be promoted by alcohols, urea, and guanidine hydrochloride. On the other hand, protein stabilization may be induced by the addition of sucrose, glycerol, certain amino acids, and salts [8, 9]. Mixed aqueous solvents are commonly used to investigate macromolecular processes, such as protein thermal stability, kinetics of protein unfolding and/or refolding, self-assembly, and aggregation.

Substances added to water in an aqueous solution are referred to as cosolutes because they coexist with the solute of interest, here a macromolecule. Cosolutes generally consist of inorganic ions or low-molecular-weight organic molecules such as glycerol, sorbitol, sucrose, glucose, single amino acids, and so on. When studying denaturation, stabilization, or precipitation, cosolutes are usually present at very high concentrations (up to ~50 wt% is not uncommon). Under these conditions, the cosolutes behave like another solvent coexisting with water, so they are also referred to as cosolvents. Throughout this chapter the word "cosolute" will be retained, although ions and simple salts will be explicitly differentiated from other cosolutes.

In an aqueous solution, the change in free energy of, for example, a folding–unfolding process is given by $\Delta G = -RT \ln K = -RT \ln(a_U/a_F)$, where a_U and a_F are the activities of the protein in the unfolded and folded states, respectively, and K is the equilibrium constant [10]. If the corresponding change in pure water is denoted

---

Modeling Solvent Environments. Applications to Simulations of Biomolecules. Edited by Michael Feig
Copyright © 2010 WILEY-VCH Verlag GmbH & Co. KGaA, Weinheim
ISBN: 978-3-527-32421-7

by ΔG^0, then the thermodynamic effects of the cosolutes on the thermodynamic equilibrium can be described by $\delta\Delta G = \Delta G - \Delta G^0 = \Delta G_U - \Delta G_F$, where ΔG_U and ΔG_F are the transfer free energies of the unfolded and folded proteins from pure water to the solution. Other thermoclynamic processes, such as self-assembly, aggregation, and precipitation can be similarly described. The presence of cosolutes affects the thermodynamic equilibrium as well as the kinetics of reactions. Biochemical pathways involve a number of steps, each characterized by its own rates and thermodynamic equilibrium. For example, protein–protein association has been described by a two-step mechanism, where a weakly bound complex is first formed as a result of protein encounters that subsequently may evolve toward a tighter complex in the second step [11]. The presence of cosolutes may affect these steps differently, or change the enthalpic and entropic contributions independently, possibly shifting the equilibrium in opposite directions [12]. A basic goal of a thermodynamic or kinetic study of reactions in solution, including computer simulations and theoretical modeling, is to estimate these transfer free energies and reaction rates, and understand how ions and cosolutes modulate these quantities.

To date, no clear picture has emerged that allows prediction of how one particular cosolute affects the properties of a given protein, and its dependence on thermodynamic descriptors such as concentration, temperature, and pH. A few trends have nonetheless been identified [1, 13–16]. In the case of denaturation, cosolutes (called denaturants in this case) appear to bind directly to the protein molecule, thus excluding water from its vicinity (preferential binding), while protein stabilization is usually accompanied by exclusion of the cosolutes (stabilizers in this case) from the immediate region surrounding the protein, thus creating an enriched water environment at the protein–liquid interface (preferential hydration). Preferential hydration appears to be the primary mechanism driving precipitation and self-assembly. Cosolutes also affect the properties of the water molecules, such as their translational and rotational motion, and induce and mediate long-range hydrogen bonding networks. These changes modify the water viscosity, self-diffusion, and dielectric response, all of which may indirectly affect macromolecular interactions, thermodynamics, and transport properties.

At the cellular level, changes in the composition of the medium can perturb the molecular machinery of the cell and compromise cell survival and growth. As a result, natural selection has evolved highly regulated control mechanisms to keep the subcellular compartments under strict thermodynamic control [1, 16]. Certain microorganisms can rapidly adapt to osmotic stress, such as desiccation, freezing, or sudden variations of cosolute concentration outside the cell. These processes affect water concentration in the cytoplasm, which may lead to protein denaturation, aggregation, changes in enzyme reaction rates, inhibition of cell growth, or even cell death. Cells have adapted by transiently inducing expression of protective genes that lead to biosynthesis of osmolyte or cryoprotectant such as glycerol [17–21]. The subtle thermodynamic balance demanded by nature is illustrated in the case of glycerol, which has been shown to inhibit self-assembly of collagen [22] and to enhance self-assembly of actin [23] and tubulin [24], all essential proteins in cells and tissues.

Interestingly, despite the variety of chemical compounds that cells could use, the number of naturally occurring osmolytes and cryoprotectants is rather small. Polyols, certain free amino acids and their derivatives, and specific urea–methylamine combinations are the only types of osmolytes found in all organisms, from bacteria, to plants, to vertebrates [16]. The apparent reason for this is the maintenance of the integrity of cellular processes, which requires either that cosolutes do not interfere with critical biomolecular processes, or that they act favorably on certain proteins when needed (for example, as structure stabilizers). From an evolutionary perspective, then, genes have evolved so that osmolytes and cryoprotectants are compatible with macromolecular structure and function. This tradeoff is rather stringent, and only a few chemical groups are suitable for the task, typically small, net-neutral, polar molecules at physiological pH. Ions, for example, are not used in general for intracellular thermodynamic control because critical cellular processes are highly sensitive to ion concentrations.

The preceding discussion illustrates the impact of the composition of the aqueous medium across all biological scales. For realistic applications in biology and biochemistry, a solvent model must describe all the physical effects that control the biomolecular processes discussed above. This includes the effects of ions and cosolutes, crowding, pH, and temperature (clearly, pure water is an unrealistic environment). The objectives of this chapter are: (i) to identify the relevant forces that operate in aqueous media, particularly in pure water and in ionic solutions, and to describe these forces quantitatively (Section 5.2); and (ii) to use the insight from (i) to guide the development of a water model for aqueous solutions of arbitrary compositions (Sections 5.3 and 5.4). Point (ii) implies that all the components of the system except water will be treated on equal footing.

5.2
Forces Induced by Water in Aqueous Solutions

From a thermodynamic point of view, the action of an aqueous solution on a protein is determined by a balance between the affinities of the proteins for the water molecules and for the cosolutes. In principle, this balance defines a broad range of cosolute actions, from strong stabilizers to strong denaturants; and from strong solubility enhancers to strong precipitants. Prediction of these effects, for example through computer simulations, may help to design agents that elicit specific actions. A number of experimental techniques are used to study preferential interactions with the solvent components [1, 15], but these techniques can only provide a thermodynamic (that is, phenomenological) description of the interactions. Microscopic details, such as the number of cosolute binding sites or hydration numbers, usually require assumptions and simplified models to interpret the data [25]. However, understanding the effects of water and cosolutes at the atomic level is ultimately needed for the development of a physically realistic model of solvent effects. Because of the experimental difficulties to produce microscopic information, computer simulations have emerged as an ideal approach to investigate molecular interactions in solution, such as short-time water dynamics, water structure and spatial reconfiguration, orientational relaxation, cooperative effects, local dielectric properties, hydrogen bond dynamics, and intermolecular forces induced by the bulk and by the hydration shells. In this section, molecular dynamic simulations are used to investigate the balance of forces operating in aqueous solutions, both in pure water and in the presence of ions.

5.2.1
Interactions in Water-Accessible Regions of Proteins

Hydrogen-bonded molecules provide a convenient framework to illustrate the effects of water on intermolecular forces and their modulations by salts and cosolutes. Hydrogen bonds (HB) are key determinants of structure, dynamics, and thermodynamics of proteins and nucleic acids, and play a role in the specificity of ligand binding and rates of enzymatic reactions [26, 27]. Many of the cosolutes discussed in the preceding section contain HB-forming groups.

Two important characteristics of HB are their geometries and energies [26, 28]. In the gas phase, the HB geometries are intrinsic properties of the interacting molecules, free from the perturbations introduced by the medium in the condensed state. Experimental studies in the gas phase have shown that the shared proton is generally oriented toward the electron lone-pair orbitals in the acceptor atom, and that the acceptor–proton–donor angle is close to 180° [29, 30]. These rules are confirmed by gas-phase quantum-chemical geometry optimization [28] and are consistent with the donor and acceptor spatial distributions in crystal structures of small molecules [30] and proteins [31]. Unlike the gas phase, an aqueous medium generates forces and torques that perturb the geometry determined by the electronic structure alone. Thus, HB geometries in liquids are modulated by several intermolecular forces.

It is generally assumed that HB energies in proteins are ~15 kcal mol^{-1} or less [26, 27], although larger values are expected between charged groups. Hydrogen bonding energies are substantially reduced in water-accessible regions because donor and acceptor groups can form HB with water molecules. Experimental studies in enzyme–substrate complexes have shown that deleting a side chain that forms a hydrogen bond between the enzyme and the substrate weakens the binding free energy by ~0.5 to ~4.5 kcal mol^{-1}, depending on the charge of the group left unpaired (in a few cases binding is slightly strengthened) [32]. In these studies, the donor donor atom and the acceptor atoms must break their HB with water molecules during substrate binding, so the measured energies contain information about the competing binding with water at the protein–liquid interface.

5.2.2
Insight from Atomistic Dynamic Simulations

An accurate description of HB energies and geometries is needed for realistic simulations of biomolecules. Failure to correctly represent HB will most likely lead

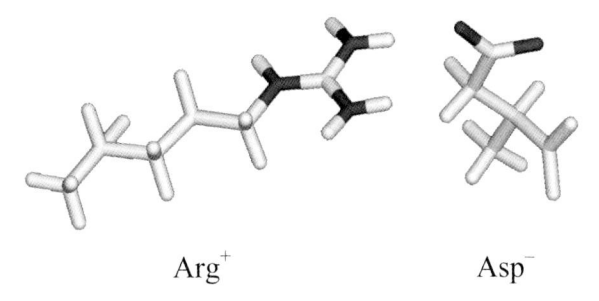

Figure 5.1 Relative orientations of protonated arginine (Arg⁺) and unprotonated aspartate (Asp⁻) used in the calculation of the intermolecular potential of mean force.

to unreliable predictions of biophysical properties, such as peptide and protein structures, protein–protein interactions, strength and modes of ligand binding, and kinetics. Given the difficulties in measuring the strength of HB interactions in solution, atomistic computer simulations can be used to address the problem systematically. Such a study was reported in [33], where HB interactions between groups commonly found at the protein–water interface were considered. The HB energies were calculated from the intermolecular potentials of mean force V between pairs of polar and charged hydrogen-bonded amino acids in pure water. The potential is given by $V(r) = V_w(r) + \phi(r)$, where ϕ is the intermolecular potential energy (bare Coulomb plus van der Waals) in the gas phase, V_w is the intermolecular potential energy associated with the mean forces exerted by the water molecules, and r is a suitable reaction coordinate, here taken as the distance between the acceptor atom and the shared proton. The relative orientations of the amino acids in the dimers were fixed at the geometry indicated by the gas-phase rules mentioned in the previous section. Figure 5.1 shows the relative orientations in one of the dimers, Asp⁻–Arg⁺. Although the stable geometries are not expected to be the same in water, this prescription nonetheless allows reproducibility of the computational setup since the precise form of V depends on the relative configurations of the monomers.

The potentials V_w are calculated as the mechanical work required to bringing the two amino acids to an intermolecular distance r, starting from an infinite separation [33]:

$$V_w(r) = -\frac{1}{2}\int_\infty^r [\langle F_A(r_A)\rangle_t - \langle F_D(r_D)\rangle_t]\cdot\hat{r}\,dr \qquad (5.1)$$

Here $\langle F_i\rangle_t$ is the time-averaged force exerted by the water molecules on the acceptor (A) and donor (D) monomers evaluated at their centers of mass with positions r_i; and $\hat{r}$ is the unit vector along the direction of movement, taken here as a straight line connecting the acceptor atom, the shared proton, and the donor atom, and oriented from donor to acceptor. The relative orientations of the monomers are kept fixed during the calculations. This is reasonable in proteins, since the backbone tends to impose constraints on the free rotation of the side chains [27]; the

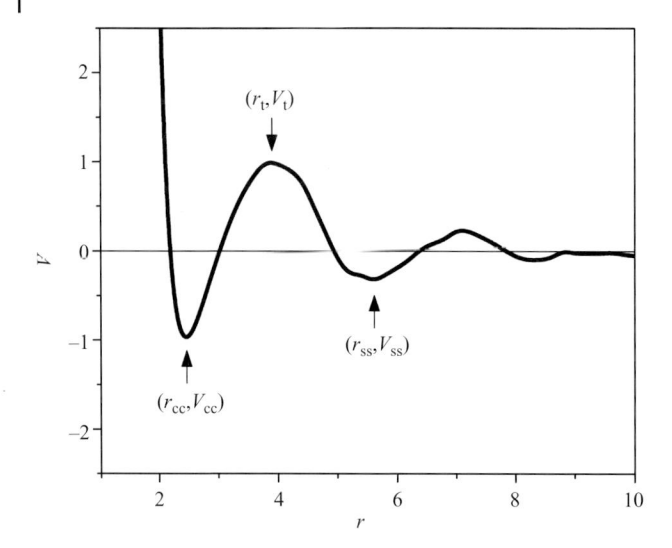

Figure 5.2 Typical intermolecular potential of mean force (kcal mol^{-1}) calculated using Equation 5.1, showing the close-contact minimum $V_{cc} \equiv V(r_{cc})$ at the close-contact distance r_{cc} (Å), the desolvation barrier $V_t \equiv V(r_t)$, and the solvent-separated minimum $V_{ss} \equiv V(r_{ss})$. Other (shallower) minima and maxima are observed at $r > r_{ss}$.

same can be said of nucleotide base pairs in nucleic acids. Moreover, the same interacting pair may appear with different relative orientations in different (or the same) proteins, so the unambiguous setup adopted here provides a standard for the calculation of the HB energy of a given pair. The results can then be used as benchmarks for simplified water models, as discussed in [34] (cf. Section 5.4).

A typical plot of V versus r is displayed in Figure 5.2. The potential shows a close-contact minimum V_{cc} and a solvent-separated minimum V_{ss}, both separated by a desolvation barrier V_t. The HB energy is defined by V_{cc}, which is generally the global minimum of the potential, and is typically within ~6 kcal mol^{-1} in all of the dimers studied. This range is similar to the estimates from the mutagenesis study described above [32], although the relationship between the HB energies and the charge/polarity of the monomers is less apparent. The potentials show additional barriers and minima beyond the water-separated distance, which appear to originate from forces induced by further restructuring of water in the intervening space between the monomers [35]. This restructuring can be visualized in Figure 5.3, which shows the peaks of water density as Asp$^-$–Arg$^+$ dissociates from close contact. The peaks are located at HB distances from either the donor or the acceptor atoms in functional groups, or from other peaks. Secondary peaks are also observed at HB distances from the first peaks (not shown), although they are smaller as water becomes less structured with increasing distance from the monomers. The sequence in Figure 5.3 shows that new peaks appear between the monomers as they dissociate from each other [35–37]. At sufficiently large intermolecular separation, the interconnections of the peaks break down and the monomers become independently hydrated. This "rupture point" is in the $r \sim 7$–12 Å range

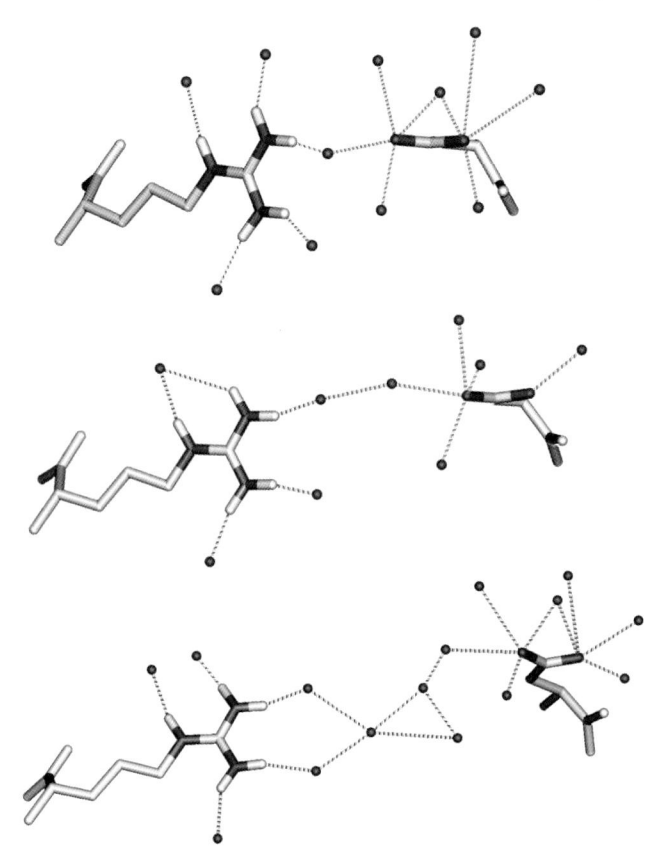

Figure 5.3 Restructuring of water around Asp⁻ and Arg⁺ as both amino acids dissociate from close contact (upper panel corresponds to $r = r_{ss}$). The dots represent peaks of water density at hydrogen bonding distances from either the acceptor or donor groups of the amino acids, or from other peaks. Secondary peaks can be observed at hydrogen bonding distances from the first peaks (not shown).

depending on the interacting monomers, being larger in general for charge–charge interactions [35]. The high degree of structural water observed in these model systems is likely to impose forces and torques on similar groups at the surface of real proteins. The long-range interconnection of the peaks also suggests the existence of transient water chains [37] that may facilitate water-mediated proton transfer between donor and acceptor groups [38].

5.2.3
Bulk and Non-Bulk Contributions of the Water Forces

The water potential V_w is determined by the forces exerted by water molecules in the bulk and in the non-bulk regions closer to the monomers. In pure water, bulk forces are mostly electrostatic, originating in the polarization and orientation of the water molecules as a response to the electric field created by a solute [39, 40]. Besides the bulk forces, other water-induced forces (generally known as "solvent-induced forces" or SIF) originate in the restructuring of water molecules surrounding a solute as a consequence of their exclusion from the region occupied by the solute itself [41, 42]. Compared to the bulk phase, this rearrangement modifies the HB network of water, which creates net forces and torques that affect the solute's equilibrium structure, dynamics, and interactions. SIF operate independently of the polar character of the solute. For example, hydrophobic forces [43] are SIF between uncharged, non-polar molecules. For polar or charged solutes, the rearrangement of the excluded water and its HB network is locally perturbed by the solute's electric field. In this case the microscopic nature of SIF is more complex, resulting in hydrophilic forces [44] that may affect a number of biophysical properties, such as protein–ligand interactions and protein folding [42, 45, 46].

The magnitude of the bulk forces and SIF are illustrated here for the dimers discussed in the preceding section. To this end, the potential V_w is partitioned into two terms, $V_w(r) = V_b(r) + V_s(r)$, where V_b is the bulk contribution and V_s is the non-bulk contribution. A practical definition of bulk and non-bulk regions is adopted here: for each value of r, the non-bulk region is defined as the first and second hydration shells of the dimers, while the bulk region is the remaining water. Figure 5.4 shows V_w, V_b, and V_s in four representative cases: (a) charged acceptor hydrogen bonded to charged donor, (b) charged acceptor and polar donor, (c) polar acceptor and charged donor, and (d) polar acceptor and polar donor. The components V_b and V_s are calculated with Equation 5.1 using in each case the forces exerted by the water molecules in the corresponding regions. In the convention used here, $dV/dr > 0$ (or < 0) means that water induces an attractive (or repulsive) force between the monomers.

The dimer Asp^-–Arg^+ (Figure 5.4a) represents the extreme case of ionic interaction. In this case the water molecules induce a strong repulsive force at all intermolecular distances, and substantial mechanical work is needed to bring the acceptor and donor groups to close contact. The partial contributions show that the bulk forces are consistently stronger than the non-bulk forces for $r > 6$ Å, but the two become comparable at shorter distances, yielding similar contributions to V_w at close contact r_{cc} (indicated by the vertical line). The bulk forces are almost constant and repulsive at all intermolecular distances (although they vanish asymptotically at sufficiently large separation). In contrast to the bulk forces, the non-bulk forces increase sharply at short distances, becoming strongly repulsive as the hydration shells of the monomers start overlapping. This is due to the fact that water molecules in the first hydration shell are more strongly hydrogen bonded to the donor and acceptor groups than they are to neighboring water molecules, so substantial work is needed to dehydrate these groups and bring them together.

The balance of forces just discussed determines the form of V_w in all of the other dimers, and can be discussed in similar terms. However, the relative weights of V_b and V_s change as the charge of the interacting monomers decreases from Figure

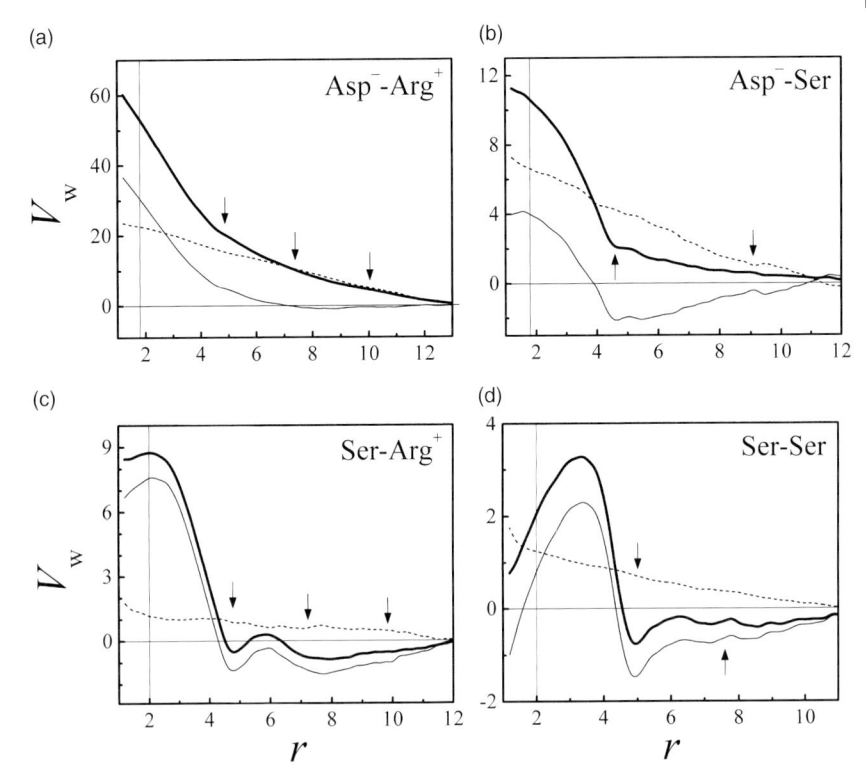

5.2 Forces Induced by Water in Aqueous Solutions | 101

Figure 5.4 Total water contribution V_w (thick solid line) to the total intermolecular potentials of mean force $V = V_w + \phi$, where ϕ is the direct gas-phase intermolecular potential energy, for four representative dimers. The bulk V_b (thin dashed line) and non-bulk V_s (thin solid line) partial contributions to $V_w = V_b + V_s$ are also shown. Arrows indicate the approximate distances where major restructuring of water occurs (all energies are in kcal mol^{-1} and distances in Å). The vertical lines indicate the close-contact minima of the potential: $r_{cc} = 1.8$ Å in (a) and (b), and $r_{cc} = 2$ Å in (c) and (d).

5.4a to d, leading to qualitatively different behavior. For example, the non-bulk forces play an increasingly dominant role as the charge decreases, and are actually the most important contribution to V_w in the case of polar–polar interactions (Figure 5.4d). The bulk forces, however, are not negligible and still contribute substantially (~$2RT$) to V_w at close intermolecular distances. Reducing the charge of one (Figure 5.4b and c) or both (Figure 5.4d) monomers tends to decrease proportionately the magnitude of the bulk forces and associated potentials. This is due to the fact that water molecules in the bulk are less affected by the field generated by polar groups than they are by charged groups. For example, V_b at r_{cc} is ~20 kcal mol^{-1} in Figure 5.4a, ~7 kcal mol^{-1} in Figure 5.4b, and only ~1 kcal mol^{-1} in Figure 5.4c and d. As in Figure 5.4a, however, the bulk forces are always repulsive and practically constant for all intermolecular distances. The magnitude of the non-bulk potentials V_s also tends to decrease proportionately with the charges of the interacting groups. Unlike ionic interactions, water molecules in the first hydration shell are more weakly hydrogen bonded to polar groups, so water can be more easily dissociated as the monomers are brought together. For example, V_s at r_{cc} is ~30 kcal mol^{-1} in Figure 5.4a, ~4 kcal mol^{-1} in Figure 5.4b, ~7 kcal mol^{-1} in Figure 5.4c, and ~1 kcal mol^{-1} in Figure 5.4d. From a thermodynamic point of view, the work required to dissociate the water molecules that are hydrogen bonded to the acceptor and the donor groups is determined by a competition between their affinities for these groups and for other water molecules available for HB. In the case of polar–polar interaction (Figure 5.4d), this balance is such that the perturbation of the hydration shells that result from bringing the monomers together actually leads to an effective attraction below $r \sim 4$ Å. In this case, the water molecules that hydrate the polar groups are easily pulled away by the surrounding water molecules, so a negative mechanical work is needed to associate the monomers. This attraction occurs once a repulsive barrier located at $r \sim 4–5$ Å is overcome. A weak attraction can already be seen in Figure 5.4b and c below $r \sim 2$ Å, reflecting the increasing imbalance of water affinities just mentioned.

In the limit of two uncharged non-polar monomers, the affinity imbalance favors water much more than the acceptor or the donor groups, so a depletion of water between the monomers is expected at sufficiently short intermolecular distances, thus resulting in hydrophobic attraction [43, 47, 48]. To illustrate and emphasize the difference between hydrophobic and hydrophilic SIF, the water component of the potential of mean force of Asp$^-$–Arg$^+$ is recalculated with all the partial charges on the monomers switched off. The potential V_w, redefined in this case as V_H for convenience in the following discussion, is plotted in Figure 5.5. The bulk (electrostatic) forces are negligible (now shown), while the non-bulk

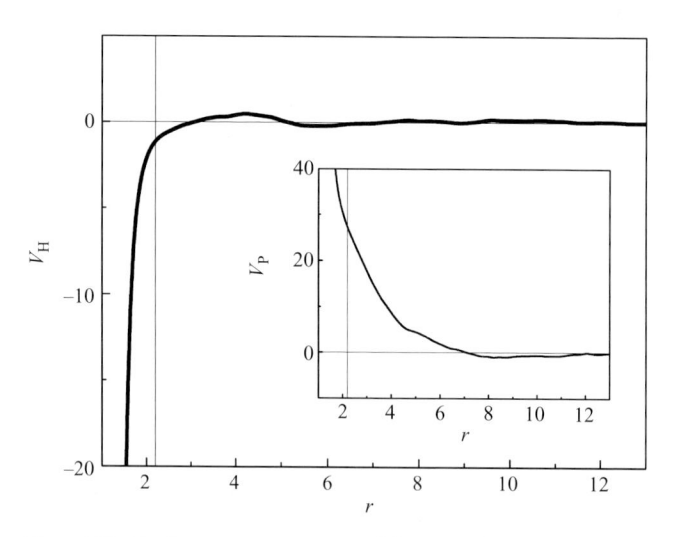

Figure 5.5 Total water component V_w of the intermolecular potential of mean force for the dimer Asp$^-$–Arg$^+$ obtained with all the partial charges on the dimers artificially switched off (in this case the potential V_w is called V_H, that is, $V = V_H + \phi$; see caption to Figure 5.4 and text for details). For the charged dimer, the non-bulk contribution to the potential is $V_s = V_H + V_P$, where V_P is shown in the inset; V_H is a purely (hypothetical) hydrophobic potential, while V_P is the "correction" to the (real) hydrophilic contribution.

forces are strongly attractive at short intermolecular distances; a weak repulsive force is observed in the $r \sim 5$–6 Å range, which can be overcome easily at room temperature. The density peaks around the dimer (not shown) are too shallow and lack the symmetry observed in the presence of polar/charged functional groups (Figure 5.3). Likewise, the water between the monomers lacks long-range structure because this can only emerge in the presence of polar/charged groups, which appear to function as anchoring points.

Any real solute contains some charge distribution that affects the local structure of water to varying degrees. For polar/charged monomers, it is instructive, then, to divide the non-bulk potential as $V_s(r) = V_H(r) + V_P(r)$, where V_H is the (hypothetical) hydrophobic contribution obtained by artificially switching the charges off (as done above for Asp$^-$–Arg$^+$), and V_P is the "correction" due to the real charge distribution on the solute. The latter potential is plotted in the inset of Figure 5.5 for Asp$^-$–Arg$^+$, showing that, in this case, V_P is larger in magnitude than V_H, so the "correction" should not be understood as being "small." Solvent models commonly used in simulations of biomolecules either ignore V_P or introduce it empirically into V_b as part of the electrostatic potential, usually through suitable parameterization of the model. However, V_P is not purely electrostatic in nature since it is determined mainly by water–water and solute–water HB interactions. These specific local interactions confer granularity and directionality to the non-bulk water, and are the origin of the cooperativity and non-pairwise additivity that characterizes the SIF.

5.2.4
Effects of Salts on the Intermolecular Interactions

Ions affect biological processes in a broad range of size scales, from single molecules to multicellular organisms [49]. These processes occur at ambient or physiological conditions, from highly diluted to near-saturating concentrations. For example, the concentrations of the most abundant ions in the extracellular fluids of the body (Na$^+$ and Cl$^-$) are in the ~100–150 mM range, while the concentrations of other common ions (K$^+$, Ca^{2+}, Mg^{2+}, HCO$_3^-$, and PO$_4^{3-}$) are much lower. The subcellular compartments also contain different types and concentrations of ions. For example, the concentrations of K$^+$ and Cl$^-$ in the cytosol are in the ~100–150 mM range, while Na$^+$ is much less abundant than in the extracellular fluids. Much higher concentrations can be found in other biological environments and *in vitro* experiments. For example, certain types of bacteria require high concentrations (above 1 M) of salts for survival, although high concentrations may kill or inhibit the growth of most other microorganisms.

The results of Section 5.2.3 already illustrate the competition of forces that underlie preferential binding and hydration discussed in Section 5.1. This balance of forces must be properly described in a solvent model to provide a good description of macromolecular thermodynamics, and to draw sensible mechanistic conclusions on biological function or to rationalize biochemical data. Interactions of cosolutes with proteins are usually weak and non-specific, but ions may also interact strongly and with high specificity to elicit particular biological responses. It is then convenient to treat ions as a special case and understand their effects at the atomic level of detail.

The treatment of ions in a solvent model is challenging because the model must capture the complex and sometimes specific nature of their interactions with macromolecules. Mean-field approximations (e.g. the Debye-Hückel model) cannot capture the microscopic aspects of these interactions, so for realistic simulations it is convenient to retain their atomistic identity. The presence of ions modifies both the bulk and the non-bulk forces with respect to those of pure water. Bulk forces are affected because the rotation and polarization of the water molecules are determined by both the solute and ion electric fields. The water molecules in the vicinity of an ion or an ion cluster are more structured than in pure water [49–51], so their response to an external field is altered, manifestation of saturation effects discussed in 5.3. This results in measurable changes in the dielectric properties of the bulk solution [52], which affect the intermolecular forces induced by the bulk. In addition, SIF are modified because ions may directly perturb the structure of the hydration shells and the long-range structure of water connecting functional groups. Finally, the direct interactions of ions with the solute generate forces and torques in addition to those of water. All these partial effects cooperate in a concentration-dependent fashion to shape the intermolecular potentials of mean force, as discussed below.

A systematic study of the effects of Na^+ and Cl^- on polar/charged dimers has been reported in [36] for NaCl salt concentrations in the 0.1–2 M range. Only two dimers, Asp^-–Arg^+ and Asp^-–Lys^+, are discussed here to illustrate the complex interplay of forces that should be described in a solvent model. Figure 5.6 shows the effects of ions on the HB energies, given in terms of $\Delta V_{cc}(c) = V_{cc}(c) - V_{cc}$,

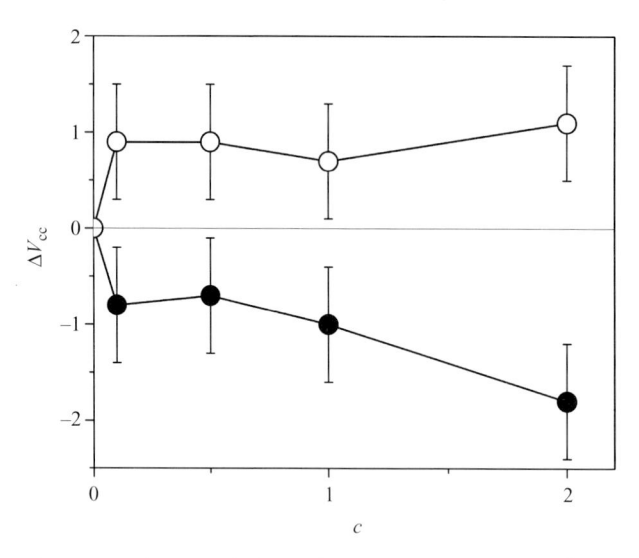

Figure 5.6 Changes in hydrogen bond energies (V_{cc}, kcal mol^{-1}; statistical errors shown) with the addition of NaCl (concentrations c, mol l^{-1}) to the aqueous solution, indicating stabilization of Asp^-–Arg^+ (solid circles) and destabilization of Asp^-–Lys^+ (open circles).

where $V_{cc}(c)$ is the potential of mean force at r_{cc} at a concentration c, and V_{cc} is the corresponding potential in pure water (Section 5.2.2). Positive (negative) values of ΔV_{cc} indicate destabilization (stabilization) of the dimer. Despite their similarities, the dimers display opposite behaviors. Out of eight dimers studied in [36], some showed a clear tendency to being stabilized and others to being destabilized. In a few cases the HB energies remained unchanged within the statistical uncertainty of the calculations. No obvious correlation was observed between the degree of (de)stabilization and the polar/charged nature of the interacting monomers.

To identify the origin of the (de)stabilizations, the quantity $\Delta V(r) = V_c(r) - V(r)$ is analyzed, where V_c is the potential of mean force at a concentration c, and V is the corresponding potential in pure water. Values of ΔV are plotted in Figure 5.7 at 2 M, showing abrupt changes within specific intermolecular distances. As r decreases, downward shifts of ΔV dominate in Asp$^-$–Arg$^+$, leading to the overall stabilization of the dimer, while upward shifts dominate in Asp$^-$–Lys$^+$, leading to its destabilization. An analysis of the forces shows that the stabilization of Asp$^-$–Arg$^+$ results from a weaker (compared to pure water) repulsive force opposing the bare Coulomb attraction between the monomers. In contrast, the repulsive forces in Asp$^-$–Lys$^+$ are generally stronger in solution than in water. The locations of the sharp changes of ΔV and an analysis of the density peaks distributions suggest that, besides the smooth, slowly varying, electrostatic effects of the bulk solution, other forces originating in the restructuring of the liquid play a decisive role.

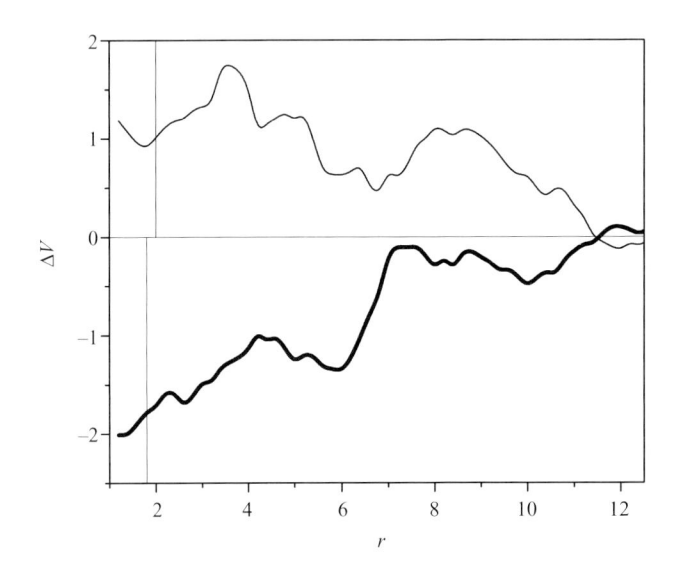

Figure 5.7 Changes in the total intermolecular potentials of mean force $\Delta V(r) = V_c(r) - V(r)$ (kcal mol^{-1}) in aqueous NaCl solutions at concentration c (mol l^{-1}) with respect to pure water, for Asp$^-$–Arg$^+$ (thick line) and Asp$^-$–Lys$^+$ (thin line). Upward and downward shifts in ΔV are apparent at specific intermolecular separations r (Å) that lead to the overall stabilization or destabilization of the dimers by the ions. Vertical lines indicate the close-contact distances r_{cc}.

A qualitative analysis of the spatial redistribution of ions as the monomers separate from close contact was reported in [36]. At certain intermolecular distances, ions relocate in pairs between Asp$^-$ and Arg$^+$. For example, at $r \sim 7$ Å, a Asp$^-$–Na$^+$–Cl$^-$–Arg$^+$ bridge is formed (not shown; see [36]) that contributes to a sudden decrease of the repulsive force. This ion-pair-mediated interaction is not observed in Asp$^-$–Lys$^+$, in spite of the fact that ion pairs do form in the vicinity of the charged groups. For example, at $r \sim 7$ Å, a Na$^+$ ion is attracted to Asp$^-$ but Lys$^+$ fails to attract the Cl$^-$ paired to Na$^+$, suggesting that, unlike Arg$^+$, the affinity of Lys$^+$ for water is greater than its affinity for the anion, at least in the particular configuration used in this study.

In an aqueous solution, the potentials of mean force are given by $V_c(r) = V_L(r) + \phi(r)$, where V_L is the potential generated by the solution at concentration c (equal to V_w in pure water; see Section 5.2.3), which can be divided into the following partial contributions: V_b (bulk solution), V_s (non-bulk solution), V_w (water), V_i (ions), V_{wb} (bulk water), V_{ws} (non-bulk water), V_{ib} (bulk ions), and V_{is} (non-bulk ions). Three partitions are considered here: (I) $V_L = V_b + V_s$; (II) $V_L = V_w + V_i$; and (III) $V_L = V_{wb} + V_{ws} + V_{ib} + V_{is}$. These partial potentials are plotted in Figure 5.8 for Asp$^-$–Arg$^+$ at 2 M; the contributions of pure water are also shown for comparison (thick line). The partition (I) (Figure 5.8a) shows that adding salt weakens the repulsive force of the bulk solution with respect to pure water, as indicated by V_b. This is consistent with the experimental observation of a reduction of the static dielectric constant of aqueous NaCl solutions, as water becomes less responsive to the field [52]. In contrast, the non-bulk contribution increases with respect to pure water. This increase is driven mainly by stronger repulsive forces at long intermolecular distances. The partition (II) (Figure 5.8b) shows that the mechanical work required to associate the monomers is facilitated by the water molecules in the solution when compared to pure water, especially below ~8 Å, as indicated by V_w. In contrast, the ions work mostly against the association of the monomers, as indicated by V_i. As a result both V_w and V_i contribute similarly to V_L at short intermolecular distances.

The partition (III) (Figure 5.8c and d) shows that ions in the bulk contribute much less to the potentials than ions in the non-bulk region, as indicated by V_{ib} and V_{is}, although the former cannot be neglected at short intermolecular distances ($V_{ib} > RT$ at r_{cc}). On the other hand, the repulsive force exerted by the water molecules in the bulk is weaker in solution than in pure water, as indicated by V_{wb}, which suggests also a decrease of the dielectric response of water; a comparison with V_b indicates that this drop is driven mainly by water. The behavior of the non-bulk water is the most complex of all the partial terms, as indicated by V_{ws}, which is larger in solution than in pure water above $r \sim 4$ Å, but reverses behavior at shorter distances. A physically correct model of water forces should, ideally, describe the behavior of V_{wb} and V_{ws} independently (this is the goal in Sections 5.3 and 5.4), while a less demanding model may describe the behavior of V_w as a whole. At any rate, the SIF elicited by the structure and restructuring of water cannot be neglected, especially in solutes with size scales comparable to those of the water molecules (or their correlation lengths), such as drugs, ions, osmolytes, and so on.

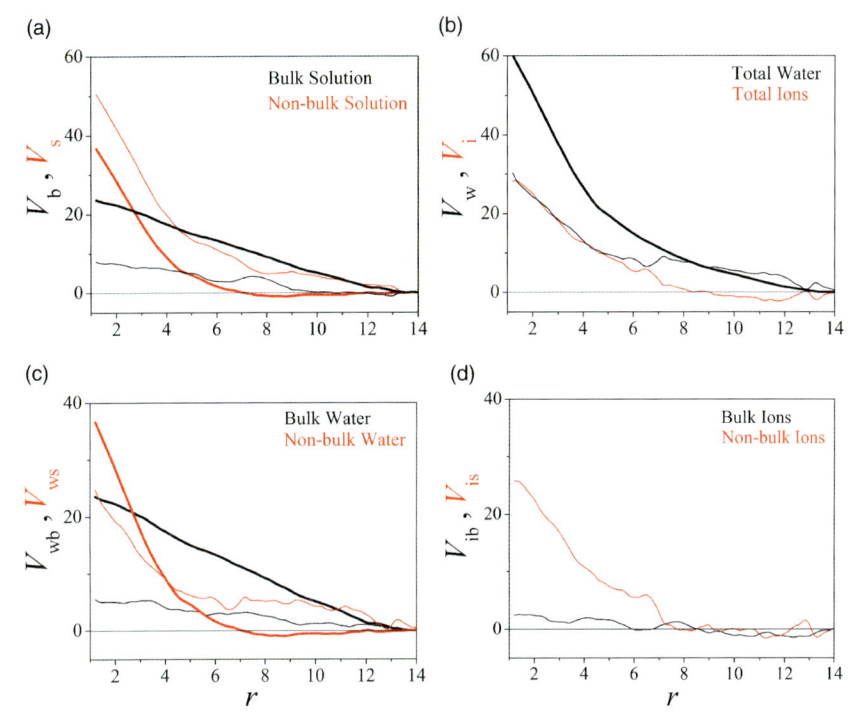

Figure 5.8 Different partitions of the total aqueous solution contribution V_L to the intermolecular potentials of mean force $V = V_L + \phi$ of the Asp$^-$–Arg$^+$ dimer in 2 mol l^{-1} NaCl (see Figure 5.4a for pure water): (a) partitioning $V_L = V_b + V_s$ between bulk V_b and non-bulk V_s partial contributions; (b) partitioning $V_L = V_w + V_i$ between water V_w and ion V_i contributions; (c) and (d) partitioning $V_L = V_{wb} + V_{ws} + V_{ib} + V_{is}$ between bulk V_{wb} and non-bulk V_{ws} water, and between bulk V_{ib} and non-bulk V_{is} ions. All energies are in kcal mol^{-1} and distances in Å. Thick lines, pure water; thin lines, solution.

5.3
Continuum Representation of Water

Solvent models for simulations of proteins and nucleic acids have focused mainly on electrostatics, and have largely neglected SIF, except for simple treatments of hydrophobic forces. In practice, to address the electrostatic problem, a compromise has been sought between physical rigor and computational feasibility. A common approach is to treat the solute as a set of point charges and the solvent as a strict continuum, and solve the Poisson equation numerically; salt effects are usually introduced in a mean-field approximation by solving, for example, the Poisson–Boltzmann [53, 54] or related equations. Another common approach is semi-microscopic, in which both the solute and the liquid are treated as a lattice of point dipoles and the resulting equations are solved self-consistently

[55, 56]. Both approaches are simplifications of the electrostatic problem in liquids, and have conceptual and practical limitations. Theoretically robust approaches to describe both electrostatics and the structure of liquids were developed during a large part of the last century, including earlier and modern theories of dielectric media (the literature on this topic is extensive, but representative publications that illustrate different aspects of the electrostatic problem in polar/polarizable liquids can be found in [40, 57–64]), and theories of liquid structure [65–69]. These approaches, however, have been largely ignored in computer simulations of proteins and nucleic acids, although they form the basis of common approaches in quantum chemistry [70] and most of the theoretical work in the physics of liquids. These approaches are ideal as a starting point for the development of simplified solvent models that retain both physical relevance and computational efficiency.

5.3.1
Dielectric and Structural Response of Water

In a polar and polarizable medium composed of molecules with isotropic polarizability α and permanent dipole moments of magnitude μ, the static polarization field P at a position r can be separated into two contributions [35, 57], one from the induced dipoles (P_α), and another one from the permanent dipoles (P_μ), that is,

$$P(r) = P_\alpha + P_\mu = \rho(r)\alpha E_i(r) + \rho(r)\langle\mu\rangle \tag{5.2}$$

where $\rho(r)$ is the number of liquid molecules per unit volume locally contributing to P, and $\langle\mu\rangle$ is the statistical average of the permanent dipole moments at position r. The microscopic field $E_i(r)$ polarizes the liquid molecules, while the directing field $E_d(r)$ orients the molecules at position r. One of the basic problems in electrostatics is to relate these fields to the macroscopic field $E(r)$ of Maxwell's equations in condensed matter. In general, the link between P and E is non-local, and given by $P(r) = \int \chi(r, r') \cdot E(r') dr'$, where $\chi(r, r')$ is the susceptibility tensor. If the field varies smoothly with the distance, a local form can be used, which is given by

$$P(r) \approx \chi(r)E(r) = \frac{\varepsilon(r)-1}{4\pi} E(r) \tag{5.3}$$

and defines $\varepsilon(r)$ as the local static dielectric permittivity of the medium, assumed here to be a scalar quantity. In the Debye theory E_i and E_d were both taken equal to the Lorentz field E_L, which is related to the macroscopic field by $3E_L(r) = [\varepsilon(r) + 2]E(r)$. From the definition of electric displacement $D(r) = E(r) + 4\pi P(r)$, and assuming spatial homogeneity of the liquid density ρ_0, a closed expression for $\varepsilon(r)$ has been obtained for a point charge within the Lorentz–Debye–Sack (LDS) approximation [58, 71], and given by

$$\varepsilon(r) = 1 + \frac{4\pi}{3}\alpha[\varepsilon(r)+2]\rho_0 + \frac{4\pi\mu}{E(r)}\rho_0 L\left(\frac{\mu[\varepsilon(r)+2]}{3k_B T\varepsilon(r)} E(r)\right) \tag{5.4}$$

where L is the Langevin function, $E(r) = E_0(r)/\varepsilon(r)$ is the magnitude of the macroscopic field at a distance r from the charge q, $E_0 = q/r^2$ is the vacuum field, k_B is Boltzmann's constant, and T is the absolute temperature. As shown by Onsager [60], however, only part of E_i contributes to E_d, that is, $E_i - E_d = R$, where R is the average value of the reaction field at position r. A corrected equation has been proposed (the Lorentz–Debye–Sack–Onsager (LDSO) approximation) [71], which is formally identical to Equation 5.4 but replacing the factor $3k_B T$ by $[3k_B + \mu R(r)]$, where R is the magnitude of the local reaction field at a distance r from the charge. When the source of the field is a point dipole with moment μ_s, an expression can be obtained for $\varepsilon(r)$ that depends not only on the distance r but also on the azimuthal angle θ (see [71]). The spherical symmetry can be recovered under certain circumstances, in which case an expression similar to Equation 5.4 is obtained that depends explicitly on μ_s.

For a spatially homogeneous distribution of the liquid, the density ρ_0 is given by $1/v$, where v is the average volume available to each liquid molecules. The most obvious source of spatial inhomogeneities is the exclusion of liquid from certain regions of space, for example, by a molecule, a lipid bilayer, or an idealized confining boundary, such as the walls of a container or a solid surface of arbitrary shape and size. The density ρ may also change within the liquid phase proper, either by changes in the volume v (a contraction, for example, leads to electrostriction), or by changes of the average number $\delta N(r)$ of liquid molecules that visit an element of volume δV per unit time at position r, that is, $\rho(r) = \delta N(r)$. Whatever the case, the number of water molecules contributing to P changes, thus affecting the local value of the static dielectric response and the component of E that is generated by P. To introduce these local inhomogeneities into the formalism, a more general expression has been derived where the reaction field was incorporated directly into E_i and E_d (see [35]), leading to

$$\frac{[\varepsilon(r)-1][1-\alpha f(r)]}{4\pi\rho(r)} = \frac{3\alpha\varepsilon(r)}{[2\varepsilon(r)+1]} + \frac{\mu}{E(r)} L\left(\frac{3\varepsilon(r)\mu}{k_B T[1-\alpha f(r)][2\varepsilon(r)+1]} E(r)\right) \quad (5.5)$$

where f is given by $R(r) = \mu f(r)/[1 - \alpha f(r)]$ or, explicitly as a function of ρ and ε, by

$$f(r) = \frac{8\pi}{3}\rho(r)\frac{\varepsilon(r)-1}{2\varepsilon(r)+1} \quad (5.6)$$

For a point charge q, the local dielectric permittivity $\varepsilon(r)$ can be calculated numerically once T, α, μ, and $\rho(r)$ are known. Assuming spatial homogeneity of the density, $\varepsilon(r)$ calculated with any of the above equations shows a sigmoidal behavior with r. The permittivity increases smoothly from low values close to the source, to the value corresponding to bulk medium far from the source. Similar equations for $\varepsilon(r)$ have been derived in the past [52, 72] and, more recently, yet another expression has been proposed based on different assumptions, which also shows a similar behavior [73]. The difference between these various approaches is quantitative, particularly in the rate of increase of ε with r, which determines the minimum distance where ε can be assumed to have reached the value of the bulk liquid, typically in the ~5–15 Å range. The decrease of ε close to the charge is known as dielectric saturation. As the strength of the field increases, the molecules contributing locally to the polarization P become increasingly less able to respond to the field. That saturation leads to a drop of ε can already be seen from Equation 5.3: as E increases and P starts saturating, ε must get closer to unity for this relationship to remain valid.

Dielectric saturation is one of several effects associated with solute–liquid interfaces. The properties of an interface are not immediately apparent from the properties of the bulk. Interfacial phenomena have been studied extensively in solid–liquid, liquid–gas, and liquid–liquid systems [67, 74], and valuable physical insight has been gained from simple geometries and model systems. However, the interface between a molecule and a liquid is among the most difficult to characterize due to the inhomogeneities of the solute's topology, and its charge and dipole distributions. Yet, the most important molecular processes in biology occur at the surface of proteins and nucleic acids. The reliability of calculated biophysical properties is likely to be dependent on the theoretical description of the solute–liquid interface, requiring that the treatment of the interface in a solvent model must be given particular consideration. For example, as shown from atomistic simulations [75] and continuum electrostatics calculations [35], the total hydration free energies of alkali and halide ions are partitioned almost equally between contributions from the bulk and from the ion–water interface. Thus, changes in the functional form of ε versus r within the critical region of the interface may introduce unacceptable errors of several RT (this problem is most apparent with simplistic applications of Poisson equation solvers; see discussion in [76]). This functional form depends on the physical information incorporated into the formalism. For example, the Onsager approximation [60, 71], which corrects the Debye model [59], brings the bulk medium closer to the interface; but the Kirkwood–Fröhlich approximation [61, 72], which is a correction to the Onsager model, takes it away from the interface, as short-range dipole–dipole correlations between the liquid molecules reinforce saturation effects. Also, the more structured liquid close to a solute may further reinforce dielectric saturation due to short-range dipole–solute correlations. The effect of the liquid density, discussed below, is yet another factor that modulates the local dielectric response. Moreover, corrections for nonlocality in the calculation of P tend to decrease ε at the interface, a process unrelated to dielectric saturation [77] (non-local effects may also induce oscillations of the permittivity around small solutes [75, 78, 79]). The coexistence of these various competing effects makes it difficult to predict the behavior of ε in the vicinity of a molecule, and the distances where properties of the bulk liquid are recovered (Section 5.4).

The local density $\rho(r)$ of a liquid far from a solute converges to the homogeneous density ρ_0 of the neat, bulk liquid in equilibrium. Closer to the solute, substantial changes in the local density are expected [35, 36], as illustrated in Figure 5.3. In the classical theory of liquid structure [68], a variational principle can be used to obtain an expression for $\rho(r)$, in which a free-energy functional $G[\rho'(r)]$ is minimized with respect to $\rho'(r)$ and the function that satisfies the extreme condition $\delta G/\delta\rho' = 0$ is the density $\rho(r)$ sought. This procedure yields [65]

$$\rho(r) = \eta \exp\{-\beta U(r) + c[\rho(r); r]\} \tag{5.7}$$

where η depends on T and the mass of the liquid molecules, and $U(r)$ is the interaction potential energy between the solute and the liquid molecules at position r. The functional $c[\rho(r); r]$ is the single-particle direct correlation function, related to the two-particle direct correlation function $c^{(2)}$ of the Ornstein–Zernike formalism through the integral equation $c[\rho(r); r] = \int \rho(r') c^{(2)}(r,r') dr'$. The term $-c[\rho(r); r]/\beta$ is an effective potential determined by the interactions between the liquid molecules at r and r'. If these pair correlations are neglected, the density can be expressed in a local form,

$$\rho(r) \cong \rho_0 \exp[-\beta V_{\text{eff}}(r)] \tag{5.8}$$

where V_{eff} is an effective interaction energy between the solute and the liquid molecules at r. Assuming that the solute is composed of N particles at positions r_i, a pairwise function of the form $V_{\text{eff}}(r) = \sum_{i=1}^{N} \zeta_i(|r - r_i|)$ can be used for simplicity. A Lennard-Jones type function has been shown to reproduce the density peaks in the vicinity of polar and charged groups in amino acids (cf. Section 5.4.1.2) [35].

5.3.2
Electrostatic and Liquid-Structure Forces on Solutes and Cosolutes

Only for spherical symmetry are the macroscopic E and the vacuum E_0 fields linearly related by $E(r) = E_0(r)/\varepsilon(r)$. For an arbitrary charge distribution, the relationship between the two fields is given by $E(r) = E_0(r) - \nabla \int P(r') \cdot \nabla' |r - r'|^{-1} dr'$ [80], which is valid for all r and r' (care must be taken in practice to avoid the singularity at $r' = r$ (see [39])). Solving this equation for $E(r)$ is the main goal of a dielectric theory, but a difficult task in general. In the general case, an effective local dielectric $\varepsilon_{\text{eff}}(r)$ can be defined such that the magnitudes of the fields satisfy the same linear relation, that is, $E(r) = E_0(r)/\varepsilon_{\text{eff}}(r)$. As a first-order approximation, it is possible to set $\varepsilon_{\text{eff}}(r) \approx \varepsilon(r)$, so Equation 5.5 can be solved for $\varepsilon(r)$. With this local dielectric permittivity, the above equation for $E(r)$ (or the Poisson equation for the associated potential) can then be solved numerically. The magnitude $E(r)$ of the new field may then be introduced into Equation 5.5 to obtain a corrected $\varepsilon(r)$; the process can be repeated. The question of convergence is not discussed here; however, the first-order approximation is expected already to provide a reasonable description of the dielectric response of the liquid to the particular charge distribution on the solute, including dielectric saturation and reaction field effects. It also provides a physically meaningful alternative to purely macroscopic models commonly used in continuum electrostatics whereby internal and external dielectric constants are defined for the solute and the solvent, and a sharp boundary is introduced to separate the two media. The conceptual and practical problems introduced by sharp boundaries when dealing with microscopic or mesoscopic length scales have been discussed [76, 81, 82]. This procedure also accounts for the local dielectric relaxation to possible charge redistributions in the system. Such flexibility in a solvent model is needed in quantum-mechanical calculations (for example, in proton or electron transfer or chemical reactions), or in molecular mechanics simulation with polarizable force fields [76] (see Section 5.4.1.1).

Once $\varepsilon(r)$ is available, the electrostatic contributions to the hydration free energies and to the energies and forces of the system can be calculated using standard techniques. A discussion involving two interacting ions is given in [35], which can be generalized to molecular solutes. The focus here is on the forces induced by the liquid structure defined by $\rho(r)$. The force $dF(r_i, r)$ exerted by the liquid molecules within a differential volume dv centered at position r on an atom i of a solute at r_i is given by [42, 44]

$$dF(r_i, r) = -\nabla_{r_i} \zeta_i(|r - r_i|) \rho(r) dv \tag{5.9}$$

where $\rho(r) = \rho_0 \exp\left[-\beta \sum_{i=1}^{N} \zeta_i(|r - r_i|)\right]$. The total liquid-structure force $F_{s,i}$ on the atom i is then calculated by integrating $dF(r_i, r)$ over r,

$$F_{s,i} = \int dF(r_i, r) = -\int \nabla_{r_i} \zeta_i(|r - r_i|) \rho(r) dv \tag{5.10}$$

and a summation over all of the atoms yields the force on the solute's center of mass, that is, $F_{CM} = \sum_{i=1}^{N} F_{s,i}$. This force is zero in equilibrium, which follows from Equation 5.10 after $-\nabla_{r_i} \to \nabla_r$, using $\int_{\mathscr{V}} \nabla f dv = \int_{\Omega} f ds$, where Ω is the surface enclosing a volume $\mathscr{V}$ containing the solute, taking the surface away from the solute as $\mathscr{V} \to \infty$, and noting that $\exp(-\beta V_{\text{eff}}) \to 1$ for all points on the surface and that $\int_{\Omega} ds = 0$. This means that the liquid-structure forces can be introduced formally as a random force in Langevin dynamics [83].

5.4
Modeling Water Effects on Proteins and Nucleic Acids

Size and time scales of biomolecular processes span several orders of magnitude. To speed up computation, simplifying approximations have long been sought to remove all or part of the solvent molecules, and incorporate their effects on the solute through suitable modifications of its Hamiltonian. A variety of approaches have been proposed, from statistical potentials, to empirical models with marginal connection to physical theories, to more rigorous physics-based continuum models. One of the difficulties in developing continuum models stems from the mesoscopic nature of biomolecules. In this size-scale regime, purely macroscopic concepts (such as the sharp solute–solvent boundaries discussed above) are not justified on physical grounds, but a complete atomistic description is not required either. Theory and simulations have shown that electrostatic [84–86] and hydrophobic [47] effects change regimes in the nanometer length scale. The development of a continuum model for biomolecules should then be formulated with consideration for this change in size regime.

Another difficulty stems from the lack of suitable benchmarks. The development of atomistic models of water follows well-established procedures, where the properties of water itself are used as benchmarks (for example, pair correlation functions, static and dynamic dielectric constants, compressibility, self-diffusion, and expansion coefficients, and so on). Because these tests are stringent, over 50 models have been reported to perform under specific thermodynamic conditions [87]. By contrast, probing the performance of a continuum model is less straightforward because the properties of solutes are used as benchmarks. For proteins and nucleic acids, this shift in focus complicates the validation processes, for example, by introducing uncertainties arising from the limitations of the macromolecular force field. Cancellations of errors are also common in larger systems, which may hide possible shortcomings.

Finally, it is not always clear what kind of tests (structure, dynamics, kinetics, pK_a shifts, and so on) may be more meaningful and convenient for benchmarking (see Section 5.4.2). Some tests may be insufficiently stringent to challenge the physical content of a model, while others may be too demanding for a particular stage of sophistication. For small molecules, these problems are less severe, and relatively stringent benchmarks can still be found, as is the case of continuum models for quantum-chemical calculations [70]. These quantum-mechanical treatments are closely related to the dielectric theories discussed in Section 5.3. Using similar ideas in macromolecules would bridge the gap between the quantum and classical levels of theory, and reduce empiricism in the latter, as discussed in [76]. In an attempt to address some of these issues, a computationally efficient continuum model has been proposed [35, 84, 88] (reviewed below) based on the theory of polar/polarizable liquids discussed in Section 5.3, which accounts for the balance of forces discussed in Section 5.2.

5.4.1
Calculation of Water Forces in Solutions of Arbitrary Compositions

The simplest task demanded from a solvent model is to provide estimates of the free energy of hydration ΔG or of the total energy E of a molecule in a fixed conformation. More challenging, however, is the description of solvent forces, which affect the dynamics and kinetics of molecules and may have biological implications. In the model described here, the total non-covalent force F_i on an atom i of a molecule, whether a macromolecule, an osmolyte, an ion, or even a single water molecule, is divided into three terms [35],

$$F_i = F_{0,i} + F_{e,i} + F_{s,i} \tag{5.11}$$

where $F_{0,i}$ is the total gas-phase force on atom i due to all the other atoms in the molecule, $F_{e,i}$ is the total electrostatic force exerted by the water phase assumed to be a structureless continuum, and $F_{s,i}$ is the added correction due to the forces exerted by the structured liquid surrounding the molecule. In turn, $F_{s,i}$ can be divided as $F_{s,i} = F_{H,i} + F_{P,i}$, where $F_{H,i}$ is a purely hydrophobic force and $F_{P,i}$ is the correction due to the charges on the molecule. The forces F_i and $F_{0,i}$ are directly related to the potentials V and ϕ introduced in Section 5.2.2, while $F_{e,i}$, $F_{s,i}$, $F_{H,i}$, and $F_{P,i}$ are closely related to V_b, V_s, V_H, and V_P discussed in Section 5.2.3. A pairwise model for $F_{e,i}$ and a non-pairwise model for $F_{s,i}$ have been developed and is reviewed below. Expressions for these terms have been obtained incrementally [35], whereby the forces and free energies of a single particle and of two interacting particles were first derived, and then generalized to N interacting particles.

5.4.1.1 Electrostatic Forces

Single charge The electrostatic potential in the liquid is given by $U(r) = U_0(r)/D(r)$, where U_0 is the gas-phase potential at a distance r from the charge q. The scalar function $D(r)$ is called the screening function, and is related to $\varepsilon(r)$ through the definition $\mathbf{E}(r) = -\nabla U(r)$. For a single charge, the relationship between the two quantities is $\varepsilon(r) = d[rD(r)]^{-1}/dr = D/[1 + (r/D) \, dD/dr]$, which can be solved for $D(r)$ once $\varepsilon(r)$ is known from theory or experiment. The dielectric permittivity ε is given by Equation 5.4 in the homogeneous LDS model (or the corresponding correction in the LDSO model), or by Equation 5.5 in a spatially inhomogeneous liquid. The *electrostatic* contribution to the free energy of hydration of the charge in the homogeneous theory is given by

$$\Delta G_{\text{elec}} = \Delta G_{\text{self}} = \frac{1}{8\pi} \int_{\Re} \mathbf{E}(r) \cdot \mathbf{D}(r) dv$$

$$= \frac{q^2}{2} \lim_{R_u \to \infty} \int_{R_B}^{R_u} \frac{1}{r^2} \left[\frac{1}{\varepsilon(r)} - 1 \right] dr = \frac{q^2}{2R_B} \left[\frac{1}{D(R_B)} - 1 \right] \quad (5.12)$$

where R_B is a lower limit of integration introduced to avoid the divergence of the integral at the origin (this limit is not needed if inhomogeneities are introduced because the integral does converge in that case; see discussion in [35]). In the form of Equation 5.12, ΔG_{self} can be evaluated with little computational cost provided that $D(r)$ and R_B are known. A sigmoidal functional form of $D(r)$ has been proposed that reproduces $\varepsilon(r)$ in various polar/polarizable liquids [76, 89], such as water, acetone, formamide, and so on, and given by $D(r) = (\varepsilon_0 + 1)/\{1 + [(\varepsilon_0 - 1)/2] \exp[-\lambda(\varepsilon_0 + 1)r]\} - 1$, where ε_0 is the static dielectric permittivity of the bulk liquid, and the parameter λ controls the rate of increase of $D(r)$ with r. This single parameter embeds the physics of the system, that is, ε_0, ε_∞, T, α, μ, ν, and q. By solving Equation 5.4 or Equation 5.5, a simple relationship between λ and q can be obtained [76], which allows fast on-the-fly adjustments of D (or ε) as q changes, which is convenient when using polarizable force fields or in quantum-chemical calculations.

Equation 5.12 is an integral form of the equation originally proposed by Born [90], although here it includes dielectric saturation and reaction field effects. The relation between R_B and the chemical properties of mono- and multivalent ions have been addressed by several authors [91–93]. It has been shown that R_B is empirically related to the ionic radius R_q through $R_B = R_q + \delta(q)$, where $\delta > 0$ is a charge-dependent extension, usually smaller for anions than for cations. These ideas have been exploited in the continuum model reviewed here, which allows efficient computation of the self-energy terms in macromolecules (see below) [84].

5.4 Modeling Water Effects on Proteins and Nucleic Acids

Two interacting charges For two charges q_1 and q_2 separated by a distance r in the liquid, the electrostatic contribution to the energy of hydration is given by

$$\Delta G_{elec} = \Delta G_{int} + \Delta G_{self} = \frac{q_1 q_2}{r}\left[\frac{1}{D(r)} - 1\right] + \frac{1}{2}\sum_{i=1}^{2}\frac{q_i^2}{R_{B,i}(r)}\left[\frac{1}{D(R_{B,i}(r))} - 1\right] \quad (5.13)$$

where the parameters λ that define D in the interaction and in the self-energy terms are different. The dependence of $\Delta G_{self,i}$ with r, i.e., with the "conformation" of the system, is introduced through the effective Born radius $R_{B,i}(r)$ as [84]

$$R_{B,i}(r) = R_{iB}\xi_i(r) + R_{ij}[1 - \xi_i(r)] \quad (5.14)$$

where R_{iB} is the Born radius of the isolated charge q_i defined in Equation 5.12; ξ_i is the fraction of water-accessible surface area of q_i, so $1 - \xi_i$ is the excluded fraction buried in q_j; and R_{ij} accounts for the effects on q_i of the absence of polar/polarizable medium in the region occupied by q_j. As $r \to \infty$, the effective radius $R_{B,i}$ converges to the Born radius R_{iB} of the isolated charge in the liquid, while R_{ij} corrects the hydration energy at short inter-particle distances. The linear combination of Equation 5.14 is the simplest way to account for the dependence of $\Delta G_{self,i}$ with r obtained from a numerical solution of $\int E \cdot D dv$ (see [35]).

N interacting charges For a molecular solute composed of N atoms with partial charges $\{q_1, q_2, ..., q_N\}$, Equation 5.13 can be generalized as [84]

$$\Delta G_{elec} = \Delta G_{int} + \Delta G_{self} = \frac{1}{2}\sum_{i \neq j}^{N}\frac{q_i q_j}{r_{ij}}\left[\frac{1}{D(r_{ij})} - \frac{1}{D'(r_{ij})}\right]$$
$$+ \frac{1}{2}\sum_{i=1}^{N}q_i^2\left\{\frac{1}{R_{B,i}}\left[\frac{1}{D(R_{B,i})} - 1\right] - \frac{1}{R'_{B,i}}\left[\frac{1}{D'(R'_{B,i})} - 1\right]\right\} \quad (5.15)$$

where D and D' are the screening functions for the molecule in the liquid and in the vacuum, respectively. In general, $D' > 1$ since a molecule in vacuum is itself a dielectric medium, so a work of polarization is needed even when the molecule is being assembled in the vacuum. Moreover, the dielectric response of a protein varies locally [94], so D' may change substantially with the local environment of the atoms (see Section 5.4.2). If $D' = 1$, Equation 5.15 is a trivial generalization of Equation 5.13. As written, Equation 5.15 describes the free energy of a general transfer process between two media, in which case both D and D' are sigmoidal with different λ, ε_0, ε_∞, α, μ, and ν. Suitable generalizations of Equation 5.14 have been proposed for the effective radii $R_{B,i}$ and $R'_{B,i}$ of atoms in a macromolecule [84]. It has been shown that both the interaction and self-energy terms in Equation 5.15 correlate well with the corresponding terms obtained from a numerical solution of the Poisson equation in peptides of varying lengths and net charges [84].

The total *electrostatic* energy of the molecule in the liquid is then given by [84]

$$E_{elec} = \frac{1}{2}\sum_{i \neq j}^{N}\frac{q_i q_j}{r_{ij}D(r_{ij})} + \frac{1}{2}\sum_{i=1}^{N}\frac{q_i^2}{R_{B,i}}\left[\frac{1}{D(R_{B,i})} - 1\right] \quad (5.16)$$

which is measured with respect to the charges at infinite separation of one another in vacuum. The dependence of E_{elec} with the conformation of the molecule stems from the explicit pairwise interaction terms through r_{ij}, and implicitly through D and R, since both quantities depend on the arrangement of atoms surrounding a given atom i. For D, the dependence is through the parameters λ, that is, $\lambda_i = \lambda_{0,i} + \delta\lambda_i$, where λ_0 is the value corresponding to the isolated atom i (same as in Equation 5.12), and $\delta\lambda$ is the correction due to the presence of the surrounding atoms. For R, the dependence is through the shielding from the liquid due to the surrounding atoms, which is incorporated into ξ_i through [88]

$$\xi_i = A_i - \sum_{j \neq i}^{N} B_{ij} \exp(-C_{ij} r_{ij}) \tag{5.17}$$

where A, B, and C are suitable geometric parameters that characterize the position of the atom i relative to all other atoms in the molecules. As an atom is brought from a distant location to its final position r_i in the molecule, ξ_i varies smoothly from $A_i \sim 1$ to its final value, with sharp changes observed as the atom approaches and crosses the protein–liquid interface. The electrostatic component of the water force on the atom i, as defined in Equation 5.11, is given by $F_{e,i} = -\nabla_{r_i} E_{elec} - F_{C,i}$, where $F_{C,i}$ is the total bare Coulomb force on i.

5.4.1.2 Liquid-Structure Forces

The only requirements for V_{eff} in Equation 5.8 are: (i) to reproduce the short-range structure of the liquid surrounding a solute; and (ii) to provide for the short-range non-pairwise additivity of the resulting forces. A Lennard-Jones function has been used for V_{eff}, and shown to reproduce the density peaks in the vicinity of polar/charged groups of single amino acids [35]. It also provides a good representation of the rearrangement of the density peaks upon pair dissociation (Figure 5.3). The force $F_{H,i}$ is obtained directly from a hydrophobic potential V_H given by $V_H = a + bS$, where a and b are positive parameters, and S is the total solvent-accessible surface area of the molecule. This is the simplest functional form commonly used in molecular mechanics calculations, and is justified by elementary thermodynamic considerations. No attempts are made here to improve upon this, so the hydrophobic force on an atom i is simply given by $F_{H,i} = -b\nabla_{r_i} S$.

To derive a computationally efficient expression for $F_{P,i}$, an incremental procedure similar to that described above for $F_{e,i}$ is followed [35]. Unlike $F_{e,i}$ and $F_{H,i}$, the forces $F_{P,i}$ are here modeled directly, and the potential can then be obtained as a mechanical work.

Two interacting charges For two charges separated by a distance r, Equation 5.10 can be solved numerically to obtain the forces exerted by the liquid structure [35]. The forces and associated potentials have been studied in detail for a Na^+ and Cl^- pair [35], and are shown in Figure 5.9; positive (negative) force implies repulsion (attraction) between the ions. The restructuring of the liquid as the inter-ionic distance changes generates potentials with a close-contact and a solvent-separated minimum, and a barrier in between. These curves, however, are the sole result of

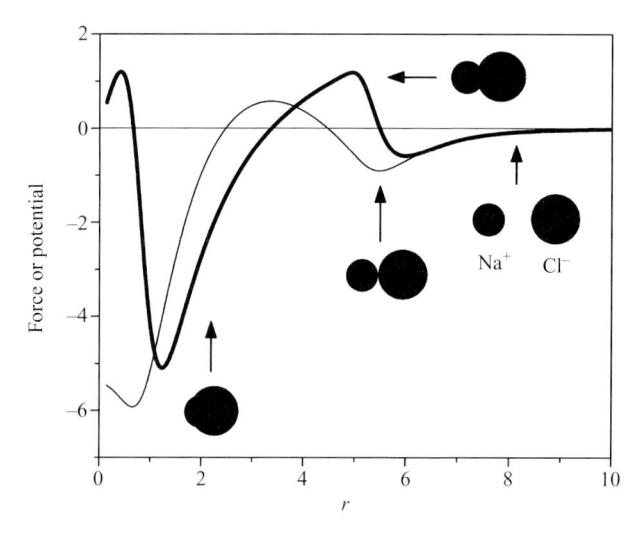

5.4 Modeling Water Effects on Proteins and Nucleic Acids | 117

Figure 5.9 Liquid-structure force (kcal mol^{-1} Å^{-1}) calculated through a numerical integration of Equation 5.10 (thick line), and the corresponding potential (kcal mol^{-1}) calculated through Equation 5.1 (thin line) for a Na$^+$–Cl$^-$ ion pair at infinite dilution in pure water (see [35] for details). The label on the vertical axis is either force or potential, in the corresponding units, and the numerical scale is the same for both. The inset black circles show the distances where major restructuring of the liquid occurs; ions are plotted to scale according to their ionic radii in water.

the action of the liquid structure on the ions, so they contain no information about the direct inter-ionic forces. The exact form of the potential depends on the form of the term V_{eff}. The overall shape of the potential is preserved if the integration domain in Equation 5.10 is restricted only to spatial regions where the liquid density is higher than a certain threshold ρ_c. As ρ_c increases, the forces become smaller, but both minima and the barrier can still be identified. This suggests that the relevant information about the forces is contained in the regions of higher density, which are located near the solute and, then, provide short-range character to the SIF. Thus, if the regions of high density are identified, the integration domain in Equation 5.10 shrinks, and the calculation of the forces becomes computationally tractable, even in complex molecular solutes.

As introduced in Section 5.3, Equation 5.10 yields the total liquid-structure force $F_s = F_H + F_P$, not only F_P as assumed above. However, as discussed in [35], ignoring liquid–liquid interactions or introducing such interactions through an effective solute–liquid potential V_{eff} is insufficient to correctly represent F_H and higher-order corrections in F_P (for example, the modulation of V beyond r_{ss}, as discussed in Section 5.2.2). The effective V_{eff} used here is more appropriate to describe the forces F_P induced by the liquid in the immediate vicinity of polar and charged groups, where solute–liquid correlations play a greater role than liquid–liquid correlations. In the limit of a pure hydrophobic solute, this approximation is weak, thus the hydrophobic force F_H is calculated separately, as indicated above, while F_P vanishes.

N interacting charges The observations above can be generalized to a molecular solute. Two systems have been considered [35]: (i) single amino acids; and (ii) amino acid dimers. Taken together, the two systems provide good benchmarks for the model's capability to represent the liquid structure and restructuring upon dimer dissociation, and their effects on the intermolecular forces. Based on the results of the two interacting charges, Equation 5.10 can be simplified as

$$F_{P,i} = -\sum_{j=1}^{n} \nabla_{r_i} \zeta_i(|r_j - r_i|) N(r_j) \tag{5.18}$$

where r_j is the position of the *j*th density peak, n is the number of peaks, and $N(r_j)$ is the number of water molecules that make up the *j*th peak. Two peaks may have similar heights but different spatial extensions, so $N(r_j) = \mathscr{V}(r_j)\rho(r_j)$, where ρ is the height of the peak and $\mathscr{V}$ is a characteristic volume of the peak. The simplest approximation is $\mathscr{V} = 1 Å^3$ for all of the peaks, which provides a qualitative representation of the SIF observed in atomistic simulations. For example, for Asp⁻–Arg⁺ the forces indeed show two attractive minima separated by a repulsive barrier [35]. Quantitative agreement, however, requires a better treatment of the local volumes $\mathscr{V}$, or an *ad hoc* pairwise correction (not discussed here; see [35]). Equation 5.18 introduces information on the microscopic nature of the liquid in the vicinity of polar/charged groups and on the non-pairwise character of the liquid-structure forces. The non-pairwise additivity of the SIF is represented by centers of force located at specific points, $\{r_i\}$, in the liquid phase, as defined by the peaks of the liquid density. These out-of-the-solute centers exert local, external forces on the atoms of the solute. It is easy to show that the total F_P on the center of mass of the solute is zero, that is, $F_P = \sum_i F_{P,i} = 0$, where i runs over all of the atoms in the solute. Thus, the location of the peaks can be used as centers of random forces in addition to the random-force background in plain Langevin dynamics.

The approach described in Section 5.4.1.1 for F_e and in Section 5.4.1.2 for F_H and F_P defines the screened Coulomb potentials (continuum) implicit solvent model (SCPISM) [34, 82, 84, 88, 95]. In the current implementation, two simplifications were made that allow fast computation but still retain important aspects of the physics of the interactions discussed throughout this chapter. First, the non-pairwise term F_P was explicitly set to zero, but its effects on the intermolecular interactions were partially retained through a suitable modification of F_e. This modification aimed at reproducing the values V_{cc}, that is, the HB energies of all the polar/charged amino acid pairs discussed in Section 5.2 [33, 34]. This provides an efficient algorithm that accounts for the strength of the HB interactions at the protein–water interface, which was shown to be essential to reproduce the structure, dynamics, and thermodynamics of peptides and proteins.

Second, the correction $\delta\lambda$ of the parameters defining D in Equation 5.16 were explicitly set to zero, but the effects of the environment on each atom were partially incorporated into λ_0 as the environments of fully hydrated amino acid side-chain analogs. Both these simplifications should be relaxed to improve the quality of the model, at the expense of computer time. The current implementation of the

SCPISM is computationally efficient, being only two to three times slower than calculations in the gas phase.

5.4.2
Calculation of pK_a in Proteins

Because of the well-documented difficulty of reliably calculating the pK_a of titratable residues (TRs) in proteins [96], the ability to do so provides a stringent test for assessing the reliability of a continuum model. Although a fully *ab initio* approach could be used starting from the calculated gas-phase proton affinities of the TR, the calculation is greatly simplified by considering the thermodynamic cycle

$$\begin{array}{ccc} AH(s) & \xleftrightarrow{2.303RT\,pK_a(s)} & A^-(s)H^+(s) \\ \Delta\mu_{s,p}(AH) \updownarrow & & \updownarrow \Delta\mu_{s,p}(A^-) \\ AH(p) & \xleftrightarrow{2.303RT\,pK_a(p)} & A^-(p)H^+(s) \end{array} \quad (5.19)$$

where p and s refer to the protein and the solvent, respectively. Thus the pK_a of a titratable group A in the protein can be calculated from its pK_a in the solvent and the additional changes in free energy that arise when the group is transferred from the solvent to the protein. Including only electrostatic contributions, the pK_a(p) is calculated as

$$pK_a(p) = pK_a(s) + (w_A^{int} + \alpha_A \Delta w_A^{tr})/2.303RT \quad (5.20)$$

where α_A is a linear scaling factor to be discussed below, w_A^{int} is the interaction free electrostatic energy of the charged group in the field of *all* the other groups in the protein, and Δw_A^{tr} is the change in self-energy on transferring the group from the solvent into the protein. The terms w_A^{int} and Δw_A^{tr} are defined from the LDS theory [58] in such a way that the contribution of the M ionizable groups to the total electrostatic free energy w of the system can be expressed as

$$w = \sum_{A=1}^{M} (w_A^{int} + \alpha_A \Delta w_A^{tr}) \quad (5.21)$$

In this method the interaction energies between non-titratable groups are constant, and can be set to zero. In most cases, the interaction energy contributes to stabilizing the system, whereas the transfer energy is positive in almost all cases. The latter is the energy penalty for transferring a charge from a hydrophilic environment (the aqueous solvent) to a more hydrophobic environment (the protein). In this section the terms "hydrophobic" and "hydrophilic" are used in their classical sense as implied, for example, in [97, 98]. They refer to the varying degrees of dielectric response of an embedded TR to the modulating effects of the surrounding local environment.

Analysis of local environments around TRs known to have large shifts favoring the neutral state shows that many of the nearest-neighbor residues around a given TR are hydrophobic, and the greater the shift in pK_a value, the more hydrophobic these residues tend to be [99]. To account for this hydrophobicity in the calculation of pK_a values, it was necessary to develop a quantitative representation of the degree of hydrophobicity or hydrophilicity (Hpy value) of the local region surrounding the ionizable group of each TR in the protein. This approach has been described elsewhere [100] and makes use of the Rekker fragmental hydrophobic constants [101], which were developed to calculate partition coefficients for potential drug candidates. Rekker's approach assigns hydrophobicity parameters to small chemical functionalities (in some cases individual atoms), and Hpy values are calculated by summing the Rekker fragmental constants for all fragments in the nearest-neighbor shell (the microenvironment) around the titratable residue in the protein. Rekker fragmental constants are completely independent of any protein properties. Moreover, by using fragmental parameters, it is possible to extend the method to coenzymes, prosthetic groups, ligands, and so on. Recently, the method has been extended to calculate the microenvironments around all fragments comprising amino acid residues [97]. The results of this generalization exhibit the high variability of these local regions in any given protein, verifying the heterogeneity of these macromolecules.

In most pK_a calculations, the equilibrium charge state is calculated from the distribution of the charge over the 2^N ionization microstates. The method developed in [100, 102] proceeds in a different way: the assignment of the ionization charge over the atoms of the protonatable moiety is determined through a variational optimization of the total free energy, thus allowing the TR to respond to the protein–solvent environment. The total variational ionization charge of each group is coupled to the pH via the Henderson–Hasselbach equation.

To derive the variational equations for assigning the titration charge, the partial charge of an atom a in a group A is defined as

$$q_a = (1-\theta_A)q_a^n + f_a q_a^0 \tag{5.22}$$

where q_a^0 are fixed initial partial charges ($\neq 0$ for titratable groups), and f_a are scaling factors that will be optimized to determine a stationary point of w. Henceforth, the group subscript A will be omitted, thus $q_A = \sum_a q_a$, where a runs over all the atoms in A; q_a^n is the partial charge from the neutral group and $\sum q_a^n = 0$. The fraction of A in the charged state, θ_A, is zero for neutral groups since $q_a^0 = 0$ for all a. Thus the first term on the right-hand side of Equation 5.22 is the contribution to the charge q_a from the neutral form of the protonatable group, while the second term represents the contribution from the ionization charge. Comparison of Equation 5.21 with Equation 5.16 shows that the electrostatic interaction free energy of the ionization charge of atom a in A (with $w_A = \sum_a w_a$) is defined by

$$w_a^{int} = \frac{1}{2}f_a q_a^0 \Phi_a(r_a) = \frac{1}{2}f_a q_a^0 \sum_{I \neq A, i} \frac{(1-\theta_I)q_i^n + f_i q_i^0}{D(r_{ai})r_{ai}} \tag{5.23}$$

where $\Phi_a(r_a)$ is the potential at r_a due to all the other groups in the system, and I runs over *all* the groups in the system. The transfer energy term is expressed as the difference of two self-energy terms (see Equation 5.12)

5.4 Modeling Water Effects on Proteins and Nucleic Acids

$$\Delta w_a^{tr} = \frac{1}{2}(f_a q_a^0)^2 \left[\frac{1}{D_p(R_a^p)R_a^p} - \frac{1}{D_s(R_a^s)R_a^s} \right](1-\xi_A) \tag{5.24}$$

where the radii R_a have been evaluated using the Born approximation so that $R_a \equiv R_a^p = R_a^s$, and are obtained from solvation free energies as described [100, 102]. The variational problem is expressed in the form

$$\delta w = \sum \delta w_A = \sum \delta w_A^{int} + \alpha_A \delta \Delta w_A^{tr} = 0 \tag{5.25}$$

and, to couple the titration charge to the Henderson–Hasselbach equation, the q_A are constrained to be constant at a given pH, that is, $\delta q_A = \sum_{A,a} q_a^0 \delta f_a = 0$. Incorporating the constraints into the variational equations yields

$$w_a^{int} + 2\Delta w_a^{tr} = \lambda_A f_a q_a^0 \tag{5.26}$$

where λ are the Lagrange multipliers that can be obtained from Equation 5.26 by summing both sides over a, yielding $\lambda_A = (w_A^{int} + 2\Delta w_A^{tr})/q_A$. Substituting into Equation 5.26 yields

$$f_a q_a^0 = q_A \frac{w_a^{int} + 2\alpha_A w_a^{tr}}{w_A^{int} + 2\alpha_A w_A^{tr}} \tag{5.27}$$

for all titratable groups A. Equation 5.27 states that the optimal distribution of the ionization charge on any atom is given by the equilibrium ionization charge scaled by a factor related to the fractional energy contribution from the charge $f_a q_a^0$ on atom a to group A. Equation 5.27 is solved iteratively until self-consistency is achieved. The pH dependence of the equilibrium charge state is coupled to the variational equation (Equation 5.27) through $\theta_A = (1+[H^+]/K_a^A)^{-1}$ (for acids) or $\theta_A = (1+K_a^A/[H^+])^{-1}$ (for bases), and requiring that at self-consistency $q_A = \sum_a q_a^0 f_a = Z_A \theta_A$, where Z_A is the formal charge on group A.

The pK_a values are calculated by first evaluating the microenvironment parameters, and then solving Equation 5.27 for the scaling factors f_a. The Hpy values are the protein's contribution to the total Hpy (THpy) of the microenvironment. For partially buried residues, the region exposed to the solvent must also be taken into account, so THpy is defined by

$$THpy_A = (1-\xi_A)Hpy_A + \xi_A Hpy_A^0 \tag{5.28}$$

where Hpy (0, A) is the value of Hpy (A) in pure solvent. Since THpy is an extensive quantity, it is normalized in the form $rHpy_A = THpy_A/Hpy_A^0$, where rHpy values range from about 1 (the value for pure water) to about −0.4. This shows that increasing hydrophobicity correlates with decreasing rHpy, with negative values indicating extremely hydrophobic microenvironments. Once all the $rHpy_A$ have been evaluated, the optimal screening function is assigned to each titratable group, as well as the scaling factor α_A. In most cases α_A is 1, but for hydrophobic microenvironments this is insufficient to yield the correct energy penalty for transferring a charged group from water into the protein. These microenvironments are very different from the typical hydrophilic to slightly hydrophobic microenvironments that surround most titratable groups, and tend to have little effect on their pK_a values. These microenvironments, however, were used to determine the values of R_a and cannot account for the unusual local environments that tend to induce large pK_a shifts.

Calibration of the scaling factor was based on highly shifted experimental pK_a values of several proteins [99]. The results indicated that: $\alpha_A = 1$ for rHpy > 0.1; $\alpha_A = 2$ for 0.02 < rHpy < 0.1; and $\alpha_A = 3$ for rHpy < 0.02. Using these values as well as the optimal screening function leads to an overall decrease of about 0.1 in the root-mean-square error of pK_a values. More importantly, the errors of the pK_a values exhibiting large shifts were significantly reduced. A good example is Gly 35 in lysozyme, where the experimental pK_a value is 6.1. Without the corrected microenvironment included, the calculated pK_a value was 4.9, but including the correction with $\alpha_A = 2$, the pK_a increased to 6.2 [100].

The complex mosaic of variable dielectric regions in proteins (reflected in the Hpy values) provides nature with a tool to evolve the specialized protein architectures that engender very large pK_a shifts of the embedded titratable residue. The reasonable accuracy of the approach is partly due to the development of a quantitative descriptor that incorporates into the theory the modulating effects of the local environment on the electrostatic interaction and transfer energies. This empirical correction would not be needed if an ideal protein force field was available and formal statistical averages are performed from the protein dynamics to compute these thermodynamic quantities. Unfortunately, this is computationally inefficient, so the method described here provides a reliable and fast alternative to the pK_a calculation of many titratable groups simultaneously present in proteins. The second feature of the method that contributes to its reliability is the use of a variational procedure to optimally distribute the ionization charge to the atoms of the titratable group, because it emulates structural relaxation of the system. As seen from Equations 5.23 and 5.24, the quantities w_a^{int} and Δw_a^{tr} have been derived from the LDS theory and provide a physically correct description of the solvent effects on the energy of the system. Nevertheless, to get reasonable results, considerable parameterization was necessary, which partially reduces the theoretical content of the results. To improve the theoretical description, it would be desirable to reduce the number of parameters in the model, for example, by calculating D and R values with the more robust formalism described in earlier sections.

Acknowledgments

This study utilized the high-performance computer capabilities of the Biowulf PC/Linux cluster at the U.S. National Institutes of Health. This work was supported by the NIH Intramural Research Program through the Center for Information Technology, U.S. Department of Health and Human Services. ELM acknowledges financial support from the NIH R01 grant DA015170, and to the NSF Terascale Computing System at the Texas Advanced Computing Center.

References

1. Timasheff, S.N. (1993) The control of protein stability and association by weak interactions with water: how do solvents affect these processes? *Annu. Rev. Biophys. Biomol. Struct.*, **22**, 67.
2. Reichardt, C. (2002) *Solvents and Solvent Effects in Organic Chemistry*, 3rd edn, Wiley-VCH Verlag GmbH, Weinheim.
3. Gregory, R.B. (ed.) (1995) *Protein–Solvent Interactions*, Marcel Dekker, New York.
4. Green, A.A. (1932) Studies on the physical chemistry of the proteins: X. The solubility of hemoglobyn in solutions of chlorides and sulfates of varying concentrations. *J. Biol. Chem.*, **95**, 47.
5. Arakawa, T., Kita, Y., and Carpenter, J.F. (1991) Protein–solvent interactions in pharmaceutical formulations. *Pharm. Res.*, **8**, 285.
6. Schellman, J.A. (2002) Fifty years of solvent denaturation. *Biophys. Chem.*, **96**, 91.
7. Shifrin, S. and Parrott, C.L. (1975) Influence of glycerol and other polyhydric alcohols on the quaternary structure of an oligomeric protein. *Arch. Biochem. Biophys.*, **166**, 426.
8. von Hippel, P.H. and Wong, K.Y. (1965) On the conformational stability of globular proteins. The effects of various electrolytes and nonelectrolytes on the thermal ribonuclease transition. *J. Biol. Chem.*, **240**, 3909.
9. Back, J.F., Oakenfull, D., and Smith, M.B. (1979) Increased thermal stability of proteins in the presence of sugars and polyols. *J. Am. Chem. Soc.*, **18**, 5191.
10. Tanford, C. (1964) Isothermal unfolding of globular proteins in aqueous urea solutions. *J. Am. Chem. Soc.*, **86**, 2050.
11. Schreiber, G. (2002) Kinetic studies of protein–protein interactions. *Curr. Opin. Struct. Biol.*, **12**, 41.
12. Zou, Q. et al. (2002) The molecular mechanism of stabilization of proteins by TMAO and its ability to counteract the effects of urea. *J. Am. Chem. Soc.*, **124**, 1192.
13. Arakawa, T. and Timasheff, S.M. (1985) The stability of proteins by osmolytes. *Biophys. J.*, **47**, 411.
14. Harries, D., Rau, D.C., and Parsegian, V.A. (2005) Solutes probe hydration in specific association of cyclodextrin and adamantane. *J. Am. Chem. Soc.*, **127**, 2184.
15. Parsegian, V.A., Rand, R.P., and Rau, D.C. (2000) Osmotic stress, crowding, preferential hydration, and binding: a comparison of perspectives. *Proc. Natl. Acad. Sci. U.S.A.*, **97**, 3897.
16. Yancey, P.H., Clark, M.E., and Hand, S.C. (1982) Living with water stress: evolution of osmolyte systems. *Science*, **217**, 1214.
17. Wood, J.M. et al. (2001) Osmosensing and osmoregulatory compatible solute accumulation by bacteria. *Compar. Biochem. Physiol. A*, **130**, 437.
18. Go, W.Y. et al. (2004) NFAT5/TonEBP mutant mice define osmotic stress as a critical feature of the lymphoid microenvironment. *Proc. Natl. Acad. Sci. U.S.A.*, **101**, 10673.
19. Woo, S.K., Lee, S.D., and Kwon, H.M. (2002) TonEBP transcriptional activator in the cellular response to increased osmolality. *Pflugers Arch.*, **444**, 579.
20. Rep, M. et al. (1999) Osmotic stress-induced gene expression in *Saccharomyces cerevisiae* requires Msn1p and the novel nuclear factor Hot1p. *Mol. Cell. Biol.*, **19**, 5474.
21. Santoro, M.M. et al. (1992) Increased thermal stability of proteins in the presence of naturally occurring osmolytes. *Biochemistry*, **31**, 5278.
22. Na, G.C. (1986) Interaction of calf skin collagen with glycerol: linked function analysis. *Biochemistry*, **25**, 967.
23. Shelanski, M.L., Gaskin, F., and Cantor, C.R. (1973) Microtubule assembly in the absence of added nucleotide. *Proc. Natl. Acad. Sci. U.S.A.*, **70**, 765.
24. Lee, J.C. and Timasheff, S.M. (1977) In vitro reconstitution of calf brain microtubules: effects of solution variables. *Biochemistry*, **16**, 1754.
25. Timasheff, S.N. (1998) In disperse solution, "osmotic stress" is a restricted case of preferential interactions. *Proc. Natl. Acad. Sci. U.S.A.*, **95**, 7363.

26 Jeffrey, G.A. (1997) *An Introduction to Hydrogen Bonding*, in Topics in Physical Chemistry (ed. D.G. Truhlar), Oxford University Press, New York.

27 Jeffrey, G.A. and Saenger, W. (1991) *Hydrogen Bonding in Biological Structures*, Springer, Heidelberg.

28 Scheiner, S. (1997) *Hydrogen Bonding: A Theoretical Perspective*, in Topics in Physical Chemistry (ed. D.G. Truhlar), Oxford University Press, New York.

29 Legon, A.C. and Millen, D.J. (1987) Directional character, strength, and nature of the hydrogen bond in gas-phase dimers. *Accts. Chem. Res.*, **20**, 39.

30 Murray-Rust, P. and Glusker, J.P. (1984) Directional hydrogen bonding to sp^2- and sp^3-hybridized oxygen atoms and its relevance to ligand–macromolecule interactions. *J. Am. Chem. Soc.*, **106**, 1018.

31 Baker, E.N. and Hubbard, R.E. (1984) Hydrogen bonding in globular proteins. *Prog. Biophys. Mol. Biol.*, **44**, 97.

32 Fersht, A.R. et al. (1985) Hydrogen bonding and biological specificity analysed by protein engineering. *Nature*, **314**, 235.

33 Hassan, S.A. (2004) Intermolecular potentials of mean force of amino acid side chain interactions in aqueous medium. *J. Phys. Chem. B*, **108**, 19501.

34 Hassan, S.A., Guarnieri, F., and Mehler, E.L. (2000) Characterization of hydrogen bonding in a continuum solvent model. *J. Phys. Chem. B*, **104**, 6490.

35 Hassan, S.A. (2007) Liquid-structure forces and electrostatic modulation of biomolecular interactions in solution. *J. Phys. Chem. B*, **111**, 227.

36 Hassan, S.A. (2005) Amino acid side chain interactions in the presence of salts. *J. Phys. Chem. B*, **109**, 21989.

37 Zeidler, M.D. (1973) Water in crystalline hydrates; aqueous solutions of simple nonelectrolytes, in *Water: A Comprehensive Treatise*, vol. 2 (ed. F. Franks), Plenum, New York.

38 Hassan, S.A., Hummer, G., and Lee, Y.S. (2006) Effects of electric fields on proton transport through water chains. *J. Chem. Phys.*, **124**, 204510.

39 Logan, D.E. (1981) On the dielectric theory of fluids. I. Fields. *Mol. Phys.*, **44**, 1271.

40 Wertheim, M.S. (1973) Theory of polar fluids. I. *Mol. Phys.*, **26**, 1425.

41 Bruge, F. et al. (1996) Solvent-induced forces on a molecular scale: non-additivity, modulation and causal relation to hydration. *Chem. Phys. Lett.*, **254**, 283.

42 Ben-Naim, A. (1990) Solvent-induced forces in protein folding. *J. Phys. Chem.*, **94**, 6893.

43 Chandler, D. (2002) Two faces of water. *Nature*, **417**, 491.

44 Durell, S.R., Brooks, B.R., and Ben-Naim, A. (1994) Solvent-induced forces between two hydrophilic groups. *J. Phys. Chem.*, **98**, 2198.

45 Ben-Naim, A. (1990) Solvent effects on protein association and protein folding. *Biopolymers*, **29**, 567.

46 Ben-Naim, A. (1990) Strong forces between hydrophilic macromolecules: implications in biological systems. *J. Chem. Phys.*, **93**, 8196.

47 Lum, K., Chandler, D., and Weeks, J.D. (1999) Hydrophobicity at small and large length scales. *J. Phys. Chem. B*, **103**, 4570.

48 Wallqvist, A. and Berne, B.J. (1995) Computer simulation of hydrophobic hydration forces on stacked plates at short range. *J. Phys. Chem.*, **99**, 2893.

49 Hassan, S.A. (2008) Computer simulation of ion cluster speciation in concentrated aqueous solutions at ambient conditions. *J. Phys. Chem. B*, **112**, 10573.

50 Hribar, B. et al. (2002) How ions affect the structure of water. *J. Am. Chem. Soc.*, **124**, 12302.

51 Mancinelli, R. et al. (2007) Perturbation of water structure due to monovalent ions in solution. *Phys. Chem. Chem. Phys.*, **9**, 2959.

52 Hasted, J.B., Ritson, D.M., and Collie, C.H. (1948) Dielectric properties of aqueous ionic solutions. Parts I and II. *J. Chem. Phys.*, **16**, 1.

53 Honig, B. and Nicholls, A. (1995) Classical electrostatics in biology and chemistry. *Science*, **268**, 1144.

54 Madura, J.D. et al. (1995) Electrostatics and diffusion of molecules in solution: simulations with the University of Houston Brownian Dynamics program. *Comput. Phys. Commun.*, **91**, 57.

55 Warshel, A. and Aqvist, J. (1991) Electrostatic energy and macromolecular function. *Annu. Rev. Biophys. Biophys. Chem.*, **20**, 267.

56 Warshel, A. and Russell, S.T. (1984) Calculation of electrostatic interactions in biological systems and in solutions. *Q. Rev. Biophys.*, **17**, 283.

57 Bottcher, C.J.F. (1952) *Theory of Electric Polarisation*, Elsevier, Amsterdam.

58 Mehler, E.L. (1996) The Lorentz–Debye–Sack theory and dielectric screening of electrostatic effects in proteins and nucleic acids, in *Molecular Electrostatic Potential: Concepts and Applications* (eds J.S. Murray and K. Sen), Elsevier Science, Amsterdam, p. 371.

59 Debye, P. (1929) *Polar Molecules*, Dover, New York.

60 Onsager, L. (1936) Electric moments of molecules in liquids. *J. Am. Chem. Soc.*, **58**, 1486.

61 Fröhlich, H. (1958) *Theory of Dielectrics*, Clarendon, Oxford.

62 Kirkwood, J.G. (1939) The dielectric polarization of polar liquids. *J. Chem. Phys.*, **7**, 911.

63 Madden, P. and Kivelson, D. (1984) A consistent molecular treatment of dielectric phenomena. *Adv. Chem. Phys.*, **56**, 467.

64 Dogonadze, R.R. and Kornyshev, A.A. (1974) Polar solvent structure in the theory of ionic solvation. *J. Chem. Soc. Faraday Trans.*, **70**, 1121.

65 Hansen, J.-P. and McDonald, I.R. (1986) *Theory of Simple Liquids*, 2nd edn, Elsevier, London.

66 Nienhuis, G. and Deutch, J.M. (1971) Structure of dielectric fluids. I. The two-particle distribution function of polar fluids. *J. Chem. Phys.*, **55**, 4213.

67 Evans, R. (1979) The nature of the liquid–vapor interface and other topics in the statistical mechanics of non-uniform, classic fluids. *Adv. Phys.*, **28**, 143.

68 Lebowitz, J.L. and Percus, J.K. (1963) Statistical thermodynamics of nonuniform fluids. *J. Math. Phys.*, **4**, 116.

69 Chandler, D., McCoy, J.D., and Singer, S.J. (1986) Density functional theory of nonuniform polyatomic systems. I. General formulation. *J. Chem. Phys.*, **85**, 5971.

70 Tomasi, J., Mennucci, B., and Cammi, R. (2005) Quantum mechanical continuum solvation models. *Chem. Rev.*, **105**, 2999.

71 Ehrenson, S. (1989) Continuum radial dielectric functions for ion and dipole solution systems. *J. Comput. Chem.*, **10**, 77.

72 Booth, F. (1951) The dielectric constant of water and the saturation effect. *J. Chem. Phys.*, **19**, 391. [Erratum in *J. Chem. Phys.* (1951) **19**, 1327.]

73 Gong, H., Hocky, G., and Freed, K.F. (2008) Influence of nonlinear electrostatics on transfer energies between liquid phases: charge burial is far less expensive than born model. *Proc. Natl. Acad. Sci. U.S.A.*, **105**, 11146.

74 Volkov, A.G. et al. (1997) *Liquid Interfaces in Chemistry and Biology*, John Wiley & Sons, Inc., New York.

75 Hyun, J.-K. and Ichiye, T. (1997) Understanding the Born radius via computer simulations and theory. *J. Phys. Chem. B*, **101**, 3596.

76 Hassan, S.A. and Mehler, E.L. (2005) From quantum chemistry and the classical theory of polar liquids to continuum approximations in molecular mechanics calculations. *Int. J. Quantum Chem.*, **102**, 986.

77 Rubinstein, A. and Sherman, S. (2007) Evaluation of the influence of the internal aqueous solvent structure on electrostatic interactions at the protein–solvent interface by nonlocal continuum electrostatic approach. *Biopolymers*, **87**, 149.

78 Chandra, A. and Bagchi, B. (1989) Molecular theory of solvation and solvation dynamics of a classical ion in a dipolar liquid. *J. Phys. Chem.*, **93**, 6996.

79 Hyun, J.-K., Babu, C.S., and Ichiye, T. (1995) Apparent local dielectric response around ions in water: a method for its determination and its applications. *J. Phys. Chem.*, **99**, 5187.

80 Jackson, J.D. (1975) *Classical Electrodynamics*, 2nd edn, John Wiley & Sons, Inc., New York.
81 Davis, M.E. and McCammon, J.A. (1991) Dielectric boundary smoothing in finite difference solutions of the Poisson equation: an approach to improve accuracy and convergence. *J. Comput. Chem.*, **12**, 909.
82 Hassan, S.A, and Mehler, E.L. (2002) A critical analysis of continuum electrostatics: the screened Coulomb potential–implicit solvent model and the study of the alanine dipeptide and discrimination of misfolded structures of proteins. *Proteins*, **47**, 45.
83 Allen, M.P. and Tildesley, D.J. (1987) *Computer Simulation of Liquids*, Clarendon, Oxford.
84 Hassan, S.A., Guarnieri, F., and Mehler, E.L. (2000) A general treatment of solvent effects based on screened Coulomb potentials. *J. Phys. Chem. B*, **104**, 6478.
85 Pollock, E.L., Alder, B.J., and Pratt, L.R. (1980) Relation between the local field at large distances from a charge or dipole and the dielectric constant. *Proc. Natl. Acad. Sci. U.S.A.*, **77**, 49.
86 Pollock, E.L., Alder, B.J., and Pratt, L.R. (1981) Remarks on "The field far from a charge and dipole". *Chem. Phys. Lett.*, **78**, 201.
87 Guillot, B. (2002) A reappraisal of what we have learnt during three decades of computer simulations on water. *J. Mol. Liq.*, **101**, 219.
88 Hassan, S.A. et al. (2003) Molecular dynamics simulations of peptides and proteins with a continuum electrostatic model based on screened Coulomb potentials. *Proteins*, **51**, 109.
89 Mehler, E.L. and Eichele, E. (1984) Electrostatic effects in water-accessible regions of proteins. *Biochemistry*, **23**, 3887.
90 Born, M. (1920) Volumen und Hydrationswärme der Ionen. *Z. Phys.*, **1**, 45.
91 Latimer, W.M, and Rodebush, W.H. (1920) Polarity and ionization from the standpoint of the Lewis theory of valence. *J. Am. Chem. Soc.*, **42**, 1419.
92 Bucher, M. and Porter, T.L. (1986) Analysis of the Born model for hydration of ions. *J. Phys. Chem.*, **90**, 3406.
93 Rashin, A. and Honig, B. (1985) Reevaluation of the Born model of ion hydration. *J. Phys. Chem.*, **89**, 5588.
94 Cohen, B.E. et al. (2002) Probing protein electrostatics with a synthetic fluorescent amino acid. *Science*, **296**, 1700.
95 Li, X., Hassan, S.A., and Mehler, E.L. (2005) Long dynamics simulations of proteins using atomistic forcefields and a continuum representation of solvent effects: calculation of structural and dynamic properties. *Proteins*, **60**, 464.
96 Stanton, C.L. and Houk, K.N. (2008) Benchmarking pK_a prediction methods for residues in proteins. *J. Chem. Theory Comput.*, **4**, 951.
97 Bandyopadhyay, D. and Mehler, E.L. (2008) Quantitative expression of protein heterogeneity: response of amino acid side chains to their local environment. *Proteins*, **72**, 646.
98 Cornette, J.L. et al. (1987) Hydrophobicity scale and computational techniques for detecting amphipathic structures in proteins. *J. Mol. Biol.*, **195**, 659.
99 Mehler, E.L. et al. (2002) The role of hydrophobic microenvironment in modulating pK_a shifts in proteins. *Proteins*, **48**, 283.
100 Mehler, E.L. and Guarnieri, F. (1999) A self-consistent, microenvironment modulated screened Coulomb potential approximation to calculate pH dependent electrostatic effects in proteins. *Biophys. J.*, **77**, 3.
101 Rekker, R.F. (1977) *The Hydrophobic Fragmental Constant, its Derivation and Application*, in Pharmacochemistry Library, vol. **1** (eds W.T. Nauta and R.F. Rekker), Elsevier, Amsterdam.
102 Mehler, E.L. (1996) A self-consistent, free energy based approximation to calculate pH dependent electrostatic effects in proteins. *J. Phys. Chem.*, **100**, 16006.

6
Continuum Electrostatics Solvent Modeling with the Generalized Born Model
Alexey Onufriev

6.1
Introduction: the Implicit Solvent Framework

An accurate description of the solvent environment is essential for realistic biomolecular modeling. Within the explicit solvent framework, the movements of individual water molecules are explicitly calculated. While arguably the most realistic of the current theoretical approaches, this methodology suffers from considerable computational costs, which often become prohibitive, especially for molecular systems undergoing significant structural transitions, such as those involved in protein folding. Other problems with the approach include the difficulty, and often complete inability, to calculate relative free energies of molecular conformations; problems arise due to the need to account for the very large number of solvent degrees of freedom.

An alternative that is becoming more and more popular, the implicit solvent framework [1–7], is based on replacing the explicit water environment consisting of discrete molecules by an infinite continuum with the dielectric and "hydrophobic" properties of water. The implicit solvent framework has several advantages over explicit water representations, especially in molecular dynamics simulations. These include the following:

1) Lower direct computational (CPU) costs for many molecular systems.
2) Enhanced sampling of conformational space. In contrast to explicit solvent models, solvent viscosity, which often slows down conformational transitions, can be drastically reduced or even turned off completely within implicit representations.
3) Effective ways to estimate (free) energies. Since solvent degrees of freedom are taken into account implicitly, estimating free energies of solvated structures is much more straightforward than with explicit water models.
4) Instantaneous dielectric response from the solvent. It eliminates the need for lengthy equilibration of water that is typically necessary in explicit water simulations. This feature of implicit solvent models becomes key when the

charge state of the system changes many times during the course of a simulation, as, for example, in constant pH simulations.

5) Implicit averaging over solvent degrees of freedom. It eliminates the "noise" – an astronomical number of local minima arising from small variations in solvent structure. Energy landscapes of molecular structures are no longer dominated by the noise and start to make sense.

The availability of all of these advantages depends critically on the availability of practical computational models based on the implicit solvent framework. One such model that has become particularly popular recently is the generalized Born (GB) approximation. The model provides a relatively simple, computationally robust, and effective way to estimate the long-range electrostatic interactions in biomolecular structures – currently the bottleneck for calculations of energy and force estimates in classical all-atom simulations [8].

The main goal of this chapter is to introduce the GB model and to present an overview of its current use in molecular modeling. Special emphasis will be given to a discussion of the approximations upon which the model rests. This chapter is organized as follows. After a brief introduction into the key approximation of the general implicit solvent framework, I will review the Poisson–Boltzmann model of continuum electrostatics as the foundation of the GB approximation. A derivation of the GB approximation will then be presented, followed by a discussion of the various "flavors" of the model. I will then give examples of some recent uses of the model in molecular modeling. Several issues pertaining to practical aspects of the model will be touched upon in some detail, including relative computational speedup and enhancement of conformational sampling that can be achieved via the use of the GB approximation. Limitations of the model will also be discussed, followed by concluding remarks and outlook.

6.1.1
Key Approximations of the Implicit Solvent Framework

In many molecular modeling applications, one needs to compute the total energy of the molecule in the presence of solvent. This energy is a function of molecular configuration, and its gradients with respect to atomic positions determine the forces on the atoms. The total energy of a solvated molecule can be conveniently written as $E_{tot} = E_{vac} + \Delta G_{solv}$, where E_{vac} represents the molecule's potential energy in vacuum (gas phase), and ΔG_{solv} is defined as the free energy of transferring the molecule from vacuum into solvent, that is, the solvation free energy. The above decomposition is already an approximation made by most classical (non-polarizable) force fields, as it assumes this specific separability of the Hamiltonian. In practice, once the choice of the gas-phase potential function, or force field, E_{vac} is made, the computation of E_{vac} is relatively straightforward [9]. The difficulty comes from the need to estimate the effects of solvent, encapsulated by the ΔG_{solv} term in the above equation. At present, the implicit solvent framework makes the following simplifying approximation to estimate ΔG_{solv}:

$$\Delta G_{solv} = \Delta G_{el} + \Delta G_{nonpolar} \tag{6.1}$$

Here $\Delta G_{\text{nonpolar}}$ is the free energy of solvating the molecule from which all charges have been removed (that is, the partial charges of every atom are set to zero), and ΔG_{el} is the free energy of first removing all the charges in vacuum, and then adding them back in the presence of a continuum solvent environment.

That Equation 6.1 holds only approximately can be seen from the fact that the absolute values of the solvation energies of ions of the same size and opposite charge (and the same magnitude) are not identical [10]. It should be noted that Equation 6.1 is not an absolutely necessary assumption within the general implicit solvent framework [11, 12], but it is the one that most current practical models make. Within this approximation, one needs practical methods of computing both ΔG_{el} and $\Delta G_{\text{nonpolar}}$. Computing the non-polar term has not so far been the computational bottleneck of molecular modeling, perhaps in part due to the simplistic nature of the approximations used to compute it. We will briefly touch upon some of these issues below. Our main focus will be ΔG_{el}, which is presently the most time-consuming part. The accuracy of ΔG_{el} estimates is of paramount concern since the underlying long-range interactions are critical to the function and stability of many classes of biological and chemical structures. To understand the physical basis of *analytical* approximations to ΔG_{el} such as the GB model, one needs to start with the more fundamental underlying approximation, the Poisson–Boltzmann (PB) model. Below is a brief introduction to the PB theory, geared toward our present goal of deriving the GB approximation from it.

6.1.2
The Poisson–Boltzmann model

If one accepts the continuum, linear-response dielectric approximation for the solvent, then, in the absence of mobile ions, the Poisson equation (PE) of classical electrostatics provides an exact formalism for computing the electrostatic potential $\phi(\mathbf{r})$ produced by an arbitrary charge distribution $\rho(\mathbf{r})$:

$$\nabla[\varepsilon(\mathbf{r})\nabla\phi(\mathbf{r})] = -4\pi\rho(\mathbf{r}) \tag{6.2}$$

Here, $\varepsilon(r)$ represents the position-dependent dielectric constant, which equals that of the bulk solvent far away from the molecule, and is expected to decrease fairly rapidly across the solute–solvent boundary. For now, we only consider $\rho(\mathbf{r})$ produced by a set of "fixed" atomic charges q_i at positions $\mathbf{r}_i$ inside the dielectric boundary, $\rho_f(\mathbf{r}) = \sum_i q_i \delta(\mathbf{r} - \mathbf{r}_i)$.

A common simplification is to assume an abrupt dielectric boundary, in which case $\varepsilon(r)$ takes only two values: ε_{in} inside the dielectric boundary, and ε_{out} outside – the so-called two-dielectric model. However, even with this assumption, analytical solutions of the PE for arbitrary $\rho(\mathbf{r})$ are available only for a handful of highly symmetric geometries, such as the sphere [13]. At the same time, the PE can be solved by a variety of standard numerical methods for essentially any realistic dielectric boundary and arbitrary $\varepsilon(\mathbf{r})$. The situation becomes more complicated if one is to consider the effect of mobile ions (salt). Not only will the total charge density on the right-hand side of Equation 6.2 depend on the potential, but it will contain non-trivial correlations between ions of finite size, which are difficult (though not impossible) to take into account computationally [14]. These correlations may be particularly strong for multivalent ions. Neglecting all such correlations via a mean-field treatment leads to the Poisson–Boltzmann (PB) equation. Namely, using the Boltzmann distribution for the density of mobile ions inside the potential field $\phi(r)$, we can represent the total charge density as

$$\rho(r) = \rho_f(r) + |e|\sum_j n_j z_j \exp[-\phi(r)|e|z_j/kT] \tag{6.3}$$

where n_j and z_j are the bulk density and charge of each ion species j, and $|e|$ is the elementary charge.

Substituting $\rho(r)$ from Equation 6.3 into the right-hand side of the Poisson equation (Equation 6.2) yields the full nonlinear Poisson–Boltzmann (NLPB) equation. The nonlinear nature of the NLPB leads to several unusual properties of $\phi(r)$ derived from it, including the fact that, unlike solutions of the PE, the NLPB potential is non-additive: the potential due to a collection of charges is not, in general, the sum of the individual potentials due to each charge. This property can make some practical calculations, in particular those involved in pK estimates [15], rather cumbersome. Several more subtle consistency problems exist: a discussion of these issues, including a derivation of the corresponding expression for the electrostatic free energy, can be found elsewhere [16].

Analytical solutions of the NLPB are not available for arbitrary $\rho_f(r)$, even for a sphere. However, a variety of numerical algorithms for solving the NLPB equation exist, many of them implemented in software packages designed specifically for biomolecular modeling, for example APBS [17] or DELPHI [2, 18, 19]. In this very brief discussion, we omit the details of these numerical procedures along with all of the related technical issues, which are discussed in detail elsewhere. The interested reader is referred, for example, to Ref. [20] or Ref. [21], which also contains a comprehensive list of the popular software packages that solve the NLPB equation and its more commonly used linearized version in a variety of contexts.

The complexity of the NLPB equation can be drastically reduced, and the familiar properties of the PE restored if the exponential in Equation 6.3 is linearized. The corresponding linear PB, or simply the PB equation, is commonly used in biomolecular simulations [22]:

$$\nabla[\varepsilon(r)\nabla\phi(r)] = -4\pi\rho_f(r) + \kappa^2 \varepsilon(r)\phi(r) \tag{6.4}$$

Here the electrostatic screening effects of (monovalent) salt enter via the second term on the right-hand side of Equation 6.4, where the Debye–Huckel screening parameter $\kappa \approx 0.1\,\text{Å}^{-1}$ at physiological conditions.* Once the potential $\phi(r)$ is computed, the electrostatic part of the solvation free energy is given by the familiar expression of classical electrostatics:

$$\Delta G_{el} = \frac{1}{2}\sum_i q_i [\phi(r_i) - \phi(r_i)|_{vac}] \tag{6.5}$$

---

*It is assumed that $\kappa = 0$ inside the solute.

where $\phi(r_i)|_{vac}$ is the electrostatic potential computed for the same charge distribution in the absence of the dielectric boundary, for example, in vacuum or more generally in a uniform dielectric equal to that of the molecular interior. Note that this simple formula is only valid for the linearized PB equation.

6.2
The Generalized Born Model

The need for computationally facile approximations for ΔG_{el} requires further trade-offs between accuracy and speed. This is especially true in dynamical applications where speed, algorithmic simplicity, and numerical stability are of paramount concern. One such approximation is the generalized Born (GB) model. A particularly computationally attractive form of the GB is the so-called *analytical* GB model. Essentially, this is a simple formula designed to provide an approximate, relative to the PB model, way to calculate the electrostatic part of the molecular solvation free energy. The methodology has become particularly popular in molecular dynamics (MD) applications [23–29], and lately in many other areas that will be discussed below.

6.2.1
Theoretical Foundation of the GB Model

We will begin by recasting the Poisson equation (Equation 6.2) in an equivalent form for the Green function [30]:

$$\nabla[\varepsilon(r)\nabla G(r_i, r_j)] = -4\pi\delta(r_i - r_j) \quad (6.6)$$

Within the two-dielectric model, the solution inside the dielectric boundary is

$$G(r_i, r_j) = \frac{1}{\varepsilon_{in}|(r_i - r_j)|} + F(r_i, r_j) \quad (6.7)$$

The first term in the Green function has the familiar form of the Coulomb potential due to a single charge source inside uniform dielectric ε_{in}, while the second term satisfies the Laplace equation $\nabla^2 F(r_i, r_j) = 0$. The non-singular function $F(r_i, r_j)$ corresponds to the reaction field due to polarization charges induced at the boundary; $F(r_i, r_j) \neq 0$ only in the presence of the boundary. Substituting $\phi(r_i) = \sum_j q_j G(r_i, r_j)$ into Equation 6.5, we obtain

$$\Delta G_{el} = \frac{1}{2}\sum_{ij} F(r_i, r_j) q_i q_j \quad (6.8)$$

Of course, computing $F(r_i, r_j)$ for an arbitrary charge distribution inside an arbitrary molecular boundary is as hard as solving the original Poisson equation, and can only be done numerically, as mentioned in the previous section. To make progress toward our goal – finding a simple, closed-form formula for ΔG_{el} – we need to make further approximations. The strategy we will pursue is to start from a known analytical solution of the PE for some very simple dielectric boundary to

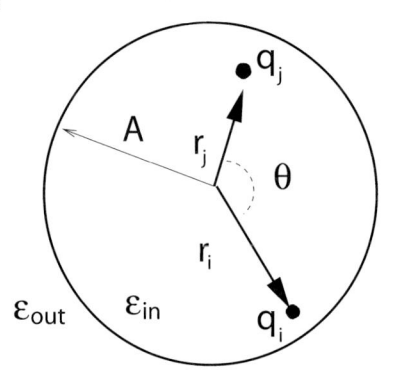

Figure 6.1 Illustration for Equations 6.9 and 6.10: a sphere of dielectric ε_{in} and radius A with two charges, q_i and q_j, at positions r_i and r_j relative to the sphere's center. The sphere is surrounded by an infinite medium of uniform dielectric ε_{out}.

get a specific *analytical* form of $F(r_i, r_j)$. Since for biomolecular modeling we need to consider an arbitrary distribution of partial charges inside the dielectric boundary, our choice is essentially limited to just one shape, the sphere, for which an exact solution of the PE exists [13]. We follow Ref. [31] and separate the self-contribution, $F(r_i, r_i)$, from the interaction part, $F(r_i, r_j)$:

$$F(r_i, r_i)^{\text{sphere}} = -\left(\frac{1}{\varepsilon_{in}} - \frac{1}{\varepsilon_{out}}\right)\frac{1}{A}\sum_{l=0}^{\infty}\frac{t_{ii}^l}{1+[l/(l+1)]\beta} \tag{6.9}$$

$$F(r_i, r_j)^{\text{sphere}} = -\left(\frac{1}{\varepsilon_{in}} - \frac{1}{\varepsilon_{out}}\right)\frac{1}{A}\sum_{l=0}^{\infty}\frac{t_{ij}^l P_l(\cos\theta)}{1+[l/(l+1)]\beta} \tag{6.10}$$

where $t_{ij} = r_i r_j / A^2$, $r_i = |r_i|$ is the position of atom i relative to the center of the sphere, A is the molecule's radius, θ is the angle between r_i and r_j, and $\beta = \varepsilon_{in}/\varepsilon_{out}$ (see Figure 6.1).

The expression for $F(r_i, r_j)$ in Equation 6.10 is extremely valuable because it is *exact*. However, its use in practical computational problems is limited because the corresponding infinite series converges slowly when t_{ij} approaches unity [32]. The latter is the typical case for biomolecules, where the largest charges are likely to be found near the molecular surface. However, under certain assumptions the series in Equation 6.10 can be summed to produce simple, finite expressions [31, 33]. Here is an outline of the derivation.

Consider the typical case of aqueous solvation $\varepsilon_{out} \gg \varepsilon_{in} \geq 1$. After making the approximation $\varepsilon_{out} \to \infty$, which in this case is equivalent to $\beta = 0$, the dependence on l in the denominators of the fractions in Equations 6.10 and 6.9 disappears, which allows us to use the well-known identity for the sum of Legendre polynomials [31] along with the geometrical identity $\cos\theta = (r_i^2 + r_j^2 - r_{ij}^2)/2r_i r_j$ (r_{ij} is the distance between the charges) to obtain for $F(r_i, r_j)^{\text{sphere}}$:

$$F(r_i, r_i)^{\text{sphere}} = -\frac{1}{\varepsilon_{in}}\frac{1}{A - r_i^2/A} \tag{6.11}$$

6.2 The Generalized Born Model

$$F(r_i, r_j)^{\text{sphere}} = -\frac{1}{\varepsilon_{\text{in}}} \frac{1}{\sqrt{r_{ij}^2 + (A - r_i^2/A)(A - r_j^2/A)}} \quad (6.12)$$

The above equations along with Equation 6.8 solve the problem of finding a simple, analytical formula for ΔG_{el}, albeit in the $\varepsilon_{\text{out}} \to \infty$ limit. However, it is not yet clear what parameters such as A, and especially "distance to center" r_i (Figure 6.1), mean in the case of realistic molecular shapes. Fortunately, the specific form of Equations 6.11 and 6.12 provides a solution that is one of the cornerstones of the GB theory. Note that both the cross-term $F(r_i, r_j)^{\text{sphere}}$ and the self-term $F(r_i, r_i)^{\text{sphere}}$ of the Green function depend only on $(A - r_i^2/A)$, which we will call the effective Born radius R_i of atom i. Thus, if we are somehow able to compute the self-term $F(r_i, r_i)$, or equivalently $\Delta G_{ii}^{\text{el}}$, for every atom in the molecule, then we can invert Equation 6.11 to calculate $(A - r_i^2/A)(A - r_j^2/A) = R_i R_j$, insert these into Equation 6.12, and hence obtain the cross-terms $F(r_i, r_j)$ as a function of R_i, R_j, and r_{ij}. For realistic biomolecular shapes, the specific form of $F(r_i, r_j)$ currently used by the generalized Born model is slightly more complicated than $F(r_i, r_j)^{\text{sphere}}$ of Equation 6.12. Namely,

$$R_i = -\frac{1}{2}\left(\frac{1}{\varepsilon_{\text{in}}} - \frac{1}{\varepsilon_{\text{out}}}\right)\frac{q_i^2}{\Delta G_{ii}^{\text{el}}} \quad (6.13)$$

$$\Delta G_{\text{el}} = \frac{1}{2}\sum_{i,j} F(r_i, r_j) q_i q_j$$
$$\approx -\frac{1}{2}\left(\frac{1}{\varepsilon_{\text{in}}} - \frac{1}{\varepsilon_{\text{out}}}\right)\sum_{i,j} \frac{q_i q_j}{\sqrt{r_{ij}^2 + R_i R_j \exp(-\gamma r_{ij}^2/R_i R_j)}} \quad (6.14)$$

The above form of $F(r_i, r_j)$ with $\gamma = 1/4$ is due to Still et al. [34] and is the most common form of $F(r_i, r_j)$ used in practice today, although alternatives such as $\gamma = 1/2$ or $\gamma = 1/10$ have also been explored. Still's formula, which we will refer to as the GB model or "canonical" GB, differs from the exact sphere-based Equations 6.11 and 6.12 in two respects: the use of $\gamma \neq 0$,[1] and slightly different dependence on the dielectrics via the prefactors $(1/\varepsilon_{\text{in}} - 1/\varepsilon_{\text{out}})$. Unless the charge is positioned exactly in the sphere's center, the specific dependence of the canonical GB on $(\varepsilon_{\text{in}}, \varepsilon_{\text{out}})$ is approximate even for a perfect sphere: the corresponding error increases with decreasing ε_{out} [31, 35]. Ways to improve the approximation are considered in Section 6.2.5. To better understand the effect of non-zero γ, note that setting $\gamma \to 0$ in Still's formula recovers the sphere limit of Equation 6.12; the use of $\gamma \neq 0$ leads to more accurate estimates of ΔG_{el} for non-spherical geometries of realistic molecules. This is likely due to the fact that, for a non-spherical molecule, many pairs of charges exist for which more of the electric field lines between the charges go through the high-dielectric region, effectively reducing the pairwise interaction compared to the purely spherical, everywhere convex, molecular boundary. Still's form of $F(r_i, r_j)$ takes this effect into account, at least to some extent, by allowing for steeper decay of the interaction with charge–charge distance.

---

1) As a consequence, $F(r_i, r_j)$ from Equation 6.14 does not satisfy the Laplace equation.

To perform practical computations based on Equation 6.14, one needs the effective Born radii R_i for every atom. The effective radius represents the atom's degree of burial within the low-dielectric solute. By definition, the effective radius can be obtained by computing ΔG_{ii}^{el} for each atom i, and then inverting Equation 6.13.[2] In fact, Equation 6.13 is simply the inverse of the famous Born formula for the solvation energy of a single ion. The idea of "generalization" of the Born formula to account for solvation energy of small molecules dates back at least 50 years [36], although the term "generalized Born" did not seem to appear in the literature until about 25 years ago [37]. Assuming that the effective Born radii can be computed efficiently for every atom in the molecule, the computational advantages of Equation 6.14 relative to the numerical PB treatment become apparent: knowledge of only N self-energy terms ΔG_{ii}^{el}, or equivalently values of R_i for each of N atoms in the molecule, gives, via the GB equation (Equation 6.14), a quick *analytical* estimate of the remaining $\sim \frac{1}{2}N^2$ charge–charge interaction contributions to the total electrostatic solvation energy. No less important is the fact that the GB formula is very simple, and its analytical derivatives with respect to atomic positions provide the forces needed in dynamical applications.

Two immediate questions arise: (i) How can the effective radii be computed *accurately* and efficiently to fully utilize these advantages? (ii) Even if the R_i are computed very accurately, will the GB formula, Equation 6.14, yield reasonably accurate estimates of the electrostatic solvation energies for *realistic molecular shapes*? A positive answer to the second question came from an analysis of the GB accuracy in the case where ΔG_{ii}^{el} and hence R_i are computed by solving the Poisson equation directly via highly accurate numerical techniques. The resulting sets of R_i – the so-called "perfect radii" – were shown to yield estimates of individual pairwise solvation cross-term energies $\Delta G_{ij}^{el} = q_i q_j F(r_i, r_j)$ in reasonable agreement with numerical results based on solving the PE directly [38] (Figure 6.2).

The PE-based perfect effective radii are very helpful in the analysis and further development of the GB model as they probe the limits of Equation 6.14 and provide a natural reference point for approximate effective radii. The perfect radii may also be useful in some practical cases, for example in MD simulations of proteins in their native states, if one assumes that the effective radii can be held constant through the simulation [38]. However, in most cases the overhead costs of performing a full PE calculation every time the radii need to be computed are prohibitive. Thus, the development of approximate methods for estimation of R_i becomes critical for the GB field. In fact, since all of the current GB models use the same basic equation (Equation 6.14), the differences between the many GB "flavors" currently available mostly reflect the differences between the particular procedures used to compute the effective Born radii. In what follows we will introduce the basic ideas behind these procedures.

---

2) In the case of a homogeneous solvent of constant dielectric, the correct procedure is to compute the radii in the $\varepsilon_{out} \to 0$ limit, which eliminates the error due to the approximate nature of the $(1/\varepsilon_{in} - 1/\varepsilon_{out})$ prefactor (see Ref. [31] for details).

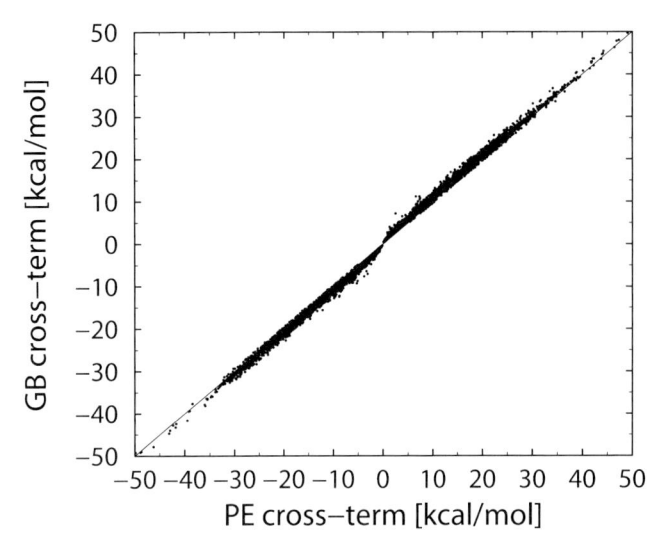

Figure 6.2 Comparison of individual cross-terms between the perfect-radii GB model (*ij* terms of Equation 6.14) to the corresponding cross-terms computed directly from the PE equation for the native state of myoglobin. The line $x = y$ represents a perfect match between the GB and PE theories.

6.2.2
Computing the Effective Born Radii

In practice, the effective radius for each atom i can be calculated by approximately evaluating the electrostatic part of the solvation free energy ΔG_{ii}^{el} in Equation 6.13 via an appropriate volume integral [39–45] (see below). Equivalent formulations based on surface integrals also exist [46, 47]. Two main challenges have to be overcome on the way to developing accurate and facile methods for computing the effective radii: (i) finding a computationally simple, physically justified integral representation for ΔG_{ii}^{el}; and (ii) developing numerical routines to perform the integration over a specific volume or surface corresponding to a physically realistic dielectric boundary between the solute and the solvent.

6.2.2.1 The Integral Approaches
Classical electrostatics provides an approach for calculating the work done to create a given charge distribution:

$$G = \frac{1}{8\pi} \int_{R^3} \frac{[D(r)]^2}{\varepsilon(r)} d^3r \tag{6.15}$$

where $D(r)$ is the dielectric displacement vector, and $\varepsilon(r)$ is the position-dependent dielectric constant. The electrostatic part of the solvation free energy ΔG_{ii}^{el} is the work of transferring the charge i from a medium of uniform dielectric constant ε_{in} equal to that of the solute into the two-dielectric solute–solvent medium, where the charge occupies its original position relative to the solute–solvent boundary. This work is given by:

$$\Delta G_{ii}^{el} = \frac{1}{8\pi\varepsilon_{out}} \int_{solvent} [D_i(r)]^2 d^3r + \frac{1}{8\pi\varepsilon_{in}} \int_{solute} [D_i(r)]^2 d^3r$$
$$- \frac{1}{8\pi\varepsilon_{in}} \int_{solvent} [D_i^0(r)]^2 d^3r - \frac{1}{8\pi\varepsilon_{in}} \int_{solute} [D_i^0(r)]^2 d^3r \quad (6.16)$$

where $D_i(r)$ is the total dielectric displacement due to charge i, ε_{out} is the dielectric constant of the solvent, and

$$D_i^0(r) \equiv \frac{q_i}{r^3} r \quad (6.17)$$

is the Coulomb field created by point charge q_i in the uniform dielectric environment. The above equation is an exact result that would lead to perfect effective radii if $D_i(r)$ were known. Since it is unknown for an arbitrary molecular shape, one has to make approximations to $D_i(r)$ to make progress. The Coulomb field approximation (CFA) was historically the first approximation of that nature. It makes what appears to be a fairly drastic assumption,

$$D_i(r) \approx D_i^0(r) \equiv \frac{q_i}{r^3} r \quad (6.18)$$

that is, the electric field generated by the atomic point charge is assumed to be unaffected by the dielectric boundary. With this assumption, the integrals over the solute volume in Equation 6.16 cancel, while the solvent volume integrals combine, to yield just one integral over the region exterior to the molecule:

$$\Delta G_{ii}^{el} = \frac{1}{8\pi} \left(\frac{1}{\varepsilon_{out}} - \frac{1}{\varepsilon_{in}} \right) \int_{ext} [D_i^0(r)]^2 d^3r$$
$$= \left(\frac{1}{\varepsilon_{out}} - \frac{1}{\varepsilon_{in}} \right) \frac{q_i^2}{8\pi} \int_{ext} \frac{1}{r^4} d^3r \quad (6.19)$$

Comparing the above with the definition of effective radius in Equation 6.13, we arrive at the following expression for the inverse effective radius α_4 in the CFA approximation:

$$\alpha_4 = R_i^{-1} = \frac{1}{4\pi} \int_{ext} \frac{1}{|r-r_i|^4} dV = \rho_i^{-1} - \frac{1}{4\pi} \int_{r>\rho_i} |r|^{-4} dV \quad (6.20)$$

where, in the first expression, the integral is taken over the region exterior to the molecule. The second expression above is often used for computational convenience [42]: the origin is moved to the atom of interest, and the integration region is the interior of the molecule outside of the atom's van der Waals (VDW) radius ρ_i. Effective ways to compute the integral will be discussed below.

The CFA is exact for a point charge at the center of a perfectly spherical solute, but it overestimates effective radii for realistic molecular geometries [48] as well as for spherical regions when the charge is off-center [33]. However, many practical routines available today still use the CFA, with a few notable exceptions discussed below. A fortuitous cancellation of errors [38] enhanced by elaborate parameterizations, plus computational efficiency of the approximation, have contributed to its success so far. Still, the limitations of the CFA have been known for quite some time, and a search for better approximations to R_i continues. In particular, a class of models exists that approximates the effective radii via integral expressions similar to Equation 6.20, but with the integrand different from the CFA's r^{-4}, namely

$$\alpha_N = \left(\frac{1}{4\pi}(N-3)\int_{\text{ext}} \frac{dV}{|r-r_i|^N}\right)^{1/(3-N)}$$
$$= \left(\rho_i^{N-3} - \frac{1}{4\pi}(N-3)\int_{r>\rho_i} \frac{dV}{|r-r_i|^N}\right)^{1/(N-3)} \tag{6.21}$$

where $N > 3$, and the additional prefactor and exponentiation are needed to preserve the dimension of α_N to be inverse length. Empirical corrections to the CFA based on a simple linear or a rational combination of α_N expressions have led to significant improvements in accuracy of the GB model [43, 44]. Several GB flavors termed "generalized Born using molecular volume" (GBMV) based on these approximations have been implemented in CHARMM, such as an expression involving α_4 and α_5 [48], and later an even more accurate expression – "GBMV2" – based on α_4 and α_7, respectively [44], $R^{-1} = \left(1-\frac{1}{2}\sqrt{2}\right)\alpha_4 + \frac{1}{2}\sqrt{2}\alpha_7$.

While all $N \neq 4$ expressions in Equation 6.21 may appear purely heuristic, at least one of them has a rigorous foundation: $N = 6$ yields the *exact* effective radius for any charge inside a perfect spherical boundary [33], not just at its center, as is the case with the CFA. More precisely, the α_6 expression is exact in the $\varepsilon_{\text{out}} \to \infty$ limit, that is, $\Delta G_{ii}^{\text{el}} = -\frac{1}{2}q_i \big/ \varepsilon_{\text{in}} \alpha_6$. Numerical tests, see Ref. [49] for important details, show that, if accurately computed "R6" radii are used in the GB equation (Equation 6.14), the resulting electrostatic solvation energies can, on average, be as accurate (relative to the PE reference) as those obtained with the use of the perfect radii. An important conclusion made in Ref. [49] was that the following expression, based on just one integral,

$$\alpha_6 = R_i^{-1} = \left(\frac{3}{4\pi}\int_{\text{ext}} \frac{dV}{|r-r_i|^6}\right)^{1/3} = \left(\rho_i^{-3} - \frac{3}{4\pi}\int_{r>\rho_i} |r|^{-6} dV\right)^{1/3} \tag{6.22}$$

represents a sufficient solution to the problem of calculating effective radii. Further attempts to increase their agreement with PB results would unlikely succeed in improving the accuracy of the GB model itself in its canonical version due to Still (Equation 6.14). The above claim makes even more sense if we recall that the "canonical" GB equation (Equation 6.14) has the same underlying physical basis as the R6 approximation for the effective radii: the Kirkwood spherical model [13]. Although the R6 radii can sometimes deviate significantly from the perfect radii, the deviations occur for those geometries where the sphere-based canonical GB model (Equation 6.14) is itself not expected to work. It remains to be seen whether the potential advantages of Equation 6.22 will translate into practical gains once its implementations, which are beginning to appear, have been extensively tested by the modeling community. Note that, in practical GB models, the accuracy of the R6 prescription may be significantly reduced by approximations made by the fast routines employed to compute the integral in Equation 6.22 or its equivalents. In general, computing the α_N integrals over a physically realistic representation of the molecular interior (or surface) is a challenge in its own right that we will discuss next.

6.2.2.2 Representations of the Dielectric Boundary

Exactly what geometrical object most closely approximates the solute–solvent dielectric boundary is still an unsettled question. Traditionally, PE or PB calculations employed the molecular (Connolly) surface for this purpose [15], although a boundary with a "smooth" transition between the two dielectrics is coming into use as well [7, 50]. The question is even more critical for the development of the GB model than for numerical PB solvers: not only the precise position of the boundary determines the numerical values of the integrals in Equations 6.20, 6.21 or 6.22, and thus the accuracy of the effective Born radii, but the mathematical form of the boundary also determines how hard it is to find analytical approximations to the integrals. Furthermore, sharper boundaries that approximate the molecular surface more closely introduce numerical instabilities that require shorter integration time steps [51].

The two limiting-case representations of the dielectric boundary commonly used for the purpose of computing the effective radii are shown in Figure 6.3. At the low end of the complexity spectrum, one has the simple and efficient model in which the solute is represented by atomic spheres of the appropriate (VDW) radii for each atom type; the dielectric boundary is taken to coincide with the surface of the spheres. In the other limiting case, the dielectric boundary is taken to be coincident with the molecular surface, and thus the integration volume is the intricate molecular volume (MV). Comparisons with explicit solvent calculations

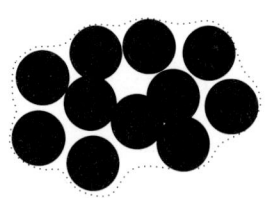

Figure 6.3 The two limiting cases of computational approximations for the solute–solvent dielectric boundary. In the simpler VDW-based approach (left), the boundary coincides with the surface of the atomic spheres, and the interstitial space is treated as high-dielectric solvent. The other approximation (right) utilizes the molecular surface (MV) for the boundary: all the space inside the surface is considered low-dielectric solute.

show that the latter is a closer approximation to physical reality [52] for macromolecules, but the higher degree of realism comes at a high price: the complexity of the molecular volume makes it very hard to approximate the corresponding volume integrals analytically. Practical routines for computing the effective radii often make additional modifications to the two basic volume definitions: the sharp dielectric boundaries in both representations can be "smoothed out," for example, via the use of atom-centered Gaussian functions. Existing GB flavors can be further classified by the specific dielectric boundary representations that each flavor is based upon. For example, the integral in Equation 6.20 can be computed numerically without further approximations over any reasonable volume representation, including the exact MV [6]. The GB flavors based on such numerical quadratures [6, 34, 44, 46, 48, 53] are in some sense similar to the "perfect" radii flavor introduced before – they play a significant role in the development of the theory, but their domain of applicability is limited, particularly in dynamics, where the GB model can be expected to offer most advantages over the PB treatment and explicit solvent.

At the other end of the "volume approximation spectrum" are GB flavors that use the VDW-based dielectric boundary to compute integrals such as Equations 6.20 or 6.21 in an efficient manner. One such flavor, "generalized Born (GB) with simple switching" (GBSW) [53] available in CHARMM, uses a smooth dielectric boundary based on the VDW volume definition, with the integrand being a rational combination of α_4 and α_7. Recently, VDW-based approximations for the R6 GB (based on α_6, Equation 6.22) have also appeared [54, 55]. Another example (still widely used) of the VDW-based approach was one of the first GB flavors – the Hawkins–Cramer–Truhlar (HCT) model [39, 40] – that employs the sharp VDW-based dielectric boundary to compute the CFA integral in Equation 6.20 in an efficient manner. Namely, the integral for each atom i is represented as a pairwise sum of integrals over spherical volumes of individual atoms $j \neq i$:

$$\alpha_4 = R_i^{-1} \approx \rho_i^{-1} - \frac{1}{4\pi} \sum_j \int_{|r_{ij}-r|<\rho_j} |r|^{-4} \, d^3r \tag{6.23}$$

The key advantage of the above approximation is that simple analytical expressions are available for it, as well as for the more general case of $\int_{|r_{ij}-r|<\rho_j} |r|^{-N} \, d^3r$ for any N, thus providing analytical expressions for α_N in Equation 6.21. Variants of the HCT are available in many modeling packages, including AMBER [56] and TINKER [57].

One problem with the VDW-based pairwise approximation such as Equation 6.23 is that it neglects possible overlaps between neighboring atoms outside of the atom of interest. In practice, the resulting over-counting of volume can be partially compensated by scaling the intrinsic radii ρ_j by a set of empirically adjusted scaling factors $\rho_j \to \rho_j f_j$, where $f_j \leq 1$ typically depend on atom type.

A more serious problem with Equation 6.23 based directly on the VDW definition of solvent volume (Figure 6.3) is the neglect of interstitial spaces between the atomic spheres in the interior of the molecule. In other words, these crevices are treated as if they belonged to the solvent space, that is, they are filled with high dielectric. As a result, the effective radii are underestimated. For molecules with little interior, the error is small, which probably explains why the original HCT model worked so well for small molecules. Also, the use of the CFA leads to a certain cancellation of errors in this case since the CFA tends to overestimate the effective radii. However, for biopolymers, such as proteins or DNA, the neglect of interstitial space leads to appreciable underestimation of the effective radii, compared to the "perfect" radii based on numerical PE estimates that use the molecular surface for the dielectric boundary [38]. Efforts to correct this deficiency while preserving the algorithmic simplicity and computational efficiency of the pairwise approximation have led to a series of GB flavors. In one of them, GB^{OBC} (available, for example, in AMBER, NAB and TINKER packages), an empirical correction is introduced [45] that modifies the pairwise integration method (Equation 6.23) to reduce the effect of interstitial high dielectrics. The procedure is designed to leave the small radii almost unaffected while larger radii are scaled up, with the scaling factor depending on the magnitude of the radius. The parameters of the rescaling function are determined based on a training set that included several proteins, in both their folded and unfolded states. Since the VDW-based pairwise HCT approach is already known to give rather accurate effective radii for surface atoms, but substantially underestimates the larger effective radii for deeply buried atoms, the rescaled radii in GB^{OBC} improve agreement with PB solvation free energies. The computational expense of the rescaling function is minimal, so that the efficiency of the HCT method is retained. In addition, effective radii are smoothly capped at about 30 Å, avoiding potential problems with numerical stability. Without the capping, stability problems may arise when the sum of the volume integrals in Equation 6.23 becomes very close to ρ_i^{-1}, making the value of the corresponding effective radius very sensitive to tiny structural variations.

However, by design, the GB^{OBC} rescaling function with tabulated parameters compensates for interstitial high dielectric only on average, in a geometry-independent manner. The problem becomes transparent in the limiting case of just two atoms that move relative to each other: while the parameters of the rescaling procedure can in principle be tabulated so as to produce the correct answer for one interatomic distance, the method will completely miss changes in molecular volume associated with the relative motion of the atoms. To address this deficiency, an additional correction to the pairwise procedure was introduced [58] that brings in elements of molecular volume, in a pairwise sense. Namely, an additional term is added to Equation 6.23 that re-introduces the molecular volume between each pair of atoms missed by the original approximation. The integral over this neck-shaped region can be approximated by a simple analytical function that carries only a small computational overhead relative to GB^{OBC}. At the same time, compared to its predecessors, the resulting GBn flavor was found to be a noticeably more faithful approximation of electrostatic solvation effects in proteins, not only by comparisons with the PE, but also with explicit solvent MD simulations [58]. The flavor is now also available in AMBER.

The "high interstitial dielectric" is not the only problem that needs to be addressed within the pairwise VDW approach. An approach to better approximate variations in the integration volume associated with changes in molecular geometry—an all-important issue in MD simulations—is presented by the "analytic generalized Born" (AGB) and "analytic generalized Born plus non-polar" (AGBNP) flavors [28] available in, for example, the IMPACT modeling package [59]. These two approximations are currently also based on the pairwise sphere-based CFA (Equation 6.23). However, unlike the original HCT, the scaling factors that multiply the VDW radii are made explicitly geometry-dependent, rather than set constant. As mentioned above, the scaling of VDW radii is designed to compensate for the over-counting of volume, which in the pairwise approximation results from multiple overlaps between atomic spheres. In AGB these overlaps are computed for each pair of atoms; the scaling factors are approximated from two-sphere overlap volumes. For computational efficiency, the atomic volumes are described by Gaussian density functions.

Yet another GB flavor based on the pairwise CFA approximation is "analytical continuum electrostatics" (ACE) [25, 60] available in, for example, the TINKER [57] package. The approach uses a set of pre-tabulated atomic volumes—for example, Voronoi volumes—to represent the total molecular volume. Each atom's contribution is described by a Gaussian density function, which leads to reasonably simple analytical expressions for the integrals in Equation 6.23.

A completely different approach, not based on the pairwise approximation and the CFA, is employed by the GBMV flavor [43, 44] and its variations [61]. Rather than augmenting the VDW representation to approximate the integrals over the molecular volume, Lee and coworkers found an analytical approximation of the appropriate integrals computed over a "smooth boundary" molecular volume that closely mimics the volume used in typical numerical PE calculations. The superior accuracy of these GBMV flavors, relative to the routines based on VDW and pairwise representations, comes at a price of noticeably higher computational costs [62].

The above examples represent several important problems and their practical solutions in the development of the GB field, but by no means give an exhaustive account of all of the variants or "flavors" of the GB model currently available. Among the relatively new developments are "residue pairwise" GB [63], the Gaussian GB [64], and FACTS [65] GB flavors. Interestingly, unlike all of the GB flavors discussed above, the FACTS approach does not rely on integral representations such as Equation 6.21 to compute the effective radii. Instead, it uses an empirical relationship between the effective radius of an atom and the distribution of other atoms around it. Work to improve the accuracy and efficiency of computational routines for estimation of effective Born radii continues.

6.2.3
Accounting for Salt Effects

When salt is present in the solvent, the GB formalism must be amended to include screening effects of the ionic atmosphere. In principle, one could envision repeating the rigorous derivations presented in Section 6.2.1 starting from the Poisson–Boltzmann equation instead of just the Poisson equation, but to the best of our knowledge the strategy has not yet been carried through. Part of the problem may be that the mathematical structures of the solution of the PB equation inside and outside the dielectric boundary are significantly more complex and substantially different from each other, unlike in the PE case. Instead, the screening effects of monovalent salt are currently introduced into the GB model via an approximate, yet very simple and computationally inexpensive, empirical correction [66] to the main formula (Equation 6.14):

$$\Delta G_{el} = -\frac{1}{2} \sum_{i,j} \left[\frac{1}{\varepsilon_{in}} - \frac{\exp(-0.73\kappa f^{GB})}{\varepsilon_{out}} \right] \frac{q_i q_j}{f^{GB}} \qquad (6.24)$$

where $f^{GB} = \sqrt{r_{ij}^2 + R_i R_j \exp(-\gamma r_{ij}^2/R_i R_j)}$ as before, and κ is the Debye–Hückel screening parameter κ (Å^{-1}) $\approx 0.316[\text{salt}]^{1/2}$ (mol$^{1/2}$l$^{-1/2}$). The above expression can be rationalized as follows. Notice that the solution of the PB equation for a single point charge has the form $\phi_i \sim [\exp(-\kappa r)/\varepsilon_{out}] \times (q_i/r)$, and the role of r in the GB formula is played by f^{GB}, suggesting a change to the above ansatz, $(1/\varepsilon_{in} - 1/\varepsilon_{out})$ $\rightarrow [1/\varepsilon_{in} - \exp(-\kappa f^{GB})/\varepsilon_{out}]$. The 0.73 prefactor was found empirically to give the best agreement with the numerical PB treatment on a set of test structures [66].

6.2.4
The Non-Polar Contribution

Although the goal of the GB model is to approximate only the electrostatic part of the solvation only, a comment is required here on how the non-polar part is currently handled in practical calculations. A common approximation widely used today [56] assumes $\Delta G_{nonpolar}$ to be proportional to the total solvent-accessible surface area (SASA) of the molecule, $\Delta G_{nonpolar} \approx \sigma \times \text{SASA}$, with the constant of proportionality derived from the experimental solvation energies of small nonpolar molecules. Substantial uncertainty exists in what appropriate value of the surface tension σ should be used in simulations, which perhaps reflects the limitations of this approximation itself. Strong arguments for the use of less drastic approximations for $\Delta G_{nonpolar}$, for example, those that treat solute–solvent van der Waals interactions ("volume term") separately from the surface area term, have been made [28, 67]. Practical models based on these ideas have already emerged. For example, the AGBNP [28] approximation mentioned above combines the basic GB framework with a model for $\Delta G_{nonpolar}$ that goes beyond the surface area approximation.

At the same time, it is clear that, at least in some cases, the $\Delta G_{nonpolar} \approx \sigma \times \text{SASA}$ approximation is not as critical to the overall accuracy of ΔG_{solv}, compared to the quality of approximating the electrostatic part ΔG_{el}. For example, in a study [25] aimed at assessing the performance of various implementations of the ACE GB flavor in MD simulations of small proteins, it was found that the overall structural deviations were insensitive to variations of the surface tension σ over a wide range from 0 to 50 cal mol^{-1}Å^{-2}. Some researchers choose to neglect the hydrophobic term altogether in MD simulations, especially if no large conformational changes are expected.

6.2.5
GB for Non-Aqueous Solvents

To the extent that solvation in non-aqueous media can be attributed to the change in the dielectric properties of the solvent, it is appropriate to seek a modification of the GB formalism to approximate the corresponding ΔG_{el}. In the simplest case of uniform solvent dielectric and arbitrary ratio $\varepsilon_{in}/\varepsilon_{out}$, one can develop a rigorous formalism similar to the one used to derive the canonical GB model valid in the $\varepsilon_{in}/\varepsilon_{out} \ll 1$ case (Section 6.2.1). Namely, the summation of the infinite series that represents the exact Green function for the sphere (Equation 6.10) can be performed for any ratio $\varepsilon_{in}/\varepsilon_{out}$, leading to the following expression [31]:

$$\Delta G_{el} \approx -\frac{1}{2}\left(\frac{1}{\varepsilon_{in}} - \frac{1}{\varepsilon_{out}}\right)\frac{1}{1+\alpha\beta}\sum_{ij} q_i q_j \left(\frac{1}{f^{GB}} + \frac{\alpha\beta}{A}\right) \tag{6.25}$$

where f^{GB} is the same as above, $\beta = \varepsilon_{in}/\varepsilon_{out}$, $\alpha = [32(3\ln 2 - 2)/(3\pi^2 - 28)] - 1 \approx 0.580127$, and A is the *electrostatic size* of the molecule. The latter provides a relationship between the molecule's global shape and its electrostatic energy [31]. Roughly speaking, A is the overall size of the structure; a rigorous definition and a way to compute it analytically are presented elsewhere [31, 35]. Whether or not Equation 6.25, termed "analytical linearized Poisson–Boltzmann" (ALPB) in Ref. [35], can be referred to as a "GB model" may be a matter of debate: unlike the canonical GB formula (Equation 6.14), Equation 6.25 contains an extra parameter A and its dependence on the solute and solvent dielectric constants is different. The model, currently available in AMBER package, is as efficient computationally as Still's formula (Equation 6.14) and can also be used in MD simulations [35] to describe solvation effects. Extensive comparisons with the PE reference on realistic biomolecular structures show [31] that the use of Equation 6.25 instead of the canonical GB to compute ΔG_{el} removes a systematic bias present in the canonical GB [35], which becomes especially pronounced outside of the $\varepsilon_{in}/\varepsilon_{out} \ll 1$ regime.

Despite the simplicity and conceptual appeal of its rigorous physical basis, the ALPB formalism of Equation 6.25 has one serious drawback: in its present form it is only applicable to the case of uniform solvent dielectric. To describe the effects of the essentially heterogeneous dielectric environment of biological membranes and the water–membrane interface, several research groups have proposed various empirical modifications to the canonical GB. The resulting approximations are currently used in practical simulations of proteins and peptides interacting with biological membranes. The key idea behind these approximations is to keep the main GB formula (Equation 6.25) intact, but to modify the effective Born radii to account for the presence of additional dielectric boundaries. In one such flavor [68], "generalized Born surface area/implicit membrane" (GBSA/IM), the membrane is modeled as a homogeneous, low-dielectric membrane "slab" that

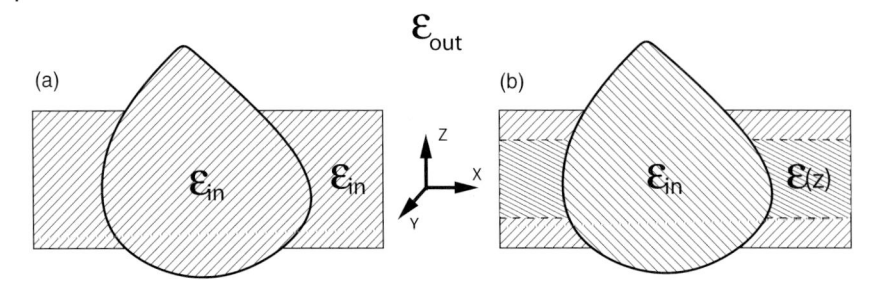

Figure 6.4 A schematic illustrating two different approaches to approximating the distributions of dielectric in the solute–membrane–solvent system used by some of the available GB flavors to compute the effective Born radii. (a) The two-dielectric model used by the GBSA/IM flavor; the low-dielectric slab is assumed infinite in the X and Y dimensions. (b) The multi-dielectric model of HDGB.

has a finite thickness in one dimension, and extends to infinity in the other two (Figure 6.4).

For each atom of the solute, the CFA pairwise summation in Equation 6.23 is split into two parts: one represents the polarization energy of the atom in the presence of the dielectric slab alone, and the other describes the contribution of the solute atoms outside of the slab. The first contribution is tabulated via a relatively simple analytical function whose parameters are set by fitting to numerical PE solutions. An obvious limitation of the approach is the assumption that the membrane environment can be represented by a single dielectric constant equal to that of the solute. Recently, the model was modified to account for heterogeneity of the membrane [69].

A different approach based on the same general idea of incorporating the electrostatic effects of a membrane into appropriately modified effective Born radii was developed in Refs. [61, 70]. The resulting heterogeneous dielectric generalized Born (HDGB) flavor is an extension of the GBMV approach discussed above. The first few terms in the infinite-series Kirkwood solution (Equation 6.9) were used to suggest a specific form for the self-energy ΔG_{ii}^{el} as a function of ε_{in} and ε_{out}, which in turn was used to modify the original (α_4, α_7)-based GBMV2 expression for the effective Born radii to include an explicit dependence on ε_{in} and ε_{out} via an analytical formula for R_i, that is, $R_i = R_i(\varepsilon_{in}, \varepsilon_{out})$. The model [70] then proceeds by partitioning the membrane "slab" into several regions of constant dielectric (Figure 6.4), approximating a realistic scenario in which the dielectric properties of the membrane vary continuously across the bilayer. The actual geometry of the slab and the variation of the dielectric constant perpendicular to the membrane plane, (z), mimic that predicted for a dipalmitoylphosphatidylcholine (DPPC) bilayer. This partition is used to define the dielectric environment for each solute atom embedded in the membrane; the corresponding effective radius is computed via the $R_i(\varepsilon_{in}, \varepsilon_{out})$ prescription. Both GBSA/IM and HDGB flavors are available in the CHARMM [71, 72] package, along with another membrane GB flavor based on the same general principle but derived from GBSW mentioned in the previous sections [73].

6.3
Applications of the GB Model

The algorithmic simplicity and reasonable accuracy of the GB approximation, combined with its availability in popular modeling packages, have made it the method of choice in many practical applications of the implicit solvent methodology. The list is expanding, and some representative examples are given below.

6.3.1
Protein Folding and Design

Exploring large conformational transitions is one of several areas where the advantages of the implicit solvent framework, and specifically of the GB model, become particularly useful. Recent molecular simulations of the protein folding process, which used all-atom physics-based potentials to obtain correctly folded structures starting from extended conformations, are arguably one of the most spectacular achievements attributable to the GB model. Examples include small proteins such as the 20-residue "Trp-cage" [74, 75], a 23-residue mixed α/β protein [76], and the 36-residue villin headpiece [77]. Successful folding simulations of even larger proteins are also beginning to appear [78, 79]. In these simulations, the folded state is typically predicted to within about 2 Å from experiment (C_α root-mean-square deviation), and in some cases [74, 80] within about 1 Å. Energy landscapes computed within the implicit solvent framework were used to gain insights into the folding mechanisms [77, 80]. The GB model can also be used to explore the influences of temperature, friction, and random forces on the folding of proteins [81].

An example of a "protein design" study in which changes of protein stability associated with point mutations were explored with the GB model can be found in Ref. [82]. Another relevant example is the use of the model in the prediction of protein loop conformations [83].

6.3.2
"Large-Scale" Motions in Macromolecules

The conformational search speedup offered by continuum solvent models such as the GB allows one to study large-scale motions in proteins and protein complexes. The use of the methodology to understand large conformational changes in proteins is exemplified by a recent study of the motions of active site flaps in HIV protease [84]: it is unlikely that a comparable explicit solvent study would currently be computationally feasible. Another relevant example is a recent work that explored conformational dynamics of avian flu virus [85].

Compared to proteins, implicit solvent MD simulations of nucleic acids are relatively new, and not as numerous. A number of methodological issues still need to be resolved. So far, the GB methodology has been employed to model free DNA in solution [86, 87], and binding between proteins and nucleic acids [88–90], as well as for energetic analysis of conformational changes such as the A $\rightarrow$ B transition [26]. A recent all-atom study of the nucleosome and its 147-base-pair DNA free in solution [91] has demonstrated the usefulness of the GB for exploring dynamics of "large" DNA fragments and protein–DNA complexes.

6.3.3
Peptides and Proteins in the Membrane Environment

Membranes are large structures, the translocation of molecular structures through membranes may involve significant molecular movements and conformational changes, and so these systems are natural candidates for implicit solvent simulations based on the GB model. The GB flavors described in Section 6.2.5 have been used in modeling of small peptides [73, 92], and membrane spanning helices in proteins [93], and in simulation of whole membrane proteins [69, 70, 92], and protein complexes, as large and complex as the bacteriorhodopsin trimer [94].

6.3.4
pK Prediction and Constant pH Simulations

Traditionally, quantitative prediction of pK values and protonation states of ionizable groups in macromolecules has been squarely in the domain of numerical PB solvers – see, for example, Refs. [15, 95–98]. While the development of PB-based approaches for pK prediction continues, GB-based calculations have begun to emerge. The first applications of the GB model to compute the energetics of proton transfer in proteins were encouraging [42], although at the time the reference PB calculations were still definitely more accurate. However, a very recent GB-based model for prediction of protein pK values was found to be competitive with the latest PB-based and empirical approaches [99].

Up until recently, physics-based pK calculations assumed limited or no coupling between protonation and conformational degrees of freedom. Likewise, the charge states of all ionizable groups were considered fixed throughout the course of a typical MD simulation, regardless of the conformational changes that the structure may undergo. In reality, changes in protonation state and conformational changes are strongly coupled. Full and consistent accounting for this coupling may be necessary for further improvement of the accuracy of pK estimates [100]; in dynamics, it may lead to non-trivial effects [101]. The GB model is an ideal candidate to introduce this type of coupling into dynamical simulations: its instantaneous dielectric response makes possible on-the-fly estimates of relative energies of protonation microstates. Several GB-based approaches have recently been developed that fully couple protonation and conformational degrees of freedom in molecular dynamics – the so-called constant-pH MD. One of the methods employs a continuous protonation state model [102], in which the equations of motion are used to time-evolve the protonation coordinate; convergence to a physical protonation state of 1 or 0 is enforced by an adjustable potential barrier. An example of recent use of the methodology is a study of the pH dependence of folding landscapes of several peptides, which provided insights into protein aggregation that occurs in Alzheimer's disease [103]. An alternative approach [104] operates directly in the physical protonation space: protonation states are accepted or rejected on the fly, according to a Metropolis criterion, during the course of the MD simulation. The approach was recently combined with the replica exchange technique to study the pH-dependent mechanism of nitric oxide release in nitrophorin proteins [105]. The accuracy of pK predictions based on the constant-pH dynamics is becoming competitive [100] with that of the more traditional PB-based models that do not fully account for structure–protonation coupling.

6.3.5
Other Uses

The use of the GB approximation in molecular modeling is not limited to the general areas outlined above. For example, applications of the model to analysis of the energetics of protein–ligand binding have been reported [45, 106–108]; several GB flavors have been implemented in the popular DOCK ligand docking program [109, 110]. Another emerging area where the GB has been found to be useful is "hybrid" explicit/implicit approaches to the treatment of solvent effects. Examples of the latter include a recent model [111] in which the immediate hydration of the solute is modeled explicitly by a layer of water molecules, and the GB model describes the electrostatics of the bulk continuum solvent outside the explicit simulation volume. A similar idea has recently been found to be very effective in the context of replica-exchange simulations [112]; a detailed account of this methodology can be found in Ref. [113]. Finally, we will mention recent application of the GB in quantum mechanics/molecular mechanics (QM/MM) simulations [114, 115].

6.4
Some Practical Considerations

The decision to use the GB approximation instead of the more rigorous PB model or the traditional explicit solvent model may depend on many factors, including the type of molecular structure, the specifics of the questions one asks of the calculation, and even the available computational resources. Which of the numerous GB flavors is an optimal choice for each task also depends on these details. Presented below is a discussion of selected aspects of the GB performance in all-atom molecular modeling that is intended to illustrate several general trends backed by specific examples.

6.4.1
The Accuracy/Speed Tradeoffs

One of the main motivations behind developing the GB model has always been its computational efficiency, relative to alternative approximations that describe solvation effects. Since the GB is just an approximation to the more fundamental PB, and both approximations share the same underlying physical framework of continuum electrostatics, performance comparison between the two models appears natural, though it is not straightforward in practice. This is because the results may depend on the type of problem, the size and shape of the molecular structure, and also on parameters of the specific algorithms and their implementations, such as grid spacing and specific convergence criteria used in numerical PB solvers. These issues were considered in detail in a recent study [62], which presents a performance comparison between several popular GB models and numerical PB solvers that were commonly available in 2004. The work used the electrostatic solvation energy as the target quantity to assess both accuracy and speed of the models. Not surprisingly, it was found that the most accurate of the GB flavors tested (for example, GBMV and GB^{OBC}) are still less accurate than the most accurate of the PB solvers, but appreciably faster. The difference in speed was up to several orders of magnitude for a small protein (36 residues, 596 atoms), but only about an order of magnitude for a much larger protein (239 residues, 3628 atoms). The trend reflects the difference in scaling behavior between the GB model, which currently scales with the number of charges as $O(N^2)$ unless further approximations are made, and the more favorable scaling of the PB-based algorithms, for example, $O(N^{3/2})$ for some algorithms that employ successive over-relaxation to solve the finite-difference matrix equations [116]. We stress that the specific trend is only applicable to compact globular structures such as proteins in their native states; a more detailed discussion of the scaling issues will be presented below in the context of MD simulations. One should also be careful not to over-interpret such comparisons between two different models, and to focus on general trends rather than precise numbers. For example, the above-mentioned comparison study [62] also found that some of the less accurate numerical PB solvers were quite competitive, speed-wise, against the more accurate among the GB flavors. Among the GB flavors, the same general tradeoffs were seen: the more accurate approximations were generally slower. For example, the most accurate of the flavors tested in Ref. [62] – GBMV – was found to be several times slower in MD simulations than the next one down the accuracy list, GB^{OBC}(II).

The algorithmic simplicity and computational speed of the GB approximation make it particularly attractive in molecular dynamics simulations of biomolecules. Relative to the traditional explicit solvent simulations, the use of the GB model to represent solvation effects can be expected to accelerate the simulation significantly in many cases. The corresponding "speedup" is the combined effect of two very different contributions: (i) the direct speedup via reduced computational (clock) time; and (ii) the indirect speedup achieved via enhanced conformational sampling. In what follows, we will consider the two contributions separately.

6.4.2
GB Computational Time Relative to Explicit Solvent

Since MD simulations in explicit solvent require tracking of a large number of water molecules and counterions placed around the solute of interest, one may expect a considerable reduction in CPU (clock) time once this need is eliminated via the use of the GB model. The exact amount of this "direct" relative speedup is not easy to quantify, for the same reasons as in the GB versus numerical PB comparison outlined above: the algorithms to be compared are very different. In explicit solvent simulations, the current practice in the field is to use the so-called particle mesh Ewald (PME) approximation [9] to speed up the computation of electrostatic interactions. Within this approach, the long-range pairwise Coulomb interaction energy is represented (via a mathematical trick that relies on imposing artificial periodic boundary conditions on the system) as a sum of rapidly converging series summed in real space, plus a rapidly converging Fourier series. Since both sums converge quickly, an accurate result can be obtained by retaining only a relatively small number of terms in the sums. An additional speedup comes from the interpolation of the potential over a regular mesh. Several adjustable parameters control the accuracy and speed of the PME. While any direct comparison of MD simulation timings between the GB and PME is bound to be implementation dependent, we can still identify several general important trends based on the computational complexity of the two algorithms and types of molecular structures used in the simulations.

The computational complexity of the canonical GB equation (Equation 6.14) is obviously $O(N_{\text{solute}}^2)$, while for the PME it is $O(N_{\text{tot}} \log N_{\text{tot}})$ (see [9]), where N_{solute} is the number of atomic charges in the solute, and N_{tot} is the total number of charges, including those of the solvent, $N_{\text{tot}} = N_{\text{solute}} + N_{\text{solvent}}$. Assuming that one keeps the same thickness of the solvent shell as the size of the solute grows, for compact globular solutes and large N_{solute} one has $N_{\text{solvent}} \sim N_{\text{solute}}^{2/3}$ and $N_{\text{tot}} \sim N_{\text{solute}}$, and thus the expense of a PME-based computation is effectively $O(N_{\text{solute}} \log N_{\text{solute}})$ for large structures. This means that, while a reasonable implementation of the "raw" GB may be expected to be faster than the PME for "small and medium-size" structures, the advantage is bound to disappear beyond a certain, implementation-dependent, crossover solute size. We stress that this result assumes no further approximations or algorithmic improvements to the general GB formalism.

The above trends for the relative speeds of GB and PME computations can be very different for non-compact structures, or simulations where transitions between compact and stretched-out conformations are expected, for example during the process of protein folding. This is because the number of solute molecules required to fill the standard simulation box in these cases will scale differently with the solute size than for the compact globular structures considered above. For example, in the limiting case of a completely stretched-out polymer chain placed inside a cubic bounding box, the total volume of the solvent in the box scales as $O(N_{\text{solute}}^3)$, and hence $N_{\text{total}} \sim N_{\text{solute}}^3$, which leads to $O(N_{\text{solute}}^3 \log N_{\text{solute}})$ for the PME computational complexity, compared to just $O(N_{\text{solute}}^2)$ for the "raw" GB. Thus, in contrast to the case of compact globular structures, GB-based simulations of extended conformations are always expected to be considerably less expensive than the corresponding explicit solvent computations that employ the PME.

We now illustrate the above general trends on specific examples. We will compare single CPU timings between the GB^{OBC} flavor and the PME, in MD simulations of a set of proteins in their compact, native states. For this illustration, we will be using the models implemented in a popular MD package, AMBER (v.8) [56]. Unless otherwise specified, we choose the default values for the input parameters such as non-bonded cutoff (9 Å) in PME. The solvent buffer size is 10 Å. No long-range cutoffs or other approximations will be applied in the GB-based simulations. The results of the comparison are as follows. For a small protein villin headpiece (36 residues, 596 atoms), a GB-based simulwation proceeds roughly 10 times faster than the corresponding explicit solvent one based on the PME. The timings become about equal for the medium-size protein ubiquitin (76 residues, 1231 atoms). For the much larger structure of the nucleosome (eight protein subunits plus 146 base-pairs of DNA, about 25 000 atoms in total), it was found earlier [91] that the ratio of computational times is about 10:1 in favor of the PME, although the parameters of that specific simulation were somewhat different from those used in the single protein examples. Thus, for these specific GB and PME implementations used to compute electrostatic interactions in all-atom MD simulations of compact globular structures, the crossover size beyond which current PME-based simulations become faster than those based on the GB is somewhere between 1000 and 2000 atoms. Note that this result applies to simulations that use a single processor. The crossover point may be effectively pushed toward larger structures if one has access to a multi-CPU (multi-core) cluster: the GB-based simulations generally tend to scale better with the number of utilized processors than the PME-based ones. Recent AMBER (v.9) benchmarks[3] provide some concrete examples: for structures of about 25 000 atoms, the maximum speedup – that is, the speedup beyond which doubling the number of CPUs does not lead to any significant increase in compute speed – is at least seven times higher for the GB-based simulations than for the PME-based runs. Thus, the relative computational speed of the GB model can be substantially increased, albeit extensively, if one has access to a large parallel machine. The use of graphics processing units (GPUs) to speed up GB-based molecular dynamics calculations may also hold considerable promise. For example, a recent study demonstrated that a two orders of magnitude acceleration versus a single CPU can be achieved for a ~5000 atom protein in an all-atom molecular dynamics running entirely on a GPU [117].

Additional approximations can also be used to reduce the associated direct simulation costs in GB-based simulations.[4] For example, the latest versions of AMBER offer at least three different strategies to speed up such simulations: some of these strategies are specific to the GB methodology, and some are generic and have been used in the explicit solvent simulations as well. Among the latter is the *multiple step* approach in which slowly varying long-range forces are not calculated at every step of MD, but only every nrespa > 1 steps. In addition, the standard long-range cutoff schemes may be very useful in GB-based simulations. The cutoff schemes have become almost obsolete in explicit solvent MD, in part due to the

---

3) See http://ambermd.org/amber9.benchl.html.
4) Note that the now standard PME approach is not applicable to GB, at least in its current form. This is because, in contrast to the explicitly pairwise Coulomb potential, the GB Green function depends, implicitly through the effective Born radii, on the positions of all the charges in the system.

success of the Ewald method and in part since it became evident that spherical cutoffs introduce artifacts into explicit solvent simulations [118] affecting particularly strongly the structure and dynamics of water. However, the very absence of explicit water in the GB-based simulations may make them more amenable to long-range cutoffs – successful long MD runs have been reported using cutoff values of 24 Å [45], 18 Å [86], or even 12 Å [89]. While for small systems the benefits of using realistic cutoffs are not very high, the speedups may become significant for larger molecules where a reasonable cutoff may be chosen to be considerably smaller than the system size. The specific numbers will necessarily be implementation dependent. Also, one should always be aware that spherical cutoffs may not be appropriate for highly charged systems or other situations where long-range electrostatic interactions play a key role, such as in nucleic acids or the nucleosome. Fortunately, yet another way to cut GB computational expense is now available, which is based on reducing the cost of computing the effective Born radii. This is achieved by setting a finite upper limit, rgbmax, in the integral (Equation 6.20), so that only the part of the solvent within rgbmax around the given atom is taken into account in computing its effective radius. Atoms whose associated spheres are farther away than rgbmax from the given atom will not contribute to that atom's effective Born radius. This is implemented in a "smooth" fashion [56, 119], so that, when part of an atom's sphere lies inside the rgbmax sphere, that part still contributes to the low-dielectric region that determines the effective Born radius. As a result, the derivatives of the total energy with respect to atomic coordinates are continuous, leading to energy conservation, and there is no large spurious force acting on the atoms coming in and out of this "reaction field cutoff." Importantly, unlike in the case of the standard cutoff, even if two charges are separated by a distance larger than rgbmax, they still interact, albeit with a somewhat altered strength.

The implicit solvent methodology is relatively new, and so studies that seek to speed up MD simulations based on it are not nearly as many or extensive as is the case with the explicit solvent MD. However, the limited available evidence is encouraging. For example [119], MD simulations of several small and medium-size proteins, 10 base-pair duplexes of B-DNA, and RNA have shown that, at least on the time scales of up to 10 ns, the use of rgbmax preserves the native structure to the same extent as do simulations in which this approximation was not used, as in Ref. [45]. Even fewer tests have been performed that examine the effects of multiple time step approximation in implicit solvent MD, or a combination of it with the traditional spherical cutoff and/or the use of rgbmax. Still, the limited experience we have is encouraging – see Table 6.1, where we have summarized the results of applying such approximations in MD simulations of a 76-residue protein.

A conclusion can be drawn that, at least for this particular protein, the use of any of the algorithmic improvements described above brings about a speedup, with only modest disruption of the native structure of the protein. When two or more of the methods are combined, their speedups combine too, though not necessarily in a linear fashion: the use of reasonable values of rgbmax, the long-range cutoff cut, and multiple time steps nrespa together (last row in Table 6.1)

Table 6.1 The effect of several additional approximations on the speed and accuracy of GB-based MD simulations. Shown are the deviations from the native structure (backbone rmsd, Å) and relative computational speedup for a set of 8 ns long MD simulations of a 76-residue protein ubiquitin (PDB 1UBQ), all done with the GB^{OBC} model (igb = 5) of AMBER. Further details of the MD protocol can be found in Ref. [45]. In simulations that employed the multiple time-step algorithm, that is, recomputed the long-range forces only at every nrespa > 1 step, Langevin dynamics was used. The collision frequency was set to a low value of gamma_ln = 0.05 ps^{-1}. The average and maximum values of the rmsd were computed over the entire trajectory in each case. The value of the long-range cutoff is specified by cut. The largest dimension of the protein is smaller than the 40 Å cutoff.

rgbmax	cut	nrespa	⟨rmsd⟩	max rmsd	Relative direct speedup
40	40	4	1.05	1.6	1
40	18	1	1.3	2.0	1.1
9	40	1	1.1	1.8	1.6
9	18	1	1.28	2.11	1.9
40	40	4	1.6	2.5	2.4
9	40	4	1.23	2.1	4.0
9	18	4	1.08	2.08	4.4

results in a four-fold increase in computational speed compared to the simulation in which none of the methods have been used (first row). The algorithmic improvements discussed above are expected to be even more efficient for larger systems. For example, our recent experience [91] with the GB-based MD simulations of the nucleosome core particle (~25 000 atoms, system size ~120 Å) shows that the use of rgbmax = 15 Å results in about three-fold increase in speed for the system, while yielding stable trajectories. Even with rgbmax = 40 Å, the speedup was still more than two-fold. Note that the use of the standard long-range cutoff would be problematic for this highly charged compound. At the same time, it is clear that the more of the additional approximations that are made, the larger are the deviations from the native structure, and so one has to be very cautious, especially in the yet unexplored regimes. It is also worth mentioning that, compared to the standard explicit solvent simulations, more careful multi-step equilibration protocols may be necessary in the regime where the solvent viscosity is considerably reduced or even set to zero [45]. Overall, based on admittedly very limited evidence, we conclude that the three approximations presented above to speed up the GB-based simulations yield encouraging results, suggesting that these are worth exploring further. Development of novel approximations that promise to bring computational complexity of the GB model to $O(N\log N)$ is under way [120].

6.4.3
Enhancement of Conformational Sampling

This is arguably one of the most significant advantages that the GB model has to offer, although a quantitative analysis of the effect and its relative contribution to the corresponding "indirect" speedup of the GB-based MD simulations is even less straightforward than for the "direct" speedup considered above. Part of the difficulty is that the sampling enhancement depends on many details of the molecular system and processes under study, and may also depend on the specifics of the GB model and MD algorithms used. Below, we will present a few semi-quantitative and qualitative observations based on the limited data available in the literature.

One can make estimates of the degree of conformational sampling enhancement by comparing the kinetics of specific conformational transitions in the implicit and the corresponding explicit solvent simulations. Generally, the enhancement of conformational sampling can be thought of as a combination of at least two effects: increased sampling due to significant reduction or complete elimination of solvent viscous forces that slow down the motion of the solute parts; and faster conformational search due to the effective smoothing of energy landscapes. The interplay of these effects in GB-based MD simulations of small model systems has recently been considered in detail by Hamelberg *et al.* [121] and Feig [122], for specific GB models in AMBER (GB^{OBC}) and CHARMM (GBMV), respectively. Relevant discussions focused on the GB-based protein folding simulations can also be found in Refs. [77, 81].

Hamelberg *et al.* [121] reported a four orders of magnitude increase in the rate of conformational sampling due to the combined effect of both the landscape smoothing and reduced solvent viscosity, as assessed by comparing the *cis* ↔ *trans* isomerization rate in a dipeptide relative to the corresponding rate in explicit water simulations. When the solvent viscosity was increased to the levels corresponding to that of water, by appropriately increasing the collision frequency in Langevin dynamics, the isomerization rate was still found to be about one to two orders of magnitude higher than that in the explicit water. Thus, in this specific model and molecular system, two to three orders of magnitude enhancement of sampling relative to explicit solvent comes from the elimination of solvent viscous forces and the remaining one to two orders of magnitude are due to the smoothing of the energy landscape. Therefore, it is reasonable to assume that the elimination and/or reduction of solvent viscosity alone should result in significant effective speedup of viscosity-controlled conformational transitions. We hypothesize that this type of transition is likely to involve large-scale relative motions of parts of the structure along a relatively smooth energy landscape. This hypothesis may explain the significant folding rate enhancement observed in the GB-based MD simulations of folding of some small proteins. For example, note that experimental folding times for even the fastest folding proteins is of the order of microseconds, whereas in some of the GB-based protein folding simulations [74] described in the previous sections the native state was reached on 10 ns time scale. Assuming that the folding rates in explicit solvent are the same order of magnitude as the experimental ones, the comparison gives a very rough idea of the magnitude of conformational search speedups – at least two orders of magnitude – that one can expect in these types of simulations through the use of the GB model. We emphasize that the specific numbers may only be applicable to the GB flavors (AMBER) used in these simulations (see below). For a relevant general discussion of the interplay of time scales and friction forces in protein folding, see, for example, Ref. [123]; a detailed analysis of the folding rate dependence on viscous effects for 20-residue protein is available in Ref. [77].

The magnitude of the conformational search speedup relative to the explicit solvent also depends on the type of conformational rearrangements. In a GB-based simulation of the A ↔ B transition in DNA, Tsui et al. [26] reported only a ~20-fold increase in the transition speed. We do not know whether the more modest transition rate enhancement observed in this system is indeed due to the fact that the transition involves relatively smaller structural rearrangements and higher barriers, but it is a plausible hypothesis. A similar amount of conformational sampling speedup was estimated from an analysis of open–close loop transition events in the avian flu virus protein [85]. On the other hand, global bending of the DNA on length scales comparable to its persistence length appears to occur about a 100 times faster in GB-based simulations [91] relative to explicit solvent [124]. This amount of conformational sampling enhancement is more consistent with viscosity-controlled dynamics.

Overall, while there is no doubt that the use of the GB model to represent solvation effects in simulations does bring about appreciable increase in the rate of conformational sampling in many cases, the precise magnitude of this increase depends strongly on the specifics of the system and processes under study. The rate increase also appears to depend on the particular implicit solvent model. For example, in contrast to the numbers discussed above appropriate for the GB flavors in AMBER, considerably smaller acceleration of conformational sampling — only about a factor of four to five — was observed [122] with a GBMV flavor available in CHARMM. Moreover, the zero solvent viscosity limit achieved in the absence of stochastic collisions with the solute (via the use of the Nosé–Hoover thermostat) actually *slowed down* conformational transitions in alanine dipeptide in that study, relative to the explicit solvent. At least some of the slowdown in this case was attributed to the lack of thermal "jolts" from stochastic collisions that help cross potential barriers.

6.5
Limitations of the GB Model

The generalized Born model is separated from reality by several layers of approximation (Figure 6.5), each of them adding its own limitations to the method. Some of these limitations directly affect the accuracy of the GB approximation, while others may simply restrict its areas of application. For example, no matter how accurate a specific flavor of the GB model may be, a continuous electrostatic potential cannot be defined within its context: at best, one can talk about potential at atomic centers only [42]. Thus, unlike the PB model, the GB approximation proper cannot be used to produce the colorful distributions of electrostatic potential that are now widely used in structural biology. To have this specific capability within an analytical model, one has to go beyond the GB [32, 125].

6.5 Limitations of the GB Model

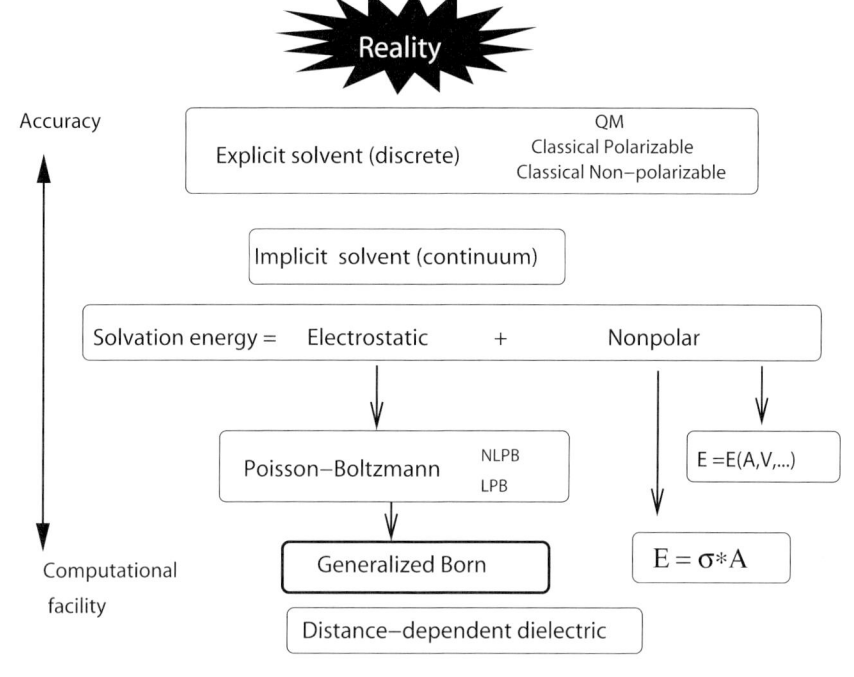

[Figure: hierarchy diagram with "Reality" at top, branching down through "Explicit solvent (discrete)" and "QM / Classical Polarizable / Classical Non–polarizable", then "Implicit solvent (continuum)", then "Solvation energy = Electrostatic + Nonpolar", with Electrostatic leading to "Poisson–Boltzmann (NLPB, LPB)" → "Generalized Born" → "Distance–dependent dielectric", and Nonpolar leading to "E = E(A,V,...)" → "E = σ∗A". Vertical axis on left shows "Accuracy" at top and "Computational facility" at bottom.]

Figure 6.5 The hierarchy of approximate representations of solvent effects in molecular modeling. The GB model is separated from reality by several layers of approximation.

More important, however, are limitations that directly affect the accuracy of the GB relative to the more fundamental descriptions of solvation, such as the explicit solvent framework or the PB model.

The most fundamental approximation step, the "discrete → continuum" approximation, obviously eliminates a number of real solvent effects that depend on the finite size of the water molecule, such as dewetting. Likewise, the implicit solvent model cannot describe effects of tightly bound water molecules, which may be a serious limitation when those are important for the function or stability of the structure of interest. For example, in protein–ligand complexes, structured water is sometimes found right at the binding interface. Also, it is not clear how well the continuum approximation works inside deep binding pockets, where solvent can hardly be considered as having the properties of the bulk. Water–solvent hydrogen bonds are present in the implicit solvent model only approximately, at a mean-field level, which may under- or over-estimate their strengths in specific cases.

The Poisson–Boltzmann approximation inherits the above generic limitations of mean-field theories and linear response approximations, and adds its own. In particular, the neglect of correlation between counterions, especially multivalent ones such as Mg^{2+}, may be a serious problem in the modeling of highly charged structures such as nucleic acids. The PB → GB step introduces several additional approximations. Earlier GB models, as well as models that used uncorrected pairwise schemes based on VDW atom spheres to compute the effective radii, could be expected to perform worse on larger structures relative to small molecules – see the discussion in Section 6.2.2.2. Fortuitous cancellation of errors often masked this problem in calculations of the total electrostatic solvation energy [38]. The latest generations of GB models have overcome some of these problems. Still, from the derivation of the GB model presented in Section 6.2.1, it can be expected that for a given structure the largest errors relative to the PB treatment will occur in regions whose local shapes deviate most from spherical. The heuristic correction in Still's formula that partially accounts for deviations of molecular shape from a perfect sphere is uncontrollable. It is unrealistic to expect that its effect on the overall accuracy, even relative to the PB, would be exactly quantifiable *a priori* for any biomolecular simulation.

An additional complication is that, while there is really only one GB model, any of its practical applications relies on a specific flavor of the model. There are now well over 10 such flavors, and the number is increasing alarmingly, especially if one counts the numerous existing parameterizations of the same basic "flavor." There is enough difference between most of these flavors and their detailed parameterizations [75] that specific results obtained with one flavor or parameterization cannot necessarily be expected to be reproducible with another. This is particularly true in dynamical applications. Note that in this case it is not only the specifics of the GB flavor that affect the outcome, but also the way the non-polar contribution is computed. The choice of underlying gas-phase force field is also important: one cannot automatically assume that a force field known to outperform its predecessor within the explicit solvent framework will also show better performance when used with a given GB model [126]. Here, by "performance," we mean agreement with the explicit solvent and/or experiment.

Even though it may not be possible to make an unambiguous choice of the best-performing combination of GB flavor and gas-phase force field, one general trend appears to emerge. The latest GB flavors that show better agreement with the underlying PB model are likely to perform better than older flavors that did not agree with the PB all that well. Since the PB is also an approximation, a natural question is then how much of the error seen in GB-based MD simulations is already present at the PB level? For example, it appears that at least some of the GB flavors do not have the right balance between intra-solute and solvent–solute charge–charge interactions, resulting in over-stabilization of solvent-exposed salt bridges [127]. However, a careful follow-up study revealed that not all of the discrepancy (with the reference explicit solvent results) came from the GB model, with some of it being inherent to the PB. Importantly, even the use of "perfect" [38] effective radii in the GB equation (Equation 6.14) does not match the accuracy of the PB in predicting the relative energies of conformational states of small peptide – see Ref. [128] for important details. In that particular study, the error of the PB itself, relative to the explicit solvent treatment, was found to be smaller, but not negligible compared to the GB error. Thus, there still appears to be room for improvement within the PB $\to$ GB approximation.

6.6
Conclusions and Outlook

Within the implicit solvent framework, solvation effects are modeled by replacing individual solvent molecules with a continuous medium that mimics the bulk properties of the solvent. Even though the framework makes several fundamental approximations to reality, it is in many respects an attractive alternative to the more conventional explicit representation, which tracks movements of discrete solvent molecules. Practical models based on the implicit solvent framework, such as the GB model considered here, offer several significant advantages over the explicit water representation, including lower computational costs, faster conformational search, and very effective ways to estimate relative free energies of conformational ensembles.

The generalized Born (GB) model provides a simple analytical formula for molecular electrostatic energy in the presence of implicit solvent. In the hierarchy of approximations that lead to the model, the GB lies below the more fundamental model based on the Poisson equation (PE) of continuum electrostatics. In fact, apart from a heuristic correction term, the general mathematical form of the GB corresponds to the exact PE result for the electrostatic part of the solvation free energy for a hypothetical perfectly spherical molecule surrounded by a uniform dielectric medium in the conductor limit (infinitely high dielectric). Heuristic corrections partially account for realistic biomolecular shapes, the screening effects of monovalent salt, and the high, but finite, dielectric of water. Non-homogeneous dielectric environments such as biological membranes require additional corrections.

The accuracy of the GB model depends critically on the accuracy of the so-called effective Born radii that characterize the positions of each partial atomic charge relative to the molecular surface of the structure. Many practical algorithms for computing the effective radii have been developed, leading to the many different "flavors" of the basic GB model available today. Expected tradeoffs between accuracy and speed apply. The accuracy of the effective radii depend largely on how realistic is the representation for the solvent–solute dielectric boundary used by the specific algorithm: simplified representations are typically more facile computationally, but lead to less accurate radii.

In several applications, such as molecular dynamics simulations, where the robustness of the algorithms and computational efficiency are of paramount concern, the generalized Born (GB) model has arguably become the most widely used approximation for the molecular electrostatic energy in the presence of implicit solvent. Perhaps one of the most spectacular achievements of the model is the successful first-principles (physics-based) simulations of the complete folding process of several small proteins at full atomic level – a fit that is probably not yet within reach for the corresponding all-atom explicit solvent simulations. Other areas where the model's effectiveness is found to be particularly useful include exploration of large-scale motions in proteins or DNA, protein design, modeling of the membrane environment, and replica-exchange simulations based on novel "hybrid" explicit or implicit approaches. For some types of calculations, for example, constant-pH molecular dynamics, models based on implicit solvation such as the GB appear to be the only ones currently available in practice. Recently, encouraging results have also been obtained in applying the GB model for prediction of pK shifts in proteins, QM/MM simulations, and in the field of protein–ligand docking.

While there is no doubt at the moment that the use of the implicit solvent in molecular simulations may bring considerable rewards, it is also associated with additional uncertainty relative to the more traditional calculations based on the explicit solvent. Compared to the latter, less is known about the domain of applicability of the implicit solvent framework, and so extra care must be taken when using practical models based on it, including the GB. In each specific case, the decision as to whether or not the potential rewards of using the GB model are likely to outweigh the risks depends on many factors. These may include the type of molecular structure, the specifics of the questions one asks of the calculation, and even the available computational resources. For example, the task of exploring large conformational changes in a protein is a good candidate for GB-based simulations, especially on a parallel machine. At the same time, if one simply needs to generate an ensemble of near-native conformations of the same protein, then the tried-and-true explicit solvent approach (with the electrostatics treated by PME) is probably a better choice. Once the decision to use the GB model is made, the researcher is typically faced with many additional practical issues, including choice of optimal GB flavor for the specific task at hand. These flavors may differ substantially in their accuracy and speed, including the speed of conformational sampling, which may be a critical factor in making the choice. An additional complication is that many of the GB flavors available in popular modeling packages deliver optimal performance only in conjunction with a specific gas-phase force field, including the associated atomic radii sets used in the calculation of the effective Born radii.

Despite the many documented successes of the GB model, situations where it clearly needs improvement are abundant. These help to establish the boundaries of applicability of the currently available GB flavors; they also suggest directions for future improvements of the model. These improvements will likely include progress in the following areas:

1) Systematic quantitative exploration of performance of the available GB flavors. Development of comprehensive consensus test sets and practices.
2) Further parameter optimization of the most promising of the existing GB flavors against the PB and explicit solvent. The challenge here is not to overparameterize the GB model beyond its natural accuracy limits. Transferability of the highly parameterized solutions will probably remain problematic.
3) Development and testing of novel ways to compute the effective Born radii, such as the "R6" prescription that yields exact effective radii in the perfect spherical case. The method is appealing from both the accuracy and computational facility standpoints, but it remains to be seen how its practical implementations will perform.

4) Development of novel approaches designed specifically to reduce the computational complexity of GB-based molecular simulations, ideally to $O(N \log N)$, without the loss of accuracy associated with the traditional spherical cutoff schemes.
5) Revision of the theoretical foundation of the GB model aimed at bringing its accuracy closer to the more fundamental PB model, while preserving the appealing simplicity of the canonical GB. It is apparent now that Still's formula has reached its accuracy limits, but developing a superior approximation that is equally simple and robust is probably the most challenging task on the list above.

Acknowledgments

The author thanks Andrew Fenley and Ramu Anandakrishnan for reading the manuscript and making helpful suggestions. Financial support from the NIH (R01 GM076121) is acknowledged.

References

1. Cramer, C.J. and Truhlar, D.G. (1999) Implicit solvation models: equilibria, structure, spectra, and dynamics. *Chem. Rev.*, **99**, 2161–2200.
2. Honig, B. and Nicholls, A. (1995) Classical electrostatics in biology and chemistry. *Science*, **268**, 1144–1149.
3. Beroza, P. and Case, D.A. (1998) Calculation of proton binding thermodynamics in proteins. *Methods Enzymol.*, **295**, 170–189.
4. Madura, J.D., Davis, M.E., Gilson, M.K., Wade, R.C., Luty, B.A., and McCammon, J.A. (1994) Biological applications of electrostatic calculations and Brownian dynamics. *Rev. Comput. Chem.*, **5**, 229–267.
5. Gilson, M.K. (1995) Theory of electrostatic interactions in macromolecules. *Curr. Opin. Struct. Biol.*, **5**, 216–223.
6. Scarsi, M., Apostolakis, J., and Caflisch, A. (1997) Continuum electrostatic energies of macromolecules in aqueous solutions. *J. Phys. Chem. A*, **101**, 8098–8106.
7. Luo, R., David, L., and Gilson, M.K. (2002) Accelerated Poisson–Boltzmann calculations for static and dynamic systems. *J. Comput. Chem.*, **23**, 1244–1253.
8. Sagui, C. and Darden, T.A. (1999) Molecular dynamics simulations of biomolecules: long-range electrostatic effects. *Annu. Rev. Biophys. Biomol. Struct.*, **28**, 155–179.
9. Schlick, T. (2002) *Molecular Modeling and Simulation*, Springer, Berlin.
10. Mobley, D.L., Ii, A.E., Fennell, C.J., and Dill, K.A. (2008) Charge asymmetries in hydration of polar solutes. *J. Phys. Chem. B*, **112** (8), 2405–2414.
11. Lazaridis, T. and Karplus, M. (1999) Effective energy function for proteins in solution. *Proteins*, **35** (2), 133–152.
12. Dzubiella, J., Swanson, J.M., and McCammon, J.A. (2006) Coupling nonpolar and polar solvation free energies in implicit solvent models. *J. Chem. Phys.*, **124** (8), 084905.
13. Kirkwood, J.G. (1934) Theory of solution of molecules containing widely separated charges with special application to zwitterions. *J. Chem. Phys.*, **2**, 351–361.
14. Frank-Kamenetskiĭ, M.D., Anshelevich, V.V., and Lukashin, A.V. (1987) Polyelectrolyte model of DNA. *Sov. Phys. Usp.*, **30** (4), 317.

15 Bashford, D. and Karplus, M. (1990) pK_as of ionizable groups in proteins: atomic detail from a continuum electrostatic model. *Biochemistry*, **29**, 10219–10225.

16 Sharp, K.A. and Honig, B. (1990) Electrostatic interactions in macromolecules. *Annu. Rev. Biophys. Biophys. Chem.*, **19**, 301–332.

17 Baker, N.A., Sept, D., Joseph, S., Holst, M.J., and McCammon, J.A. (2001) Electrostatics of nanosystems: application to microtubules and the ribosome. *Proc. Natl. Acad. Sci. U.S.A.*, **98** (18), 10037–10041.

18 Rocchia, W., Alexov, E., and Honig, B. (2001) Extending the applicability of the nonlinear Poisson–Boltzmann equation: multiple dielectric constants and multivalent ions. *J. Phys. Chem. B*, **105** (28), 6507–6514.

19 Nicholls, A. and Honig, B. (1991) A rapid finite difference algorithm, utilizing successive over relaxation to solve the Poisson–Boltzmann equation. *J. Comput. Chem.*, **12**, 435–445.

20 Simonson, T. (2003) Electrostatics and dynamics of proteins. *Rep. Prog. Phys.*, **66**, 737–787.

21 Baker, N., Bashford, D., and Case, D. (2006) Implicit solvent electrostatics in biomolecular simulation, in *New Algorithms for Macromolecular Simulation* (eds B. Leimkuhler, C. Chipot, R. Elber, A. Laaksonen, A. Mark, T. Schlick, C. Schütte, and R. Skeel), Lecture Notes in Computational Science and Engineering, vol. 49, Springer, Berlin, pp. 263–295.

22 Baker, N.A. (2005) Improving implicit solvent simulations: a Poisson-centric view. *Curr. Opin. Struct. Biol.*, **15** (2), 137–143.

23 Dominy, B.N. and Brooks, C.L. (1999) Development of a generalized Born model parametrization for proteins and nucleic acids. *J. Phys. Chem. B*, **103**, 3765–3773.

24 Bashford, D. and Case, D. (2000) Generalized Born models of macromolecular solvation effects. *Annu. Rev. Phys. Chem.*, **51**, 129–152.

25 Calimet, N., Schaefer, M., and Simonson, T. (2001) Protein molecular dynamics with the generalized Born/Ace solvent model. *Proteins*, **45** (2), 144–158.

26 Tsui, V. and Case, D. (2000) Molecular dynamics simulations of nucleic acids using a generalized Born solvation model. *J. Am. Chem. Soc.*, **122**, 2489–2498.

27 Wang, T. and Wade, R. (2003) Implicit solvent models for flexible protein–protein docking by molecular dynamics simulation. *Proteins*, **50**, 158–169.

28 Gallicchio, E. and Levy, R.M. (2004) AGBNP: an analytic implicit solvent model suitable for molecular dynamics simulations and high-resolution modeling. *J. Comput. Chem.*, **25**, 479–499.

29 Nymeyer, H. and Garcia, A.E. (2003) Simulation of the folding equilibrium of α-helical peptides: a comparison of the generalized Born approximation with explicit solvent. *Proc. Natl. Acad. Sci. U.S.A.*, **100**, 13934–13949.

30 Jackson, J.D. (1975) *Classical Electrodynamics*, John Wiley & Sons, Inc., New York.

31 Sigalov, G., Scheffel, P., and Onufriev, A. (2005) Incorporating variable dielectric environments into the generalized Born model. *J. Chem. Phys.*, **122** (9), 094511.

32 Fenley, A.T., Gordon, J.C., and Onufriev, A. (2008) An analytical approach to computing biomolecular electrostatic potential. I. Derivation and analysis. *J. Chem. Phys.*, **129** (7), 075101.

33 Grycuk, T. (2003) Deficiency of the Coulomb field approximation in the generalized Born model: an improved formula for Born radii evaluation. *J. Chem. Phys.*, **119** (9), 4817–4826.

34 Still, W.C., Tempczyk, A., Hawley, R.C., and Hendrickson, T. (1990) Semianalytical treatment of solvation for molecular mechanics and dynamics. *J. Am. Chem. Soc.*, **112**, 6127–6129.

35 Sigalov, G., Fenley, A., and Onufriev, A. (2006) Analytical electrostatics for biomolecules: beyond the generalized Born approximation. *J. Chem. Phys.*, **124** (12), 124902.

36 Hoijtink, G.J., de Boer, E., van der Meij, P.H., and Weijland, W. (1956) Reduction potentials of various aromatic

hydrocarbons and their univalent anions. *Rec. Trav. Chim. Pays Bas*, **75**, 487–503.

37 Constanciel, R. and Contreras, R. (1984) Self consistent field theory of solvent effects representation by continuum models: introduction of desolvation contribution. *Theor. Chim. Acta*, **65** (1), 1–11.

38 Onufriev, A., Case, D.A., and Bashford, D. (2002) Effective Born radii in the generalized Born approximation: the importance of being perfect. *J. Comput. Chem.*, **23** (14), 1297–1304.

39 Hawkins, G.D., Cramer, C.J., and Truhlar, D.G. (1995) Pairwise solute descreening of solute charges from a dielectric medium. *Chem. Phys. Lett.*, **246**, 122–129.

40 Hawkins, G.D., Cramer, C.J., and Truhlar, D.G. (1996) Parametrized models of aqueous free energies of solvation based on pairwise descreening of solute atomic charges from a dielectric medium. *J. Phys. Chem.*, **100**, 19824–19836.

41 Scarsi, M., Apostolakis, J., and Caflisch, A. (1997) Continuum electrostatic energies of macromolecules in aqueous solutions. *J. Phys. Chem. A*, **101** (43), 8098–8106.

42 Onufriev, A., Bashford, D., and Case, D.A. (2000) Modification of the generalized Born model suitable for macromolecules. *J. Phys. Chem. B*, **104** (15), 3712–3720.

43 Lee, M.S., Salsbury, F.R. Jr, and Brooks, C.L. III (2002) Novel generalized Born methods. *J. Chem. Phys.*, **116**, 10606–10614.

44 Lee, M.S., Feig, M., Salsbury, F.R., and Brooks, C.L. (2003) New analytic approximation to the standard molecular volume definition and its application to generalized Born calculations. *J. Comput. Chem.*, **24** (11), 1348–1356.

45 Onufriev, A., Bashford, D., and Case, D.A. (2004) Exploring protein native states and large-scale conformational changes with a modified generalized Born model. *Proteins*, **55** (2), 383–394.

46 Ghosh, A., Rapp, C.S., and Friesner, R.A. (1998) Generalized Born model based on a surface integral formulation. *J. Phys. Chem. B*, **102**, 10983–10990.

47 Romanov, A.N., Jabin, S.N., Martynov, Y.B., Sulimov, A.V., Grigoriev, F.V., and Sulimov, V.B. (2004) Surface generalized Born method: a simple, fast, and precise implicit solvent model beyond the Coulomb approximation. *J. Phys. Chem. A*, **108** (43), 9323–9327.

48 Lee, M.S., Salsbury, F.R., and Brooks, C.L. (2002) Novel generalized Born methods. *J. Chem. Phys.*, **116** (24), 10606–10614.

49 Mongan, J., Svrcek-Seiler, W.A., and Onufriev, A. (2007) Analysis of integral expressions for effective Born radii. *J. Chem. Phys.*, **127** (18), 185101.

50 Im, W., Beglov, D., and Roux, B. (1998) Continuum solvation model: computation of electrostatic forces from numerical solutions to the Poisson–Boltzmann equation. *Comput. Phys. Commun.*, **111** (1–3), 59–75.

51 Chocholousová, J. and Feig, M. (2006) Balancing an accurate representation of the molecular surface in generalized Born formalisms with integrator stability in molecular dynamics simulations. *J. Comput. Chem.*, **27**, 719–729.

52 Swanson, J.M.J., Mongan, J., and McCammon, J.A. (2005) Limitations of atom-centered dielectric functions in implicit solvent models. *J. Phys. Chem. B*, **109** (31), 14769–14772.

53 Im, W., Lee, M.S., and Brooks, C.L. (2003) Generalized Born model with a simple smoothing function. *J. Comput. Chem.*, **24** (14), 1691–1702.

54 Tjong, H. and Zhou, H.X. (2007) GBr6: a parameterization-free, accurate, analytical generalized Born method. *J. Phys. Chem. B*, **111** (11), 3055–3061.

55 Labute, P. (2008) The generalized Born/volume integral implicit solvent model: estimation of the free energy of hydration using London dispersion instead of atomic surface area. *J. Comput. Chem.*, **29**, 1693–1698.

56 Case, D.A., Cheatham, T.E., Darden, T., Gohlke, H., Luo, R., Merz, K.M., Onufriev, A., Simmerling, C., Wang, B., and Woods, R.J. (2005) The Amber biomolecular simulation programs. *J. Comput. Chem.*, **26** (16), 1668–1688.

57 Ponder, J., (2001) TINKER: Software Tools for Molecular Design, 3.9,

Washington University School of Medicine, Saint Louis, MO.

58 Mongan, J., Simmerling, C., McCammon, J., Case, D., and Onufriev, A. (2007) Generalized Born model with a simple, robust molecular volume correction. *J. Chem. Theory Comput.*, **3**, 156–169.

59 Banks, J.L., Beard, H.S., Cao, Y., Cho, A.E., Damm, W., Farid, R., Felts, A.K., Halgren, T.A., Mainz, D.T., Maple, J.R., Murphy, R., Philipp, D.M., Repasky, M.P., Zhang, L.Y., Berne, B.J., Friesner, R.A., Gallicchio, E., and Levy, R.M. (2005) Integrated modeling program, applied chemical theory (Impact). *J. Comput. Chem.*, **26** (16), 1752–1780.

60 Schaefer, M. and Karplus, M. (1996) A comprehensive analytical treatment of continuum electrostatics. *J. Phys. Chem.*, **100**, 1578–1599.

61 Feig, M., Im, W., and Brooks, C.L. (2004) Implicit solvation based on generalized Born theory in different dielectric environments. *J. Chem. Phys.*, **120** (2), 903–911.

62 Feig, M., Onufriev, A., Lee, M.S., Im, W., Case, D.A., and Brooks, C.L. (2004) Performance comparison of generalized Born and Poisson methods in the calculation of electrostatic solvation energies for protein structures. *J. Comput. Chem.*, **25**, 265–284.

63 Archontis, G. and Simonson, T. (2005) A residue-pairwise generalized Born scheme suitable for protein design calculations. *J. Phys. Chem. B*, **109** (47), 22667–22673.

64 Grant, J.A., Pickup, B.T., Sykes, M.J., Kitchen, C.A., and Nicholls, A. (2007) The Gaussian generalized Born model: application to small molecules. *Phys. Chem. Chem. Phys.*, **9** (35), 4913–4922.

65 Haberthür, U. and Caflisch, A. (2007) Facts: fast analytical continuum treatment of solvation. *J. Comput. Chem.*, **29**, 701–715.

66 Srinivasan, J., Trevathan, M., Beroza, P., and Case, D. (1999) Application of a pairwise generalized Born model to proteins and nucleic acids: inclusion of salt effects. *Theor. Chem. Accts*, **101**, 426–434.

67 Wagoner, J.A. and Baker, N.A. (2006) Assessing implicit models for nonpolar mean solvation forces: the importance of dispersion and volume terms. *Proc. Natl. Acad. Sci. U.S.A.*, **103** (22), 8331–8336.

68 Spassov, V.Z., Yan, L., and Szalma, S. (2002) Introducing an implicit membrane in generalized Born/solvent accessibility continuum solvent models. *J. Phys. Chem. B*, **106** (34), 8726–8738.

69 Ulmschneider, M.B., Ulmschneider, J.P., Sansom, M.S., and Di Nola, A. (2007) A generalized Born implicit-membrane representation compared to experimental insertion free energies. *Biophys. J.*, **92** (7), 2338–2349.

70 Tanizaki, S. and Feig, M. (2005) A generalized Born formalism for heterogeneous dielectric environments: application to the implicit modeling of biological membranes. *J. Chem. Phys.*, **122** (12), 124706.

71 Brooks, B., Bruccoleri, R., Olafson, D., States, D., Swaminathan, S., and Karplus, M. (1983) Charmm: a program for macromolecular energy, minimization, and dynamics calculations. *J. Comput. Chem.*, **4**, 187–217.

72 MacKerel A. Jr, Brooks, C. III, Nilsson, L., Roux, B., Won, Y., and Karplus, M. (1998) CHARMM: the energy function and its parameterization with an overview of the program, in *Encyclopedia of Computational Chemistry*, vol. **1** (ed. P. von Ragué Schleyer), John Wiley & Sons, Ltd, Chichester, pp. 271–277.

73 Im, W., Feig, M., and Brooks, C.L. (2003) An implicit membrane generalized Born theory for the study of structure, stability, and interactions of membrane proteins. *Biophys. J.*, **85** (5), 2900–2918.

74 Simmerling, C., Strockbine, B., and Roitberg, A.E. (2002) All-atom structure prediction and folding simulations of a stable protein. *J. Am. Chem. Soc.*, **124**, 11258–11259.

75 Chen, J., Im, W., and Brooks, C.L. (2006) Balancing solvation and intramolecular interactions: toward a consistent generalized Born force field. *J. Am. Chem. Soc.*, **128** (11), 3728–3736.

76 Jang, S., Kim, E., and Pak, Y. (2008) All-atom level direct folding simulation of a ββα miniprotein. *J. Chem. Phys.*, **128** (10), 105102.

77 Zagrovic, B., Snow, C.D., Shirts, M.R., and Pande, V.S. (2002) Simulation of folding of a small alpha-helical protein in atomistic detail using worldwide-distributed computing. *J. Mol. Biol.*, **323** (5), 927–937.

78 Jang, S., Kim, E., Shin, S., and Pak, Y. (2003) Ab initio folding of helix bundle proteins using molecular dynamics simulations. *J. Am. Chem. Soc.*, **125** (48), 14841–14846.

79 Lei, H. and Duan, Y. (2007) Ab initio folding of albumin binding domain from all-atom molecular dynamics simulation. *J. Phys. Chem. B*, **111** (19), 5458–5463.

80 Pitera, J.W. and Swope, W. (2003) Understanding folding and design: replica-exchange simulations of "trp-cage" miniproteins. *Proc. Natl. Acad. Sci. U.S.A.*, **100** (13), 7587–7592.

81 Jagielska, A. and Scheraga, H.A. (2007) Influence of temperature, friction, and random forces on folding of the B-domain of staphylococcal protein A: all-atom molecular dynamics in implicit solvent. *J. Comput. Chem.*, **28** (6), 1068–1082.

82 Lopes, A., Alexandrov, A., Bathelt, C., Archontis, G., and Simonson, T. (2007) Computational sidechain placement and protein mutagenesis with implicit solvent models. *Proteins*, **67** (4), 853–867.

83 Felts, A.K., Gallicchio, E., Chekmarev, D., Paris, K.A., Friesner, R.A., and Levy, R.M. (2008) Prediction of protein loop conformations using the AGBNP implicit solvent model and torsion angle sampling. *J. Chem. Theory Comput.*, **4** (5), 855–868.

84 Hornak, V., Okur, A., Rizzo, R.C., and Simmerling, C. (2006) HIV-1 protease flaps spontaneously open and reclose in molecular dynamics simulations. *Proc. Natl. Acad. Sci. U.S.A.*, **103** (4), 915–920.

85 Amaro, R.E., Cheng, X., Ivanov, I., Xu, D., and McCammon, A.J. (2009) Characterizing loop dynamics and ligand recognition in human- and avian-type influenza neuraminidases via generalized Born molecular dynamics and end-point free energy calculations. *J. Am. Chem. Soc.*, **131** (13), 4702–4709.

86 Tsui, V. and Case, D. (2001) Theory and applications of the generalized Born solvation model in macromolecular simulations. *Biopolymers*, **56**, 275–291.

87 Sorin, E., Rhee, Y., Nakatani, B., and Pande, V. (2003) Insights into nucleic acid conformational dynamics from massively parallel stochastic simulations. *Biophys. J.*, **85** (2), 790–803.

88 De Castro, L.F. and Zacharias, M. (2002) DAPI binding to the DNA minor groove: a continuum solvent analysis. *J. Mol. Recognit.*, **15** (4), 209–220.

89 Allawi, H., Kaiser, M., Onufriev, A., Ma, W., Brogaard, A., Case, D., Neri, B., and Lyamichev, V. (2003) Modeling of flap endonuclease interactions with DNA substrate. *J. Mol. Biol.*, **328**, 537–554.

90 Chocholousová, J. and Feig, M. (2006) Implicit solvent simulations of DNA and DNA protein complexes: agreement with explicit solvent vs experiment. *J. Phys. Chem. B*, **110** (34), 17240–17251.

91 Ruscio, J.Z. and Onufriev, A. (2006) A computational study of nucleosomal DNA flexibility. *Biophys. J.*, **91** (11), 4121–4132.

92 Spassov, V.Z., Yan, L., and Szalma, S. (2002) Introducing an implicit membrane in generalized Born/solvent accessibility continuum solvent models. *J. Phys. Chem. B*, **106**, 8726–8738.

93 Zheng, W., Spassov, V., Yan, L., Flook, P., and Szalma, S. (2004) A hidden Markov model with molecular mechanics energy scoring function for transmembrane helix prediction. *Comput. Biol. Chem.*, **28** (4), 265–274.

94 Tanizaki, S. and Feig, M. (2006) Molecular dynamics simulations of large integral membrane proteins with an implicit membrane model. *J. Phys. Chem. B*, **110** (1), 548–556.

95 Beroza, P. and Fredkin, D.R. (1996) Calculation of amino acid pK_as in a protein from a continuum electrostatic model: methods and sensitivity analysis. *J. Comput. Chem.*, **17**, 1229–1244.

96 Antosiewicz, J., McCammon, J.A., and Gilson, M.K. (1994) Prediction of pH

97 Nielson, J.E. and Vriend, G. (2001) Optimizing the hydrogen-bond network in Poisson–Boltzmann equation-based pK_a calculations. *Proteins*, **43**, 403–412.

98 Georgescu, R., Alexov, E., and Gunner, M. (2002) Combining conformational flexibility and continuum electrostatics for calculating pK_as in proteins. *Biophys. J.*, **83**, 1731–1748.

99 Spassov, V. and Yan, L. (2008) A fast and accurate computational approach to protein ionization. *Protein Sci.*, **17**, 1955–1970.

100 Khandogin, J. and Brooks, C.L. (2006) Toward the accurate first-principles prediction of ionization equilibria in proteins. *Biochemistry*, **45** (31), 9363–9373.

101 Ripoll, D.R., Vorobjev, Y.N., Liwo, A., Vila, J.A., and Scheraga, H.A. (1996) Coupling between folding and ionization equilibria: effects of pH on the conformational preferences of polypeptides. *J. Mol. Biol.*, **264**, 770–783.

102 Lee, M.S., Salsbury, F.R., and Brooks, C.L. (2004) Constant pH molecular dynamics using continuous titration coordinates. *Proteins*, **56** (4), 738–752.

103 Khandogin, J. and Brooks, C.L. (2007) Linking folding with aggregation in Alzheimer's beta-amyloid peptides. *Proc. Natl. Acad. Sci. U.S.A.*, **104** (43), 16880–16885.

104 Mongan, J., Case, D.A., and McCammon, J.A. (2004) Constant pH molecular dynamics in generalized Born implicit solvent. *J. Comput. Chem.*, **25** (16), 2038–2048.

105 Swails, J.M., Meng, Y., Walker, A.F., Marti, M.A., Estrin, D.A., and Roitberg, A.E. (2009) pH-dependent mechanism of nitric oxide release in nitrophorins 2 and 4. *J. Phys. Chem. B*, **113** (4), 1192–1201.

106 Gohlke, H. and Case, D.A. (2004) Converging free energy estimates: MM-PB(GB)SA studies on the protein–protein complex ras–raf. *J. Comput. Chem.*, **25** (2), 238–250.

107 Liu, H.Y. and Zou, X. (2006) Electrostatics of ligand binding: parametrization of the generalized Born model and comparison with the Poisson–Boltzmann approach. *J. Phys. Chem. B*, **110** (18), 9304–9313.

108 Wittayanarakul, K., Hannongbua, S., and Feig, M. (2008) Accurate prediction of protonation state as a prerequisite for reliable MM-PB(GB)SA binding free energy calculations of HIV-1 protease inhibitors. *J. Comput. Chem.*, **29** (5), 673–685.

109 Zou, X., Sun, Y., and Kuntz, I.D. (1999) Inclusion of solvation in ligand binding free energy calculations using the generalized-Born model. *J. Am. Chem. Soc.*, **121** (35), 8033–8043.

110 Ewing, T.J.A., Makino, S., Skillman, A.G., and Kuntz, I.D. (2001) Dock 4.0: search strategies for automated molecular docking of flexible molecule databases. *J. Comput. Aided Mol. Des.*, **15**, 411–428.

111 Lee, M.S., Salsbury, F.R., and Olson, M.A. (2004) An efficient hybrid explicit/implicit solvent method for biomolecular simulations. *J. Comput. Chem.*, **25** (16), 1967–1978.

112 Okur, A., Wickstrom, L., Layten, M., Geney, R., Song, K., Hornak, V., and Simmerling, C. (2006) Improved efficiency of replica exchange simulations through use of a hybrid explicit/implicit solvation model. *J. Chem. Theory Comput.*, **2** (2), 420–433.

113 Okur, A. and Simmerling, C. (2006) Hybrid explicit/implicit solvation methods, in *Annual Reports in Computational Chemistry*, vol. 2 (ed. D. Spellmeyer), Elsevier, Amsterdam, pp. 97–109.

114 Pellegrini, E. and Field, M.J. (2002) A generalized-Born solvation model for macromolecular hybrid-potential calculations. *J. Phys. Chem. A*, **106** (7), 1316–1326.

115 Walker, R.C., Crowley, M.F., and Case, D.A. (2007) The implementation of a fast and accurate qm/mm potential method in Amber. *J. Comput. Chem.*, **29** (7), 1019–1031.

116 Bashford, D. (1997) An object-oriented programming suite for electrostatic effects in biological molecules, in *Scientific Computing in Object-Oriented Parallel Environments* (eds Y. Ishikawa,

R.R. Oldehoeft, J.V.W. Reynders, and M. Tholburn), Lecture Notes in Computer Science, vol. **1343**, Springer, Berlin, pp. 233–240.

117 Friedrichs, M.S., Eastman, P., Vaidyanathan, V., Houston, M., Legrand, S., Beberg, A.L., Ensign, D.L., Bruns, C.M., and Pande, V.S. (2009) Accelerating molecular dynamic simulation on graphics processing units. *J. Comput. Chem.*, **30** (6), 864–872.

118 Loncharich, R.J. and Brooks, B.R. (1989) The effects of truncating long-range forces on protein dynamics. *Proteins*, **6**, 32–45.

119 Svrcek-Seiler, A. and Onufriev, A. Fast computation of effective Born radii for use in molecular dynamics simulations (unpublished).

120 Anandakrishnan, R. and Onufriev, A. (2009) An $N \log N$ approximation based on the natural organization of biomolecules for speeding up the computation of long range interactions. *J. Comput. Chem.* doi: 10.1002/jcc.21357.

121 Hamelberg, D., Shen, T., and McCammon, J.A. (2006) Insight into the role of hydration on protein dynamics. *J. Chem. Phys.*, **125** (9), 094905.

122 Feig, M. (2007) Kinetics from implicit solvent simulations of biomolecules as a function of viscosity. *J. Chem. Theory Comput.*, **3** (5), 1734–1748.

123 Hagen, S.J., Qiu, L., and Pabit, S.A. (2005) Diffusional limits to the speed of protein folding: fact or friction? *J. Phys. Condens. Matter*, **17** (18), S1503–S1514.

124 Lankas, F., Lavery, R., and Maddocks, J.H. (2006) Kinking occurs during molecular dynamics simulations of small dna minicircles. *Structure*, **14** (10), 1527–1534.

125 Gordon, J.C., Fenley, A.T., and Onufriev, A. (2008) An analytical approach to computing biomolecular electrostatic potential. ii. Validation and applications. *J. Chem. Phys.*, **129** (7), 075102.

126 Shell, S.M., Ritterson, R., and Dill, K.A. (2008) A test on peptide stability of Amber force fields with implicit solvation. *J. Phys. Chem. B*, **112**, 6878–6886.

127 Zhou, R. and Berne, B.J. (2002) Can a continuum solvent model reproduce the free energy landscape of a beta-hairpin folding in water? *Proc. Natl. Acad. Sci. U.S.A.*, **99** (20), 12777–12782.

128 Roe, D.R., Okur, A., Wickstrom, L., Hornak, V., and Simmerling, C. (2007) Secondary structure bias in generalized Born solvent models: comparison of conformational ensembles and free energy of solvent polarization from explicit and implicit solvation. *J. Phys. Chem. B*, **111** (7), 1846–1857.

7
Implicit Solvent Force-Field Optimization
Jianhan Chen, Wonpil Im, and Charles L. Brooks III

7.1
Introduction

Ab initio simulation of protein folding has been one of the greatest challenges in structural biology [1–3]. Such an ability is not only an ultimate and most stringent test of the protein force-field and sampling methodology, but also provides tremendous opportunities for successful application of molecular simulations to study the structure and dynamics of biomolecules in general. Widespread availability of high-performance computing and development of efficiently parallelized simulation codes in recent years have greatly expanded the sampling capability and really pushed the limit of accessible time and length scales. It is now possible to simulate systems with over a million atoms for tens of nanoseconds [4, 5] or fully solvated small proteins for several microseconds [6, 7]. However, it has been somewhat disappointing that atomistic simulations on the microsecond time scale (which is considered the protein folding speed limit [8]) have not produced the success greatly anticipated [6]. Instead, such long-time simulations actually confirm many of the limitations in the classical physics-based molecular mechanical force fields [9, 10], especially in their abilities to faithfully describe protein folding and large conformational transitions. Partially due to the original parameterization to study the folded native structures of proteins, virtually all modern protein force fields have suffered from systematic biases such as over-stabilization of intramolecular interactions with respect to solvation [11] and tendency to overestimate the helical contents [12, 13]. Addressing these limitations in a consistent and transferable fashion has proven to be a formidable task and so far only rather limited progress has been made. One of the main bottlenecks for continual improvement of the force fields is to do with the necessity for an accurate and efficient description of the solvent environment, which is critical for the structure, dynamics, and function of biological macromolecules [14].

The conventional approach of explicitly including the water molecules in the simulation arguably provides the most detailed description of the solvent, but, at the same time, dramatically increases the system size and thus the associated computational cost. Furthermore, longer simulations are necessary to obtain statistically meaningful averages for protein structure, dynamics and thermodynamic properties. The expensive computational cost, together with the highly complex nature and large parameter space of the related optimization problem, has so far prevented the translation of a high level of detail into a high level of accuracy in describing protein conformational equilibria. Instead, so-called implicit solvent models have recently emerged as a powerful alternative to explicit water for representing the solvent environment [15]. Here, the mean influence of the solvent molecules on the solute is captured via a direct estimation of the solvation free energy, defined as the reversible work required to transfer the solute from vacuum to solution in a fixed configuration. Elimination of the solvent molecules by the implicit treatment substantially reduces the number of atoms that need to be simulated. More importantly, this can be achieved with only a moderate increase in the computational cost required for calculating the solvation free energy on the fly, such as using continuum electrostatics-based methods including Poisson–Boltzmann (PB) and generalized Born (GB) theories [16–18]. Clearly, such a dramatic reduction in the system size does not come without a loss of detail and achievable accuracy. For example, implicit solvent models may yield considerable disagreement with explicit water simulations in short-range effects when the detailed interplay of a few water molecules (which are distinct from the bulk water) is important [19, 20]. Examinations of the potentials of mean force (PMFs) between model compounds also revealed a lack of fine structures in the implicit solvent PMFs due to continuum descriptions [21, 22]. The entropic contribution of solvation is in principle included in the solvation free energy, but the temperature dependence of solvation and its influence on protein stability generally cannot be modeled by implicit solvent. Nonetheless, the substantial reduction in the computational cost and extension of accessible simulation time scales with an implicit treatment of the solvent have opened a door to address many biological problems that are otherwise difficult with explicit solvent. Particularly, it has also allowed development of new simulation techniques, such as constant-pH molecular dynamics to study pH-dependent protein folding and unfolding [23, 24] and implicit membrane models to study insertion, folding, and assembly of membrane proteins [25–28].

One of the advantages of implicit solvent that might be easily overlooked is that substantial reduction in the computational cost allows careful optimization of the force field through extensive peptide simulations to obtain a highly accurate description of peptide conformational equilibria. The protein conformational equilibrium is governed by a delicate balance among sets of underlying competing interactions, that is, the solvation preference of side chains and backbones in solution versus the strength of solvent-mediated interactions between these moieties in a complex protein environment [22, 29]. The extent to which a solvent model (explicit or implicit) can capture this delicate balance is a key to its success in describing conformational equilibria. Achieving sufficient balance of the competing interactions for complex heterogeneous systems is a challenging task. To a large extent, this is due to a severe lack of direct experimental measurements or reliable high-level quantum mechanics data. In practice, one has to resort to indirect experimental observables, such as thermodynamic stability and conformation equilibria of model peptides and proteins, in order to rebalance the force field [10,

22, 29, 30]. However, reliable calculation of these thermodynamic quantities requires extensive folding and unfolding simulations, and is generally only accessible with implicit solvent. In this chapter, we will first provide a brief overview of the theoretical frameworks of implicit solvent, particularly the GB theories, and then focus on discussing our recent efforts in the optimization of implicit solvent force fields. We need to emphasize that several other groups have also made substantial progress in optimizing other implicit force fields [30–34]. Owing to space limitations, we will only briefly discuss a few aspects that are closely related to our work in this chapter.

7.2
Theoretical Foundations of Implicit Solvent

7.2.1
General Principles of Implicit Solvent

An implicit treatment of solvent captures the mean influence of solvent via direct estimation of the solvation free energy, which is defined as the reversible work required to transfer the solute in a fixed configuration from vacuum to solution. The solvation free energy as defined in principle captures *all* information on the influence of the solvent on the equilibrium properties of the solute [15]. In practice, important assumptions need to be made for computational efficiency. For example, the so-called reference interaction-site models (RISM) might be used to approximate the average solvent density from site–site radial distributions for computing the solvation energy [35, 36] (see Chapter 2). For biomolecular simulations, various semi-analytical approximations have been more effective. Recognizing that water molecules in the first solvation shell dominate the solvation energetics, fully empirical approaches estimate the solvation free energy directly from certain geometric properties of the solute, such as the atomic solvent-exposed surface area or solvent-excluded volume, using empirical free-energy scales [37, 38]. A more rigorous approach is to decompose the total solvation free energy into non-polar and electrostatic contributions, which correspond to the reversible work required (i) to insert the solute into the solvent with zero atomic partial charges, and (ii) to switch the partial charges from zero to their full values [15]:

$$\Delta G_{solv} = \Delta G_{elec} + \Delta G_{np} \tag{7.1}$$

Such a decomposition is path dependent, but nonetheless allows both components to be related to appropriate continuum models of water, and is generally more accurate than the fully empirical approaches.

7.2.2
Continuum Electrostatics

Continuum electrostatics is the most well-established theory for electrostatic solvation, where the solute is represented as a low-dielectric cavity embedded in a featureless, high-dielectric solvent medium. The corresponding electrostatic solvation free energy can be rigorously calculated by solving the PB equation using finite-difference methods [39–41]. Despite progress in fast PB computational methodologies [42–44], the computational cost of solving the PB equation remains a bottleneck in its application to molecular dynamics (MD) simulations of biomolecules. Alternatively, the GB approximation can be used to calculate the same electrostatic solvation energy using an efficient pairwise summation [45, 46]:

$$\Delta G_{elec} = -\frac{1}{2}\left(1-\frac{1}{\varepsilon}\right)\sum_{ij}\frac{q_i q_j}{\sqrt{r_{ij}^2 + R_i^{GB} R_j^{GB} \exp(-r_{ij}^2/FR_i^{GB}R_j^{GB})}} \quad (7.2)$$

Here r_{ij} is the distance between atoms i and j, q_i is the atomic charge, R_i^{GB} is the so-called "effective Born radius" of atom i, ε is the solvent (high) dielectric constant, and F is an empirical factor whose value may range from 2 to 10, with 4 being the most common value. The (low) dielectric constant of the solute interior is generally assumed to be 1 (same as vacuum), and ΔG_{elec} corresponds to the electrostatic free energy of transferring the solute from vacuum to a medium of dielectric constant ε. Several extensions have also been described to model the free energy of transfer between two (solvent) media with different dielectric constants [47–50]. GB offers higher computational efficiency and readily allows the evaluation of analytical forces. Therefore, it is often preferred over the PB methods for MD simulations.

The effective Born radius, R_i^{GB}, is a key quantity in the GB formalism. It corresponds to the distance between a particular atom and its hypothetical spherical dielectric boundary, chosen such that the self (or atomic) electrostatic solvation energy, $\Delta G_{elec,i}$, satisfies the Born equation [51],

$$\Delta G_{elec,i} = -\frac{1}{2}\left(1-\frac{1}{\varepsilon}\right)\frac{q_i^2}{R_i^{GB}} \quad (7.3)$$

In principle, the "exact" effective Born radii can be calculated from Equation 7.3 using the self electrostatic solvation energy obtained through the PB theory. The principal assumption in the GB approximation is that the solvent-shielded charge–charge interactions can be reproduced by the cross-term summation in Equation 7.2 with the effective Born radii. Indeed, Equation 7.2 has been shown to closely reproduce the PB electrostatic solvation energy, provided that the effective Born radii are accurate [52, 53]. As such, most of the literature on extensions of the GB methods has been focused on efficient and accurate evaluation of the Born radii, and $\Delta G_{elec,i}$ or R_i^{GB} from PB calculations serve as standard benchmarks for assessing various GB approximations. Many modifications, extensions, and improvements have been made over the past several years, and various implementations are now available in virtually all major molecular modeling software packages [54–69]. At present, the GB formalism has reached a mature stage and many GB models are capable of achieving a similar level of numerical accuracy as PB [53]. As such, GB is now considered a prime choice for implicit treatment of solvent. Successful applications to various biological problems have demonstrated the great potential, and, at the same time, have revealed many remaining limitations, of the GB models for studies of biomolecular structure and function [16, 18]. One of the main limitations of the GB implicit solvent force fields at present appears to lie in co-optimization of the solvation model in the context of the underlying protein force field, besides the treatment of non-polar solvation.

7.2.3
Treatment of Non-Polar Solvation

In contrast to the advances in the GB-based continuum electrostatics, a well-established continuum theory for non-polar solvation is not yet available, and this is one of the key limitations in implicit solvent. Indeed, accurate calculation of non-polar solvation energy is much more challenging for biomolecules with complex shapes. At present, the non-polar solvation free energy is either largely ignored, or estimated from the solvent-accessible surface area (SASA). With substantial improvement in the electrostatic solvation models, limitations of simple SASA models for biomolecule simulation are becoming increasingly evident [70–72]. In comparison to explicit solvent results, simple SASA models not only systematically overestimate the stabilities of pairwise interactions between non-polar side chains, but also give rise to anti-cooperativity instead of cooperativity for multi-body hydrophobic interactions [71, 72]. These limitations have been shown to originate from insufficient description of two known properties of hydrophobic solvation and association. First, there exists a length-scale dependence of hydrophobic solvation, such that the solvation free energy scales with volume for small solutes but with surface area for larger ones [73–76]. The crossover occurs around 10 Å radius, a scale relevant to protein folding. Ignoring the length-scale dependence in the current SASA model results in both over-stabilization of pairwise interactions and anti-cooperativity of three-body hydrophobic associations [71]. Second, SASA models are not able to describe the solvent screening of mid-range solute–solute dispersion interactions [70, 72], which also contributes to the systematic over-stabilization of non-polar interactions. Various models have been proposed to describe either key property [67, 71], but none has demonstrated meaningful improvement on the peptide simulation level. An integration of these proposed models to account for both critical properties of non-polar solvation in a single, consistent model seems to be necessary. Detail of developments along this line is beyond the scope of this chapter, and readers are encouraged to refer to the literature for additional information.

7.3
Optimization of Implicit Solvent Force Fields

As discussed above, the theoretical frameworks of PB or GB continuum electrostatics are well developed, and co-optimization of the solvation model with the underlying protein force field is a key area for continual development of the implicit solvent models. In the context of GB optimization, the important difference between numerical and physical parameters needs to be emphasized.

Numerical parameters refer to those in a particular GB formalism that are adjusted to maximally reproduce the "exact" results from equivalent high-resolution PB calculations. Key quantities that need to be reproduced in reference to PB include solvation free energies of small model compounds as well as proteins with various sizes and structures, and, importantly, the effective Born radii (see Equation 7.3) in a protein environment. Physical parameters include the intrinsic atomic radii, definition of the solvent boundary (if adjustable), and those associated with the non-polar solvation component (such as effective surface tension coefficients). These parameters all have defined physical meanings and are optimized to reproduce certain (experimental) physical properties (see below). Optimization of the physical parameters is meaningful when, and only when, satisfactory numerical accuracy has been achieved. Otherwise, improper cancellation of errors might occur and lead to unphysical results. We want to note that there has been some confusion in the literature concerning the optimization of GB implicit solvent, where numerical accuracy is often confused with physical accuracy, or where physical accuracy is discussed without first establishing the numerical accuracy of the method. Furthermore, the quality of the implicit solvent force field is ultimately limited by the underlying protein model, and importantly it is necessary to optimize the non-bonded parameters of the protein model with the solvation parameters in a self-consistent fashion.

The rest of this chapter focuses on optimization of the implicit solvent protein force fields using an array of physical benchmarks, with the ultimate goal of achieving a realistic description of peptide conformational equilibria. Key physical parameters to be optimized and the associated computational approaches will be discussed. These principles and approaches apply to PB force-field optimization as well, and the optimized physical parameters (for example, intrinsic radii and torsion potentials) should be directly transferable to the correspondingly equivalent PB models.

7.3.1
Solvation Free Energies of Small Molecules

The solvation free energies of amino acid backbone and side-chain analogs are among the few types of experimental data that might be directly used in protein force-field parameterization. A complication is that the electrostatic component needs to be separated from the total solvation free energy. As such, results from explicit solvent simulations are typically used instead for optimization of continuum electrostatics models. In continuum electrostatics, the extent of solvent exposure of each atom at the dielectric boundary dictates all of the electrostatic and most of the non-polar solvation energetics. Thus, it is physically appropriate to optimize the intrinsic radii, which, together with the choice of solute surface type (for example, van der Waals like or molecular surface), defines the location of the boundary between the low- and high-dielectric regions. The intrinsic atomic radii for continuum electrostatics have previously been optimized based on the radial solvent charge distribution to reproduce the electrostatic solvation energy obtained from explicit solvent charging free-energy calculations for both proteins [77, 78] (hereinafter referred to as Nina's radii) and nucleic acids [79]. The Nina's radii set has been shown to work well in several applications, including ion channel modeling [80], peptide folding [26], and protein nuclear magnetic resonance (NMR) structure refinement [81, 82]. A similar strategy has more recently been applied to identify optimal radius sets for the AMBER protein force field [83].

7.3.2
Potentials of Mean Force of Pairwise Interactions

Key to an accurate description of peptide conformational equilibria is the ability to capture the delicate balance between sets of competing interactions, that is, the solvation preference of side chains and backbones in solution versus the strength of solvent-mediated interactions between these moieties in a complex protein environment. The intramolecular Coulomb interaction energy in the protein is strongly anticorrelated with the electrostatic solvation energy [84]. These two opposing terms are both large and mostly cancel each other. As such, a small relative error in either term will result in a substantial shift in the balance, which can subsequently lead to a bias in conformational equilibria. This simple analysis suggests that optimization based on the solvation free energy alone might not be sufficient [84]. Indeed, it has been noticed previously that many existing continuum electrostatics solvation models (GB as well as PB) over-stabilize salt bridges [21, 32, 85–88], which can partially account for the observed discrepancies in the conformational equilibria and free energy surfaces for several peptides [87, 89, 90]. This over-stabilization might be amplified even more in the low-dielectric protein interior, which appears to be particularly problematic in certain applications such as protein design [91].

Potentials of mean force (PMFs) for the interactions between the polar moieties of peptides can provide a direct measure of how well an implicit solvent force field is able to capture the balance of solvation and intramolecular interactions. With a lack of experimental data on these interactions, explicit solvent simulations have been used in GB force-field optimization. We have systematically examined the PMFs between polar side chains and between side chains and the peptide backbone in various representative configurations in both GBSW (generalized Born with simple switching) implicit [65] and TIP3P (transferable intermolecular potential, three-point) explicit [92] solvents with the CHARMM param22/CMAP all-atom protein force field [93–95]. The explicit solvent PMFs were computed using an umbrella sampling protocol where the dimers were constrained to move along a reaction coordinate, that is, a straight line in specific dimer orientations (for example, see Figure 7.1) [22]. The data were analyzed using the weighted histogram analysis method (WHAM) [96, 97]. Corresponding PMFs in implicit solvent were computed directly by translating the molecules away from each other along the reaction coordinate. The resulting PMFs do not include the contribution of solute conformational entropy. However, this contribution is assumed to be

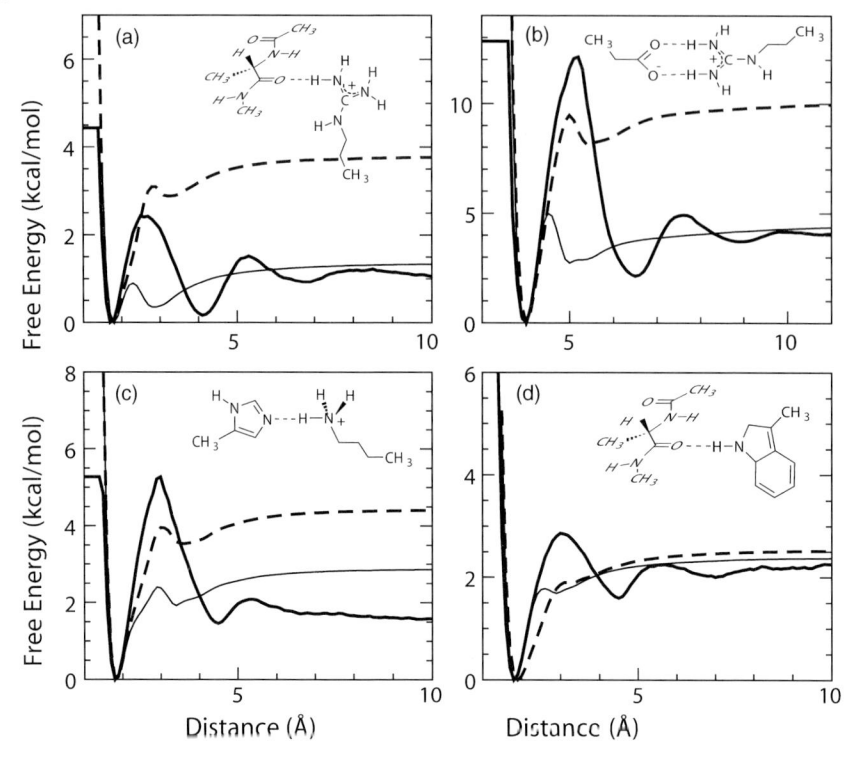

Figure 7.1 Free-energy profiles of four dimers in TIP3P water (thick lines) and GBSW implicit solvent with the Nina's radii (dashed lines) and re-optimized radii (thin lines). The dimer configurations are also shown. The reaction coordinates plotted in the x coordinates are (a) $r(O \cdots H)$, (b) $r(CZ \cdots CD)$, (c) $r(NE2 \cdots H)$ and (d) $r(O \cdots H)$. Note that the heavy atoms were constrained in two orthogonal planes for the dimers shown in panels (a) and (d). Figure originally published in Chen et al. (2006) [22].

similar in both explicit and implicit solvent models, and thus omitting it in both cases should not affect the implicit solvent optimization.

Comparison of PMFs in GBSW and TIP3P solvents demonstrates that many interactions are indeed significantly over-stabilized with the Nina's radii that were optimized based on solvation free energies alone – for example, see Figure 7.1. Through systematic adjustments of the intrinsic atomic radii, mainly those of charged atoms, agreement with the TIP3P results can be dramatically improved. Note that the solvation peaks (oscillations in the TIP3P PMFs) are mostly absent in the GBSW implicit solvent, which is due to the lack of solvent granularity and adoption of a van der Waals like surface in GBSW. The first solvation peak can be effectively reproduced by incorporating the solvent reentrant surface in the dielectric boundary such as in the GBMV model [63], without having to include any explicit water molecules in the implicit solvent. However, it is not clear whether there is any significant consequence in capturing such fine details in the interactions [98, 99]. While the folding kinetics might be altered, the absence of large

Table 7.1 Modifications to the Nina's input radii that are self-consistent with a CHARMM22/CMAP^{GBSW} (see next section) force field with the GBSW implicit solvent. Data originally published in Chen et al. (2006) [22].

Residue	Atom	Nina (Å)	New (Å)
Backbone	NH1	2.30	2.03
Lys	NZ	2.13	1.80
Arg	N$^{a)}$	2.13	1.70
	CZ	2.80	2.20
Gln/Asn	O$^{a)}$	1.42	1.60
	N	2.15	2.00
Hse	ND	1.80	1.90
Hsp	N$^{a)}$	2.30	1.90
Trp	NE	2.40	1.85
	C$^{a)}$	1.78	2.00

a) Refers to a wildcard character.

solvation peaks might actually speed up the conformational sampling without introducing any thermodynamic bias. It turned out that only a few atom types need to be adjusted and mainly involved backbone amide nitrogen (CHARMM atom type NH1) and other charged nitrogen atoms. The proposed modifications to the Nina's radii set are summarized in Table 7.1. Simmerling, Pak and their coworkers have also separately examined the strength of salt-bridge interactions of the GB/SA (generalized Born surface area) models in AMBER, and showed that empirical adjustment of the intrinsic atomic radii could improve the agreement with TIP3P results and in turn the ability to simulate peptide and protein folding [32, 33].

7.3.3
Conformational Equilibria of Model Peptides

While the PMFs provide an important guide for systematic improvement of the balance between solvation and intramolecular interactions, substantial errors and uncertainties remain in both the numerical accuracy of PMFs and the quality of the reference explicit solvent force fields. Therefore, it is necessary to directly examine the ability of the force field to reproduce the experimental conformational equilibria of model peptides. These model peptides need to be carefully selected to provide sufficient control. A set of model peptides that was used to optimize the GBSW implicit solvent force field is shown in Table 7.2. Different sets of model peptides have been used by other groups [31–34]. There are several important considerations for choosing the model peptides. First, recognizing the importance of balancing the secondary structure preferences, the set should consist of both helical peptides and β-hairpins. Second, the peptides need to display a range of thermodynamic stabilities to avoid potentially over-stabilizing the native interactions. Third, the peptides need to be small enough to make it feasible to obtain

Table 7.2 A set of model peptides for force-field optimization.

Name	Sequence	Structure (%)	REX-MD[a]
(AAQAA)$_3$	AAQAAAAQAAAAQAA	α (50%)[b]	65%
GB1p	GEWTYDDATKTFTVTE	β (42–80%)[c]	61%
GB1m1	GEWTYDDATKTATVTE	β (6%)[d]	0%
HP5A	KKYTWNPATGKATVQE	β (21%)[d]	43%
GB1m3	KKYTWNPATGKFTVQE	β (86%)[d]	88%

a) Using an optimized GBSW force field. Data originally reported in Chen et al. (2006) [22].
b) Helicity measured by NMR chemical shifts at 270 K [100].
c) Population estimated from multiple NMR chemical shift probes (~42% at 278 K [101]) and from the tryptophan fluorescence experiment (~80% at 273 K) [102].
d) Folded population estimated by NMR chemical shifts at 298 K [103].

reasonably converged conformational equilibria from folding and unfolding simulations. The set of model peptides shown in Table 7.2 appears to satisfy all these criteria. In particular, peptides GB1m1, HP5A, and GB1m3 are derived from the native sequence of the C-terminal β-hairpin (residues 41–56) of the B1 domain of protein G (GB1p) but display reduced or enhanced stabilities: (unfolded) GB1ml < HP5A < GB1p < GB1m3 (most folded) [103]. Therefore, these peptides as a series provide a particularly useful control for force-field optimization. It needs to be emphasized that reliable calculation of peptide conformational equilibria critical to peptide-based optimization remains a challenging and expensive task even with implicit solvent, and that advance sampling techniques such as the replica exchange (REX) method [104, 105] are indispensable in such efforts. It should also be stressed that much still needs to be understood about when and how REX can be most effectively used for sampling the configurational space of a given peptide with particular folding kinetics and thermodynamics [106–109] and this might lead to substantial uncertainty in the computed conformational equilibria. Accordingly, particular care needs to be paid to examining the actual convergence of the peptide simulations. Reaching a stationary steady state as the simulation progresses is far from being a sufficient proof. Instead, it is more effective to initiate REX simulations from completely different initial conformations (for example, fully folded and fully extended) and to compare the resulting structure ensembles [22].

Besides the physical parameters of the solvent models, non-bonded parameters (partial charges and Lennard-Jones parameters) and torsional energetics in the underlying molecular mechanics force field in principle need to be adjusted self-consistently for a given implicit solvent model to achieve sufficient balance. However, considering that current protein force fields have been extensively calibrated over the past decades to achieve proper solvent–solute and solute–solute interactions in explicit solvent [10], peptide backbone torsional potentials have been the main focus of co-optimization with the solvation parameters [22, 30, 31, 33]. These efforts are similar in essence to previous parameterization of the

7.3 Optimization of Implicit Solvent Force Fields

peptide backbone in the context of the TIP3P explicit solvent [110]. In particular, the ϕ/ψ CMAP torsion cross-term recently introduced in CHARMM [94, 95, 110] provides a powerful and convenient means to modulate the peptide backbone conformational preferences. An iterative procedure was previously applied to optimize the GBSW implicit solvent force field, where the intrinsic atomic radii as well as the backbone dihedral energetics were empirically adjusted based on results of folding and unfolding simulations of model peptides. Briefly, whenever the backbone energetics is changed, the backbone input radii are adjusted such that the helicity of (AAQAA)$_3$ is close to the experimental value (~50% at 270 K). Folding and unfolding simulations of the series of GB1p β-hairpins (GB1ml, GB1p, HP5A, and GB1m3) then follow to examine whether the correct folding thermodynamics is obtained. The final parameters are then further examined by folding simulations of additional small proteins as well as control simulations of a range of larger proteins with various folds. This strategy relies heavily on manual adjustments of key parameters such as backbone torsion potentials, and might sound less rigorous than a systematic optimization such as the Z-score optimization [111, 112]. However, it appears to be more effective in practice due to the under-determined nature of the optimization problem and expensive computational cost of peptide folding–unfolding simulations.

Iterative optimization of the solvation parameters with the peptide backbone torsion potentials lead to a consistent GBSW implicit solvent force field that is able to accurately describe the conformational equilibria of all the model peptides shown in Table 7.2. For example, Figure 7.2 compares the probability distributions of the number of native hydrogen bonds (N_{hb}^{nat}) at 270 K for the GB1p β-hairpin series. REX-MD simulations initiated from both fully extended conformations (folding) and folded hairpin conformations (control) demonstrate very good convergence for all hairpins except for GB1p, which might be related to the possible difference in the folding or unfolding rates of the three sequences. A recent study demonstrates that a stronger turn-promoting sequence (such as the D47P mutation in GB1m3) increases the hairpin stability primarily by increasing the folding rate, whereas a stronger hydrophobic cluster stabilizes the hairpin by decreasing the unfolding rate [113]. Furthermore, the poor convergence for the GB1p folding simulation might be related to the fact that the REX setup is suboptimal for GB1p. The melting temperature of GB1p was shown to be $T_m \sim 300$ K, compared to $T_m \sim 330$ K for Gb1m3 [103]. In the original setup with 16 replicas spanning 270 K to 550 K, only three replicas are actually simulated under the T_m of GB1p. An additional REX-MD simulation with the temperature range reduced to 270–400 K (with six temperature windows below $T_m \sim 300$ K) appears to converge faster, yielding a native hydrogen bond probability distribution very similar to that of the control simulation (see Figure 7.2b). These folding and unfolding simulations seem to correctly reproduce the experimental results that GB1m3 is the most folded and GB1m1 is largely unfolded. Furthermore, assuming conformations with $N_{hb}^{nat} \geq 4$ as native, the native populations turn out to be about 88% for GB1m3, 43% for HP5A, 61% (fold_2) for GB1p, and 0% for GB1m1 from the folding simulations, which are also in very good agreement with the experimental data (see Table 7.2).

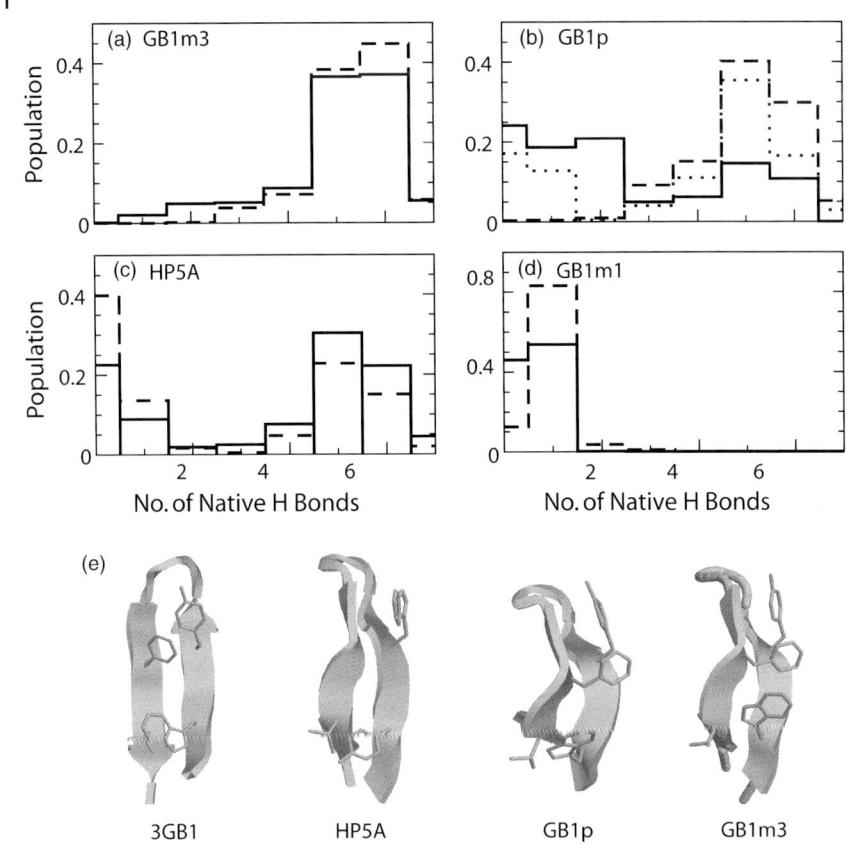

Figure 7.2 Probability distributions of the number of native hydrogen bonds for (a) GB1m3, (b) GB1p, (c) HP5A, and (d) GB1m1 at 270 K; and (e) representative folded hairpin structures of HP5A, GB1p, and GB1m3 in comparison with the experimental fragment structure (PDB ID: 3gb1). The distributions were computed from the last 10 ns of REX-MD simulations of 30–50 ns in total length. The hydrogen bonds taken as native are the same for all peptides. They are (in protein GB1 residue numbering): E42(N)-T55(O), E42(O)-T55(N), T44(N)-T53(O), T44(O)-T53(N), D46(N)-T51(O), D46(O)-T51(N), and D47(O)-K50(N). The fold_2 (dotted lines) is an additional REX-MD folding simulation for GB1p using 16 replicas at 270–400 K, carried out to improve the convergence. Both folding (dashed lines) and control (full lines) simulations of HP5A used 16 replicas spanning 270–400 K. Figure originally published in Chen et al. (2006) [22].

Representative structures from the folded structures of HP5A, GB1p, and GB1m3 are shown in Figure 7.2e, in comparison with the fragment structure from the protein GB1 domain [114]. All stand-alone hairpins show a characteristic twist observed in other stable hairpins such as the trpzip series [115]. Close packing of hydrophobic side chains is present in most folded structures. In particular, residue Phe52 in GB1m3 packs with both Tyr45 and Trp43 and thus contributes significantly to the observed stability. Mutation of Phe52 to alanine thus dramatically destabilizes the hairpin (such as in GB1m1). The ability of the optimized force field to reproduce both the experimental structures and stabilities of these model peptides is a strong indication that the underlying balance between competing solvation forces and intramolecular interactions is quite well captured.

7.3.4
Folding Simulations of Small Proteins

Folding and unfolding studies of peptides and proteins that have not been used in the course of optimization is important for understanding the robustness and transferability of the optimized implicit solvent force field. These applications also help to identify certain (fundamental) limitations of the solvation model and force field, and provide important insights for further modification and improvement. Control simulations of larger stable proteins are necessary to first verify that the native structures are sufficiently stable. Folding proteins using a first-principles approach is much more costly and is typically limited to short sequences. For this, small designed proteins that are stable and fold fast are particularly suitable. Figure 7.3 summarizes the folding simulations of two sequences, trpzip2 and Trp-cage, using the same optimized GBSW force field. Trpzip2, a 12-residue tryptophan zipper, is a designed β-hairpin with a type I′ turn and contains a characteristic structural motif of tryptophan–tryptophan cross-strand pairs [115]. It is the smallest peptide to adopt an unique tertiary, β-fold with exceptional stability. Trp-cage is a 20-residue designed mini-protein with a stable compact folded state [116]. The native structure contains a short α-helix, a single turn of 3_{10}-helix, and a rigidified polyproline C-terminal tail. The well-structured hydrophobic core consists of the indole side chain of Trp6 buried between rings of Pro12 and Pro18. The structure is further stabilized by two tertiary hydrogen bonds, one between the side chain of Trp6 ($N^{\varepsilon 1}H$) and the backbone of Arg16 (CO), and the other one between the backbone groups of Trp6 (CO) and Gly11 (NH). The results of the folding simulations demonstrate that the optimized force field is able to fold both sequences with a very high accuracy, yielding conformations about 1 Å C_α root-mean-square deviation (RMSD) from the corresponding NMR structures. Closer inspections show that the most important structural features described above are faithfully captured in the simulated structures, except for the packing of the tryptophan side chains in trpzip. These side chains appear to be versatile, and the most populated configuration, particularly for Trp4 and Trp11, is different from the NMR structures. The force field clearly has a systematic tendency to yield structures where large hydrophobic side chains form (non-specific) contacts with the peptide backbone and other nearby molecular segments (for example, see also Figure 7.2e). Further theoretical and modeling studies later identified several limitations in the current GB/SA force field, one of which is to do with the fundamental limitations of the SASA-based treatment of non-polar solvation [71, 72].

The same force field has also been applied to fold two small helical proteins, including villin headpiece subdomain (residues 41–76; Protein Data Bank (PDB) ID: 1vii) (HP36) [117], and a 46-residue segment of staphylococcal protein A fragment B (residues 10–55; PDB ID: 1bdc) [118]. The results are summarized in

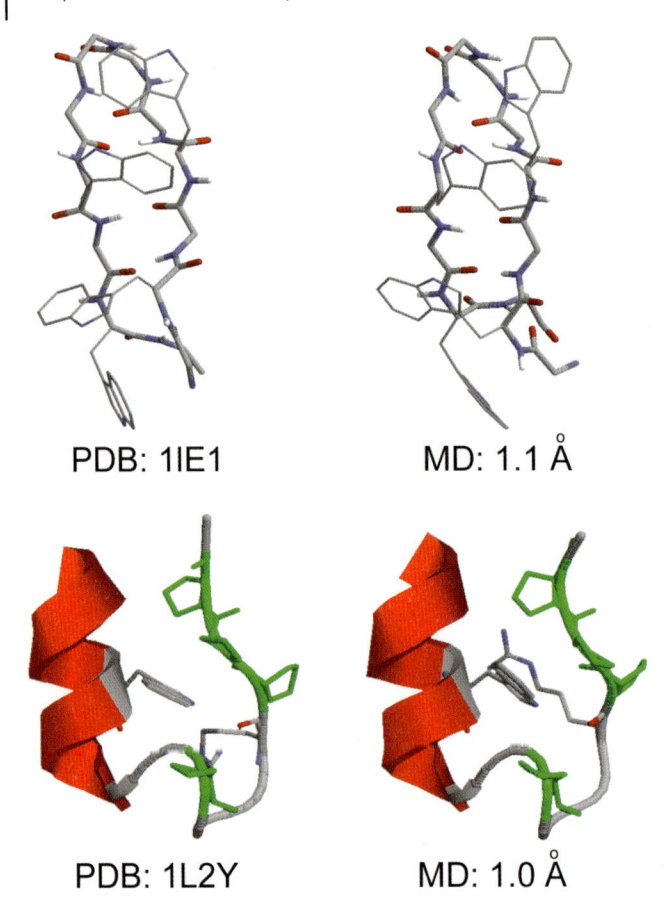

Figure 7.3 Representative folded structures of trpzip2 beta hairpin (top) and Trp-cage mini-protein (bottom), in comparison with the corresponding NMR structures. Both folded structures correspond to the average structures of the largest ensembles at 270 K sampled during the last 10 ns of REX folding simulations [22]. RMSD values shown were computed using all backbone heavy atoms. Data originally published in Chen et al. (2006) [22].

Figure 7.4. The REX simulations were initiated from fully extended conformations, and the total simulation lengths were 160 ns for villin headpiece and 100 ns for protein A. For both proteins, near-native conformations, shown in Figure 7.4b–c, were reached within the first 40 ns of REX-MD simulations, but little further progress toward the fully folded structures was made for the rest of the simulations. The near-native conformations are within 4 Å C_α RMSD from the experimental structures. The secondary elements are largely correct, with helical content of about 60% as in the native structures. The arrangements of helices are essentially native-like, but the packing is not as compact. The critical bottleneck to the fully folded conformations appears to be the formation of the hydrophobic cores, highlighted in green color in Figure 7.4b–c. For example, HP36 contains a

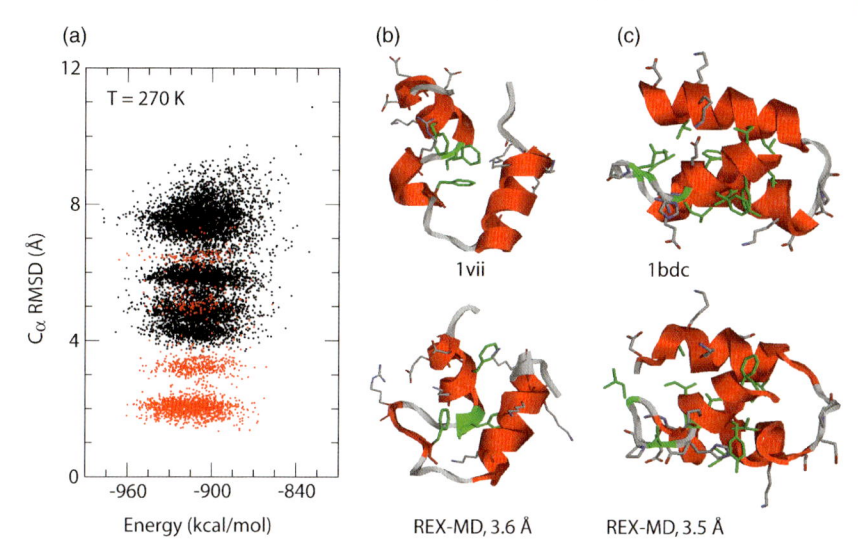

Figure 7.4 (a) Total potential energy versus C$_\alpha$ RMSD plot from REX-MD folding (black dots) and control (red dots) simulations of villin headpiece subdomain (HP36). The folding simulation was initiated from a fully extended conformation, and the control simulation from the PDB structure. Snapshots were taken every 10 REX exchange steps at 270 K. The total simulation lengths were 160 ns and 20 ns for folding and control runs, respectively. (b–c) Representative near-native conformations from folding simulations in comparison with the experimental structures for (b) HP36, and (c) protein A. The length of protein A REX-MD folding simulation is 100 ns. Representative conformations from the largest cluster at the lowest temperature (270 K) are shown, and their occupancies are 25% and 24%, respectively, during the last 20 ns of the REX-MD simulations. The helical segments (as defined in the PDB structures) are colored red, residues in the hydrophobic cores in green, and charged residues in CPK colors. Figure originally published in Chen and Brooks (2008) [72].

mini hydrophobic core that consists of three phenylalanines, which is not formed in the simulated structure even at the end of the 160 ns REX-MD simulation. It is interesting that the folding of secondary structures and the formation of tertiary hydrophobic cores are clearly coupled, which is expected, as both proteins are known to fold cooperatively.

The failure to reach fully folded states might be attributed to two reasons. First, the compact, near-native conformations are too stable in the current force field. To reach the true native fold requires searching through a large number of such compact conformations by breaking and re-forming many (non-polar) contacts. With systematic over-stabilization of non-polar interactions, this becomes very slow and cannot be accomplished within the time scales simulated (even though these simulations are two of the longest REX-MD simulations reported to date). Second, the underlying free-energy surface is significantly distorted and the true native structure no longer corresponds to the global free-energy minimum. To examine whether the native structure has lower potential energy, a 20 ns control REX-MD simulation was initiated from the PDB structure of HP36. The results are summarized in Figure 7.4a. It shows that the fully folded conformations (that is, those with C_α RMSD of about 2 Å) have lower energies on average. It indicates that the true global free-energy minimum for HP36 might not have been severely distorted. Similar control simulations of protein A show that the native structures no longer have lower potential energies on average compared to other compact structures. Therefore, both reasons are responsible for the observed inability to fully fold HP36 and protein A. Nonetheless, the fact that near-native conformations are reached in both cases is encouraging. It indicates that the force field has nearly proper balance of the underlying interactions, particularly with respect to electrostatic solvation and intramolecular interactions, and that a fine tuning of the non-polar solvation model might be sufficient to fully fold both proteins.

7.3.5
Optimization Based on Other Experimental Measurables

Several other types of experimental measurables have also been used in (mostly explicit solvent) force-field optimization and validation. For example, protein backbone and side-chain order parameters derived from NMR relaxation studies have been used to validate new force-field developments [30, 119–121]. Improved agreement between simulation and NMR experiment is generally considered an evidence of improvement in the force field. While benchmarking the force field's ability to capture both structural and dynamical properties of proteins is critical, one needs to be cautious and to recognize the potential issues with interpretation of the NMR relaxation data, especially when it comes to slower motions on the nanosecond time scales [122]. Nanosecond motions are also where the largest discrepancies between simulation and experiment are observed. Recently, the Kirkwood–Buff (KB) integrals were used to quantify solute–solute and solute–solvent interactions, and allowed direct comparison to experiments on a range of thermodynamic properties of solution mixtures [11, 123, 124]. Re-parameterization of important protein parameters (such as charges and Lennard-Jones parameters) for a given water model might allow one to construct the so-called Kirkwood–Buff based force field (KBFF) that promises to provide a much better balance between solvation and intramolecular interactions (see Chapter 3). While a full KBFF protein force field is yet to be developed, the promised substantial improvement in the quality of the underlying protein force field will naturally lead to a better ability of the implicit solvent force field to describe protein conformations. Interestingly, KB integrals can also be directly used to quantify the balance of solvation and solute–solute interactions for implicit solvent force fields (Chen and Smith, unpublished data).

7.4
Concluding Remarks and Outlook

Implicit solvent has emerged as one of the most powerful techniques for classical simulation of proteins and other biomolecules in aqueous solution, offering a favorable compromise between speed and accuracy. Recent methodological advances have continued to push the level of accuracy and complexity of continuum electrostatics based solvation theories. In particular, GB/SA theory has become a prime choice for the simulation of various conformational properties of biological assemblies that cannot be directly studied using explicit solvent. It is particularly encouraging that GB/SA-based *ab initio* folding simulations combined with advanced conformation sampling protocols are now able to reproduce native or near-native structures for peptides and mini-proteins, and to capture the thermodynamics and kinetics of folding at a quantitative level. Nonetheless, substantial challenges and opportunities remain, particularly in the treatment of non-polar solvation. As implicit solvent models are maturing for biomolecular simulations, one major goal is to achieve a force field consistent with the implicit solvent model and vice versa. The initial successes in extensive co-optimization of the GB/SA models and the protein force field, albeit limited, are very encouraging.

Continual development of implicit solvent force fields for accurate modeling of protein folding and conformational transitions are faced with a range of challenges.

First, an extremely high level of accuracy is required, considering that the average thermodynamic stability of proteins is only up to the order of 0.1 kcal mol^{-1} per residue [115].

Second, substantial limits exist in the modern protein force field underlying the implicit solvent models [9, 10]. In particular, it is clear that there is a systematic tendency to underestimate the solvation free energy of key protein functional groups [125] and to overestimate solute–solute interactions [11, 126]. The current limited success of implicit solvent force fields rely heavily on parameterization to achieve sufficient cancellation of errors. However, it seems that some aspects of the protein force fields, particularly regarding aromatic side chains [127, 128], will need to be improved along with the construction of the implicit solvent models. A potential pitfall of the peptide-based optimization that has proved to be effective so far is that it might attempt to improperly compensate for errors in the underlying protein force field and lead to unexpected consequences when the force field is applied outside of the model peptide set.

Third, the treatment of non-polar solvation clearly needs to be substantially improved. While the fundamental limitations of the simple SASA-based models are now well understood, numerically stable and efficient solutions that can describe both the length-scale dependence of hydrophobic solvation and solvent screening of solute–solute dispersion interactions are not straightforward. One of the conceptual difficulties is in fast analytical estimation of the "local" surface curvature for modeling the length-scale dependence. Furthermore, owing to the short-range nature of dispersion interactions (with an r^{-6} dependence), any model that explicitly includes the dispersion term in non-polar solvation will be highly sensitive to the definition of the surface and parameters such as solvent (probe) radius and input radii of solute atoms. Achieving sufficient (numerical) accuracy to arrive at a reliable estimation of the (small) net non-polar solvation free energy is challenging in practice.

Finally, further complications arise from breakdowns of the continuum approximation at short range. The presence of backbone and other charged atoms can induce order or disorder in the local water structures, giving rise to non-trivial secondary contributions [129]. It has also been argued that polar and non-polar solvation is coupled, and one might need to solve an optimization problem to derive the most appropriate solvent-accessible surface [130].

The implications of these secondary effects on the modeling of protein conformational equilibria are unclear, even though it appears that the current implicit solvent models are far too inaccurate to justify a meaningful attempt to incorporate these effects for protein simulations. Looking forward, these challenges are also opportunities for continual improvement of the implicit solvent force field. Together with development and application of novel implicit solvent-based modeling methodologies, such as constant-pH MD, implicit solvent force fields are expected to play even a bigger role in biomolecular simulations.

Acknowledgments

This work was supported by the National Institutes of Health (RR12255) and Alfred P. Sloan Research Fellowship (to W.I.).

References

1 Anfinsen, C.B. (1973) Principles that govern folding of protein chain. *Science*, **181**, 223–230.

2 Gnanakaran, S., Nymeyer, H., Portman, J., Sanbonmatsu, K.Y., and Garcia, A.E. (2003) Peptide folding simulations. *Curr. Opin. Struct. Biol.*, **13**, 168–174.

3 Shakhnovich, E. (2006) Protein folding thermodynamics and dynamics: where physics, chemistry, and biology meet. *Chem. Rev.*, **106**, 1559–1588.

4 Kitao, A., Yonekura, K., Maki-Yonekura, S., Samatey, F.A., Imada, K., Namba, K., and Go, N. (2006) Switch interactions control energy frustration and multiple flagellar filament structures. *Proc. Natl. Acad. Sci. U.S.A.*, **103** (13), 4894–4899.

5 Freddolino, P.L., Arkhipov, A.S., Larson, S.B., McPherson, A., and Schulten, K. (2006) Molecular dynamics simulations of the complete satellite tobacco mosaic virus. *Structure*, **14** (3), 437–449.

6 Freddolino, P.L., Liu, F., Gruebele, M., and Schulten, K. (2008) Ten-microsecond molecular dynamics simulation of a fast-folding WW domain. *Biophys. J.*, **94** (10), L75–L77.

7 Maragakis, P., Lindorff-Larsen, K., Eastwood, M.P., Dror, R.O., Klepeis, J.L., Arkin, I.T., Jensen, M.O., Xu, H., Trbovic, N., Friesner, R.A., Palmer, A.G. III, and Shaw, D.E. (2008) Microsecond molecular dynamics simulation shows effect of slow loop dynamics on backbone amide order parameters of proteins. *J. Phys. Chem. B*, **112** (19), 6155–6158.

8 Kubelka, J., Hofrichter, J., and Eaton, W.A. (2004) The protein folding "speed limit". *Curr. Opin. Struct. Biol.*, **14**, 76–88.

9 Ponder, J.W. and Case, D.A. (2003) Force fields for protein simulations. *Adv. Protein Chem.*, **66**, 27–85.

10 MacKerell, A.D. Jr (2004) Empirical force fields for biological macromolecules: overview and issues. *J. Comput. Chem.*, **25**, 1584–1604.

11 Kang, M. and Smith, P.E. (2006) A Kirkwood–Buff derived force field for

amides. *J. Comput. Chem.*, **27**, 1477–1485.

12 Yoda, T., Sugita, Y., and Okamoto, Y. (2004) Secondary-structure preferences of force fields for proteins evaluated by generalized-ensemble simulations. *Chem. Phys.*, **307**, 269–283.

13 Best, R.B., Buchete, N., and Hummer, G. (2008) Are current molecular dynamics force fields too helical? *Biophys. J.*, **95**, 4494.

14 Brooks, C.L. III, Karplus, M., and Pettitt, B.M. (1987) *Proteins: A Theoretical Perspective of Dynamics, Structure, and Thermodynamics*, John Wiley & Sons, Inc., New York.

15 Roux, B. and Simonson, T. (1999) Implicit solvent models. *Biophys. Chem.*, **78**, 1–20.

16 Feig, M. and Brooks, C.L. III (2004) Recent advances in the development and application of implicit solvent models in biomolecule simulations. *Curr. Opin. Struct. Biol.*, **14**, 217–224.

17 Baker, N.A. (2005) Improving implicit solvent simulations: a Poisson-centric view. *Curr. Opin. Struct. Biol.*, **15**, 137–143.

18 Chen, J., Brooks, C.L., and Khandogin, J. (2008) Recent advances in implicit solvent based methods for biomolecular simulations. *Curr. Opin. Struct. Biol.*, **18**, 140–148.

19 Cramer, C.J. and Truhlar, D.G. (1999) Implicit solvation models: equilibria, structure, spectra, and dynamics. *Chem. Rev.*, **99**, 2161–2200.

20 Bashford, D. and Case, D.A. (2000) Generalized Born models of macromolecular solvation effects. *Annu. Rev. Phys. Chem.*, **51**, 129–152.

21 Masunov, A. and Lazaridis, T. (2003) Potentials of mean force between ionizable amino acid side chains in water. *J. Am. Chem. Soc.*, **125**, 1722–1730.

22 Chen, J., Im, W., and Brooks, C.L. III (2006) Balancing solvation and intramolecular interactions: towards a consistent generalized Born force field. *J. Am. Chem. Soc.*, **128**, 3728–3736.

23 Khandogin, J., Chen, J., and Brooks, C.L. III (2006) Exploring atomistic details of pH-dependent peptide folding. *Proc. Natl. Acad. Sci. U.S.A.*, **103**, 18546–18550.

24 Khandogin, J., Raleigh, D.P., and Brooks, C.L. III (2007) Folding intermediate in the villin headpiece domain arises from disruption of a N-terminal hydrogen-bonded network. *J. Am. Chem. Soc.*, **129**, 3056–3057.

25 Im, W., Feig, M., and Brooks, C.L. III (2003) An implicit membrane generalized Born theory for the study of structure, stability, and interactions of membrane proteins. *Biophys. J.*, **85**, 2900–2918.

26 Im, W. and Brooks, C.L. III (2004) De novo folding of membrane proteins: an exploration of the structure and Nmr properties of the fd coat protein. *J. Mol. Biol.*, **337**, 513–519.

27 Im, W. and Brooks, C.L. III (2005) Interfacial folding and membrane insertion of designed peptides studied via molecular dynamics simulations. *Proc. Natl. Acad. Sci. U.S.A.*, **102**, 6771–6776.

28 Feig, M. (2008) Implicit membrane models for membrane protein simulation. *Methods Mol. Biol.*, **443**, 181–196.

29 Im, W., Chen, J., and Brooks, C.L. III (2005) Peptide and protein folding and conformational equilibria: theoretical treatment of electrostatics and hydrogen bonding with implicit solvent models. *Adv. Protein Chem.*, **72**, 173–198.

30 Hornak, V., Abel, R., Okur, A., Strockbine, B., Roitberg, A., and Simmerling, C. (2006) Comparison of multiple Amber force fields and development of improved protein backbone parameters. *Proteins*, **65**, 712–725.

31 Okur, A., Strockbine, B., Hornak, V., and Simmerling, C. (2003) Using PC clusters to evaluate the transferability of molecular mechanics force fields for proteins. *J. Comput. Chem.*, **24**, 21–31.

32 Geney, R., Layten, M., Gomperts, R., Hornak, V., and Simmerling, C. (2006) Investigation of salt bridge stability in a generalized Born solvent model. *J. Chem. Theory Comput.*, **2**, 115–127.

33 Jang, S., Kim, E., and Pak, Y. (2006) Free energy surfaces of miniproteins with

αββα motif: replica exchange molecular dynamics simulation with an implicit solvation model. *Proteins*, **62**, 663–671.

34 Jang, S., Kim, E., and Pak, Y. (2007) Direct folding simulation of alpha-helices and beta-hairpins based on a single all-atom force field with an implicit solvation model. *Proteins*, **66**, 53–60.

35 Chandler, D. and Andersen, H.C. (1972) Optimized cluster expansions for classical fluids. II. Theory of molecular liquids. *J. Chem. Phys.*, **57** (5), 1930–1937.

36 Beglov, D. and Roux, B. (1997) An integral equation to describe the solvation of polar molecules in liquid water. *J. Phys. Chem. B*, **101** (39), 7821–7826.

37 Ferrara, P., Apostolakis, J., and Caflisch, A. (2002) Evaluation of a fast implicit solvent model for molecular dynamics simulations. *Proteins*, **46**, 24–33.

38 Lazaridis, T. (2005) Implicit solvent simulations of peptide interactions with anionic lipid membranes. *Proteins*, **58**, 518–527.

39 Warwicker, J. and Watson, H.C. (1982) Calculation of the electric potential in the active site cleft due to alpha-helix dipoles. *J. Mol. Biol.*, **157**, 671–679.

40 Nicholls, A. and Honig, B. (1991) A rapid finite difference algorithm, utilizing successive over-relaxation to solve the Poisson–Boltzmann equation. *J. Comput. Chem.*, **12**, 435–445.

41 Im, W., Beglov, D., and Roux, B. (1998) Continuum solvation model: electrostatic forces from numerical solutions to the Poisson–Boltzmann equation. *Comput. Phys. Commun.*, **111**, 59–75.

42 David, L., Luo, R., and Gilson, M.K. (2000) Comparison of generalized Born and Poisson models: energetics and dynamics of hiv protease. *J. Comput. Chem.*, **21**, 295–309.

43 Luo, R., David, L., and Gilson, M.K. (2002) Accelerated Poisson–Boltzmann calculations for static and dynamic systems. *J. Comput. Chem.*, **23**, 1244–1253.

44 Prabhu, N.V., Zhu, P., and Sharp, K.A. (2004) Implementation and testing of stable, fast implicit solvation in molecular dynamics using the smooth-permittivity finite difference Poisson–Boltzmann method. *J. Comput. Chem.*, **25**, 2049–2064.

45 Still, W.C., Tempczyk, A., Hawley, R.C., and Hendrickson, T. (1990) Semi-analytical treatment of solvation for molecular mechanics and dynamics. *J. Am. Chem. Soc.*, **112**, 6127–6129.

46 Constanciel, R. and Contreras, R. (1984) Self-consistent field theory of solvent effects representation by continuum models – introduction of desolvation contribution. *Theor. Chim. Acta*, **65**, 1–11.

47 Feig, M., Im, W., and Brooks, C.L. III (2004) Implicit solvation based on generalized Born theory in different dielectric environments. *J. Chem. Phys.*, **120**, 903–911.

48 Sigalov, G., Scheffel, P., and Onufriev, A. (2005) Incorporating variable dielectric environments into the generalized Born model. *J. Chem. Phys.*, **122**, 094511.

49 Sigalov, G., Fenley, A., and Onufriev, A. (2006) Analytical electrostatics for biomolecules: beyond the generalized Born approximation. *J. Chem. Phys.*, **124**, 124902.

50 Tjong, H. and Zhou, H.X. (2007) GBr^6: a parameterization-free, accurate, analytical generalized Born method. *J. Phys. Chem. B*, **111**, 3055–3061.

51 Born, M. (1920) Volumen und hydratation-swarme der ionen. *Z. Phys.*, **1**, 45–48.

52 Onufriev, A., Case, D.A., and Bashford, D. (2002) Effective Born radii in the generalized Born approximation: the importance of being perfect. *J. Comput. Chem.*, **23**, 1297–1304.

53 Feig, M., Onufriev, A., Lee, M.S., Im, W., Case, D.A., and Brooks, C.L. III (2004) Performance comparison of generalized Born and Poisson methods in the calculation of electrostatic solvation energies for protein structures. *J. Comput. Chem.*, **25**, 265–284.

54 Hawkins, G.D., Cramer, C.J., and Truhlar, D.G. (1996) Parameterized models of aqueous free energies of solvation based on pairwise descreening of solute atomic charges from a

dielectric medium. *J. Phys. Chem.*, **100**, 19824–19839.
55. Schaefer, M. and Karplus, M. (1996) A comprehensive analytical treatment of continuum electrostatics. *J. Phys. Chem.*, **100**, 1578–1599.
56. Qiu, D., Shenkin, P.S., Hollinger, F.P., and Still, W.C. (1997) The gb/sa continuum model for solvation. A fast analytical method for the calculation of approximate Born radii. *J. Phys. Chem. A*, **101**, 3005–3014.
57. Scarsi, M., Apostolakis, J., and Caflisch, A. (1997) Continuum electrostatic energies of macromolecules in aqueous solutions. *J. Phys. Chem. A*, **101**, 8098–8106.
58. Ghosh, A., Rapp, C.S., and Friesner, R.A. (1998) Generalized Born model based on a surface integral formulation. *J. Phys. Chem. B*, **102**, 10983–10990.
59. Dominy, B.N. and Brooks, C.L. III (1999) Development of a generalized Born model parameterization for proteins and nucleic acids. *J. Phys. Chem. B*, **103**, 3765–3773.
60. Srinivasan, J., Trevathan, M.W., Beroza, P., and Case, D.A. (1999) Application of a pairwise generalized Born model to proteins and nucleic acids: inclusion of salt effects. *Theor. Chem. Accts.*, **101**, 426–434.
61. Tsui, V. and Case, D.A. (2000) Molecular dynamics simulations of nucleic acids with a generalized Born solvation model. *J. Am. Chem. Soc.*, **11**, 2489–2498.
62. Onufriev, A., Bashford, D., and Case, D.A. (2000) Modification of the generalized Born model suitable for macromolecules. *J. Phys. Chem. B*, **104**, 3712–3720.
63. Lee, M.S., Salsbury, F.R. Jr, and Brooks, C.L. III (2002) Novel generalized Born methods. *J. Chem. Phys.*, **116**, 10606–10614.
64. Spassov, V.Z., Yan, L., and Szalma, S. (2002) Introducing an implicit membrane in generalized Born/solvent accessibility continuum solvent models. *J. Phys. Chem. B*, **106**, 8726–8738.
65. Im, W., Lee, M.S., and Brooks, C.L. III (2003) Generalized Born model with a simple smoothing function. *J. Comput. Chem.*, **24**, 1691–1702.
66. Im, W., Feig, M., and Brooks, C.L. III (2003) An implicit membrane generalized Born theory for the study of structures, stability, and interactions of membrane proteins. *Biophys. J.*, **85**, 2900–2918.
67. Gallicchio, E. and Levy, R.M. (2004) AGBNP: an analytic implicit solvent model suitable for molecular dynamics simulations and high-resolution modeling. *J. Comput. Chem.*, **25**, 479–499.
68. Feig, M., Im, W., and Brooks, C.L. III (2004) Implicit solvation based on generalized Born theory in different dielectric environments. *J. Chem. Phys.*, **120**, 903–910.
69. Zhu, J., Alexov, E., and Honig, B. (2005) Comparative study of generalized Born models: Born radii and peptide folding. *J. Phys. Chem. B*, **109**, 3008–3022.
70. Levy, R.M., Zhang, L.Y., Gallicchio, E., and Felts, A.K. (2003) On the nonpolar hydration free energy of proteins: surface area and continuum solvent models for the solute–solvent interaction energy. *J. Am. Chem. Soc.*, **125**, 9523–9530.
71. Chen, J. and Brooks, C.L. III (2007) Critical importance of length-scale dependence in implicit modeling of hydrophobic interactions. *J. Am. Chem. Soc.*, **129**, 2444–2445.
72. Chen, J. and Brooks, C.L. (2008) Implicit modeling of nonpolar solvation for simulating protein folding and conformational transitions. *Phys. Chem. Chem. Phys.*, **10**, 471–481.
73. Tolman, R.C. (1949) The effect of droplet size on surface tension. *J. Chem. Phys.*, **17**, 333–337.
74. Reiss, H., Frisch, H.L., and Lebowitz, J.L. (1959) Statistical mechanics of rigid spheres. *J. Chem. Phys.*, **31**, 369–380.
75. Lum, K., Chandler, D., and Weeks, J.D. (1999) Hydrophobicity at small and large length scales. *J. Phys. Chem. B*, **103**, 4570–4577.
76. Chandler, D. (2005) Interfaces and the driving force of hydrophobic assembly. *Nature*, **437**, 640–647.
77. Nina, M., Beglov, D., and Roux, B. (1997) Atomic radii for continuum electrostatics calculations based on

molecular dynamics free energy simulations. *J. Phys. Chem. B*, **101**, 5239–5248.

78 Nina, M., Im, W., and Roux, B. (1999) Optimized atomic radii for protein continuum electrostatics solvation forces. *Biophys. Chem.*, **78**, 89–96.

79 Banavali, N.K. and Roux, B. (2002) Atomic radii for continuum electrostatics calculations on nucleic acids. *J. Phys. Chem. B*, **106**, 11026–11035.

80 Roux, B., Allen, T., Berneche, S., and Im, W (2004) Theoretical and computational models of biological ion channels. *Q. Rev. Biophys.*, **37** (1), 15–103.

81 Chen, J., Im, W., and Brooks, C.L. III (2004) Refinement of NMR structures using implicit solvent and advanced sampling techniques. *J. Am. Chem. Soc.*, **126**, 16038–16047.

82 Chen, J., Won, H.-S., Im, W., Dyson, H.J., and Brooks, C.L. (2005) Generation of native-like models from limited NMR data, modern force fields and advanced conformational sampling. *J. Biomol. NMR*, **31**, 59–64.

83 Swanson, J.M.J., Adcock, S.A., and McCammon, J.A. (2005) Optimized radii for Poisson–Boltzmann calculations with the AMBER force field. *J. Chem. Theory Comput.*, **1** (3), 484–493.

84 Scarsi, M. and Caflisch, A (1999) Comment on the validation of continuum electrostatics models. *J. Comput. Chem.*, **20** (14), 1533–1536.

85 Luo, R., David, L., Hung, H., Devaney, J., and Gilson, M.K. (1999) Strength of solvent-exposed salt-bridges. *J. Phys. Chem. B*, **103**, 727–736.

86 Gallicchio, E., Zhang, L.Y., and Levy, R.M. (2002) The sgb/np hydration free energy model based on the surface generalized Born solvent reaction field and novel nonpolar hydration free energy estimators. *J. Comput. Chem.*, **23** (5), 517–529.

87 Zhou, R. (2003) Free energy landscape of protein folding in water: explicit vs. implicit solvent. *Proteins*, **53**, 148–161.

88 Khandogin, J. and Brooks, C.L. III (2005) Constant pH molecular dynamics with proton tautomerism. *Biophys. J.*, **89**, 141–157.

89 Zhou, R. and Berne, B.J. (2002) Can a continuum solvent model reproduce the free energy landscape of a β-hairpin folding in water? *Proc. Natl. Acad. Sci. U.S.A.*, **99**, 12777–12782.

90 Nymeyer, H. and Garcia, A.E. (2003) Simulation of the folding equilibrium of alpha-helical peptides: a comparison of the generalized Born approximation with explicit solvent. *Proc. Natl. Acad. Sci. U.S.A.*, **100**, 13934–13939.

91 Jaramillo, A. and Wodak, S.J. (2005) Computational protein design is a challenge for implicit solvation models. *Biophys. J.*, **88**, 156–171.

92 Jorgensen, W.L., Chandrasekhar, J., Madura, J.D., Impey, R.W., and Klein, M.L. (1983) Comparison of simple potential functions for simulating liquid water. *J. Chem. Phys.*, **79**, 926–935.

93 MacKerell, A.D. Jr, Bashford, D., Bellott, M., Dunbrack, R.L., Evanseck, J.D., Field, M.J., Fischer, S., Gao, J., Guo, H., Ha, S., Joseph-McCarthy, D., Kuchnir, L., Kuczera, K., Lau, F.T.K., Mattos, C., Michnick, S., Ngo, T., Nguyen, D.T., Prodhom, B., Reiher, W.E. III, Roux, B., Schlenkrich, M., Smith, J.C., Stote, R., Straub, J., Watanabe, M., Wiorkiewicz-Kuczera, J., Yin, D., and Karplus, M. (1998) All-atom empirical potential for molecular modeling and dynamics studies of proteins. *J. Phys. Chem. B*, **102**, 3586–3616.

94 Feig, M., MacKerell, A.D. Jr, and Brooks, C.L. III (2003) Force field influence on the observation of π-helical protein structures in molecular dynamics simulations. *J. Phys. Chem. B*, **107**, 2831–2836.

95 MacKerell, A.D. Jr, Feig, M., and Brooks, C.L. III (2004) Improved treatment of the protein backbone in empirical force fields. *J. Am. Chem. Soc.*, **126**, 698–699.

96 Kumar, S., Bouzida, D., Swendsen, R.H., Kollman, P.A., and Rosenberg, J.M. (1992) The weighted histogram analysis method for free-energy calculations on biomolecules. I. The method. *J. Comput. Chem.*, **13**, 1011–1021.

97 Roux, B. (1995) The calculation of the potential of mean force using computer simulations. *Comput. Phys. Commun.*, **91**, 275–282.

98 Yu, Z.Y., Jacobson, M.P., and Friesner, R.A. (2006) What role do surfaces play in GB models? A new generation of surface-generalized Born model based on a novel Gaussian surface for biomolecules. *J. Comput. Chem.*, **27**, 72–89.

99 Mongan, J., Simmerling, C., McCammon, J.A., Case, D.A., and Onufriev, A. (2007) Generalized Born model with a simple, robust molecular volume correction. *J. Chem. Theory Comput.*, **3**, 156–169.

100 Shalongo, W., Dugad, L., and Stellwagen, E. (1994) Distribution of helicity within the model peptide acetyl (AAQAA)₃ amide. *J. Am. Chem. Soc.*, **116**, 8288–8293.

101 Blanco, G., Rivas, F.J., and Serrano, L. (1994) A short linear peptide that folds into a β-hairpin in aqueous solution. *Nat. Struct. Biol.*, **1**, 584–590.

102 Munoz, V., Thompson, P.A., Hofrichter, J., and Eaton, W.A. (1997) Folding dynamics and mechanism of β-hairpin formation. *Nature*, **390**, 196–199.

103 Fesinmeyer, R.M., Hudson, F.M., and Andersen, N.H. (2004) Enhanced hairpin stability through loop design: the case of the protein GB1 domain hairpin. *J. Am. Chem. Soc.*, **126**, 7238–7243.

104 Sugita, Y. and Okamoto, Y. (1999) Replica-exchange molecular dynamics method for protein folding. *Chem. Phys. Lett.*, **314**, 141–151.

105 Feig, M., Karanicolas, J., and Brooks, C.L. III (2004) MMTSB tool set: enhanced sampling and multiscale modeling methods for applications in structural biology. *J. Mol. Graph. Model.*, **22**, 377–395.

106 Periole, X. and Mark, A.E. (2007) Convergence and sampling efficiency in replica exchange simulations of peptide folding in explicit solvent. *J. Chem. Phys.*, **126**, 014903.

107 Zheng, W., Andrec, M., Gallicchio, E., and Levy, R.M. (2007) Simulating replica exchange simulations of protein folding with a kinetic network model. *Proc. Natl. Acad. Sci. U.S.A.*, **104**, 15340–15345.

108 Denschlag, R., Lingenheil, M., and Tavan, P. (2008) Efficiency reduction and pseudo-convergence in replica exchange sampling of peptide folding–unfolding equilibria. *Chem. Phys. Lett.*, **458**, 244–248.

109 Nymeyer, H. (2008) How efficient is replica exchange molecular dynamics? An analytic approach. *J. Chem. Theory Comput.*, **4**, 626–636.

110 MacKerell, A.D. Jr, Feig, M., and Brooks, C.L. III (2004) Extending the treatment of backbone energetics in protein force fields: limitations of gas-phase quantum mechanics in reproducing protein conformational distributions in molecular dynamics simulations. *J. Comput. Chem.*, **25**, 1400–1415.

111 Kortemme, T., Morozov, A.V., and Baker, D. (2003) An orientation-dependent hydrogen bonding potential improves prediction of specificity and structure for proteins and protein–protein complexes. *J. Mol. Biol.*, **326**, 1239–1259.

112 Liwo, A., Khalili, M., and Scheraga, H.A. (2005) Ab initio simulations of protein-folding pathways by molecular dynamics with the united-residue model of polypeptide chains. *Proc. Natl. Acad. Sci. U.S.A.*, **102**, 2362–2367.

113 Du, D., Zhu, Y., Huang, C.Y., and Gai, F. (2004) Understanding the key factors that control the rate of β-hairpin folding. *Proc. Natl. Acad. Sci. U.S.A.*, **101**, 15915–15920.

114 Gronenborn, A.M., Filpula, D.R., Essig, N.Z., Achari, A., Whitlow, M., Wingfield, P.T., and Clore, G.M. (1991) A novel, highly stable fold of the immunoglobulin binding domain of streptococcal protein G. *Science*, **253**, 657–661.

115 Cochran, A.G., Skelton, N.J., and Starovasnik, M.A. (2001) Tryptophan zippers: stable, monomeric β-hairpins. *Proc. Natl. Acad. Sci. U.S.A.*, **98**, 5578–5583.

116 Neidigh, J.W., Fesinmeyer, R.M., and Andersen, N.H. (2002) Designing a 20-residue protein. *Nat. Struct. Biol.*, **9**, 425–430.

117 McKnight, C.J., Doerin, D.S., Matsudaira, P.T., and Kim, P.S. (1997) NMR structure of the 35-residue villin headpiece subdomain. *Nat. Struct. Biol.*, **4**, 180.

118 Gouda, H., Torigoe, H., Saito, A., Sato, M., Arata, Y., and Shimada, I. (1992) Three-dimensional solution structure of the B domain of staphylococcal protein A: comparisons of the solution and crystal structures. *Biochemistry*, **31**, 9665.

119 Buck, M., Bouguet-Bonnet, S., Pastor, R.W., and MacKerell, A.D. (2006) Importance of the CMAP correction to the CHARMM22 protein force field: dynamics of hen lysozyme. *Biophy. J.*, **90** (4), L36–L38.

120 Trbovic, N., Kim, B., Friesner, R.A., and Palmer, A.G. III (2008) Structural analysis of protein dynamics by MD simulations and NMR spin-relaxation. *Proteins*, **71** (2), 684–694.

121 Kent, A., Jha, A.K., Fitzgerald, J.E., and Freed, K.F. (2008) Benchmarking implicit solvent folding simulations of the amyloid beta (10–35) fragment. *J. Phys. Chem. B*, **112** (19), 6175–6186.

122 Chen, J., Brooks, C.L., and Wright, P.E. (2004) Model-free analysis of protein dynamics: assessment of accuracy and model selection protocols based on molecular dynamics simulation. *J. Biomol. NMR*, **29**, 243–257.

123 Smith, P.E. (2008) On the Kirkwood–Buff inversion procedure. *J. Chem. Phys.*, **129** (12), 124509.

124 Pierce, V., Kang, M., Aburi, M., Weerasinghe, S., and Smith, P.E. (2008) Recent applications of Kirkwood–Buff theory to biological systems. *Cell Biochem. Biophys.*, **50** (1), 1–22.

125 Shirts, M.R. and Pande, V.S. (2003) Extremely precise free energy calculations of amino acid side chain analogs: comparison of common molecular mechanics force fields for proteins. *J. Chem. Phys.*, **119**, 5740–5761.

126 Chen, J., Brooks, C.L., and Scheraga, H.A. (2008) Revisiting the carboxylic acid dimers in aqueous solution: interplay of hydrogen bonding, hydrophobic interactions and entropy. *J. Phys. Chem. B*, **112**, 242–249.

127 Xu, Z., Luo, H.H., and Tieleman, D.P. (2007) Modifying the OPLS-AA force field to improve hydration free energies for several amino acid side chains using new atomic charges and an off-plane charge model for aromatic residues. *J. Comput. Chem.*, **28**, 689–697.

128 Baker, C.M. and Grant, G.H. (2007) Modeling aromatic liquids: toluene, phenol, and pyridine. *J. Chem. Theory Comput.*, **3**, 530–548.

129 Pratt, L.R. (2002) Molecular theory of hydrophobic effects: "she is too mean to have her name repeated". *Annu. Rev. Phys. Chem.*, **53**, 409–436.

130 Dzubiella, J., Swanson, J.M.J., and McCammon, J.A. (2006) Coupling hydrophobicity, dispersion, and electrostatics in continuum solvent models. *Phys. Rev. Lett.*, **96**, 087802.

8
Modeling Protein Solubility in Implicit Solvent
Harianto Tjong and Huan-Xiang Zhou

8.1
Introduction

Protein solubility is defined as the equilibrium concentration of the protein in the solution phase when a saturation amount of the protein is present. Protein solubility is of scientific interest because it plays important roles in protein expression and purification, structural genomics, human diseases, and industrial applications. Various factors can influence protein solubility. While the intrinsic factor is determined by the amino acid sequence and the protein structure, the extrinsic factors include pH, ionic strength, temperature, and concentration of precipitants [1, 2]. Generally, proteins are least soluble at the pH, known as the isoelectric point, or pI, where their net charges are zero.

When a protein of interest is not soluble enough for a particular application, then it is termed a low-solubility protein. Low solubility is found to enhance amyloid fibril formation [3], and is expected to play a major role in diseases associated with protein aggregation. A number of techniques have been proposed to increase protein solubility, including time-consuming screening strategies for optimal solvent conditions [4, 5], co-expression with molecular chaperones [6], attachment of a small protein (fusion tag) [7, 8] or peptide (poly-Lys or poly-Arg) [9] as a solubility enhancement tag (SET), and rational mutation of surface-exposed hydrophobic residues to polar residues [10–12]. In addition, charge mutations have been introduced to manipulate the pI [13–15]. An excellent review by Trevino *et al.* [16] provides further reading.

Theoretical methods to predict protein solubility are of great interest, as they can be used to obtain the desired levels of solubility for particular proteins, and may provide the necessary guidance to achieve optimal solvent conditions for protein solubility. Recently, we have developed a method for calculating the effects of pH and mutations on protein solubility [17]. Here, we present a summary of this method and its applications. Our method was inspired by our study of the transfer free energies of pentapeptides, AcWL-X-LL (acetyl-Trp-Leu-X-Leu-Leu, where X is any of the 20 amino acids), from an organic solvent to water. The experimental data obtained by Wimley *et al.* [18] were reproduced reasonably well

Modeling Solvent Environments. Applications to Simulations of Biomolecules. Edited by Michael Feig
Copyright © 2010 WILEY-VCH Verlag GmbH & Co. KGaA, Weinheim
ISBN: 978-3-527-32421-7

by our transfer free energy model [17]. In that model, the octanol-to-water transfer free energy was calculated as the difference in solvation free energy in octanol and in water. We sought the optimal effective dielectric constant of octanol, while the dielectric constant of water was fixed at the appropriate value (around 78–80). In predicting protein solubility, the basic tenet of our approach is the modeling of the crystalline phase as an implicit solvent; we sought the optimal effective dielectric constant for the crystalline phase.

8.2
The Models

8.2.1
Transfer Free Energy

Thermodynamically, protein solubility is defined as the protein concentration at which the chemical potentials of the protein molecules in the solution (G^s) and in the crystalline condensed phase (G^c) are equal. The chemical potential in the solution phase can be written in the form

$$G^s = G^{s0} + k_B T \ln C \tag{8.1}$$

where C is the protein concentration, k_B is the Boltzmann constant, T is the absolute temperature, and G^{s0} is the chemical potential at a "standard" concentration (for example, at $C = 1$ M). In the crystalline phase, the individual protein molecules do not mix, so there is no concentration-dependent term like in Equation 8.1 for G^c. When the two chemical potentials are equal, the concentration is then defined as the solubility, S,

$$\begin{aligned} S &= \exp[-(G^{s0} - G^c)/k_B T] \\ &= \exp[-\Delta G^{c \to s}/k_B T] \end{aligned} \tag{8.2}$$

The difference of the two potentials, $\Delta G^{c \to s}$, can be viewed as the transfer free energy from the crystalline phase to the solution phase. The more favorable it is for a protein to transfer from the crystalline phase to the solution phase, the lower the transfer free energy, and correspondingly the higher the solubility.

We model the crystalline phase as an implicit solvent and decompose the transfer free energy $\Delta G^{c \to s}$ into electrostatic and non-electrostatic components:

$$\Delta G^{c \to s} = \Delta G^{c \to s}_{el} + \Delta G^{c \to s}_{ne} \tag{8.3}$$

The electrostatic component is calculated by solving the Poisson–Boltzmann (PB) equation [19–29], and the non-electrostatic component is obtained based on the solvent-accessible surface area [21, 26, 30–33].

8.2.2
Electrostatic Free Energy

A proper theoretical description of biomolecules must consider effects coming from the environment, and one major effect comes from the solvent. Solute–solvent interactions are often described by solvation energy, the free energy of transferring the solute from a vacuum or some dielectric medium to the solvent environment of interest (for example, water at a certain ionic strength). This model treats the solvent as a high-dielectric medium interacting with charges that are embedded in solute molecules of lower dielectric. The electrostatic free energy is calculated according to

$$G_{el} = \frac{1}{2}\sum_i q_i \varphi(r_i) \tag{8.4}$$

by solving the electrostatic potential, φ, from the PB equation

$$\nabla \cdot \varepsilon(r)\nabla \varphi(r) - \varepsilon_s \kappa^2 \varphi(r) = -4\pi \sum_{i=1}^{N} q_i \delta(r - r_i) \tag{8.5}$$

in the linearized form, where $\kappa = \sqrt{8\pi e^2 I/\varepsilon_s k_B T}$ is the inverse Debye screening length, ε_s is the solvent dielectric constant, r_i are the positions of protein point charges q_i, e is the elementary charge, and I is the ionic strength (same as the salt concentration in the solution for 1:1 electrolyte).

The electrostatic free energy of a solute molecule inside an implicit solvent can be separated into a Coulomb term and the solvation term:

$$G_{el} = G_{Coul} + G_{solv} \tag{8.6}$$

The Coulomb term is given by

$$G_{Coul} = \sum_{i>j} \frac{q_i q_j}{\varepsilon_p r_{ij}} \tag{8.7}$$

where r_{ij} are inter-charge distances, and ε_p is the solute dielectric constant. The electrostatic component of the transfer free energy, $\Delta G_{el}^{c \to s}$, is given by the difference in electrostatic solvation energy of the solution phase from the crystalline phase (the Coulomb electrostatic energy of the protein is assumed to be the same in the two phases). The non-electrostatic component will be described later.

8.3
Applications

8.3.1
Transfer Free Energy from Octanol to Water

Before going to the solubility problem, we discuss a direct use of implicit solvent for the transfer free energy between two solvents, which has a similar expression to Equation 8.3. The model systems are AcWL-X-LL pentapeptides, where X is any of the 20 amino acids, dissolved in two different solvents (octanol and water). The octanol-to-water transfer free energy takes the form

$$\Delta G^{o \to w} = \Delta G_{el}^{o \to w} + \Delta G_{ne}^{o \to w} \tag{8.8}$$

where the superscripts "o" and "w" denote octanol and water, respectively. Again, the electrostatic component, $\Delta G_{el}^{o \to w}$, is calculated as the difference in electrostatic solvation energy between the two solvents. The variation of $\Delta G_{ne}^{o \to w}$ among the 20 pentapeptides was modeled based on the non-polar and polar portions of the solvent-accessible surface areas calculated over the side chain of guest residue X:

$$\Delta G_{ne}^{o \to w} = \Delta \sigma_{np}^{o \to w} A_{np}(X) + \Delta \sigma_{p}^{o \to w} A_{p}(X) + \Delta G_{ne0}^{o \to w} \tag{8.9}$$

where $\Delta G_{ne0}^{o \to w}$ represents the contribution of the common moiety of the 20 pentapeptides to the non-electrostatic component of the transfer free energy. The values for $A_{np}(X)$ and $A_p(X)$ used in this study were taken directly from the paper of Wimley et al. [18] (as listed under the heading "ASA of X in AcWL-X-LL (Å²)" of their table 1).

The 20 pentapeptides AcWL-X-LL were built with the LEAP program and then energy-minimized before molecular dynamics (MD) simulations in vacuum [17] (using the SANDER program in the AMBER package [34]). Once the pentapeptide samples were collected, we performed PB calculations to obtain $\Delta G_{el}^{o \to w}$ as the difference in electrostatic solvation energy between two implicit solvents, one modeling water with a dielectric constant of 78.5 and another modeling octanol with an adjustable dielectric constant. The pentapeptide dielectric constants were all set to 4, the value that we have always assigned for proteins in our studies [23, 27, 35–40]. The PB solver used was the UHBD program [19]. The dielectric constant of the octanol solvent was adjusted such that a multi-linear regression of the experimental data of Wimley et al. [18] for $\Delta G^{o \to w}$ (at pH 9, taken from their table 1) against $\Delta G_{el}^{o \to w}$, $A_{np}(X)$, and $A_p(X)$ resulted in a coefficient of unity for the $\Delta G_{el}^{o \to w}$ term. We fitted our calculations with the experimental data at pH 9 under the assumption that our MD simulations were appropriate for that pH, where we run fixed protonation states with R and K protonated, and D, E, H, and the C-terminal unprotonated.

The coefficient of unity was achieved when a dielectric constant of 15 was assigned to the octanol solvent, with the regression analysis reaching a correlation R^2 of 0.91 (see Figure 8.1). From the regression analysis, the non-polar ($\Delta \sigma_{np}^{o \to w}$) and polar ($\Delta \sigma_p^{o \to w}$) octanol-to-water solvation parameters are 18.1 and 13.5 cal mol⁻¹ Å⁻², respectively. The value of the non-polar solvation parameter, $\Delta \sigma_{np}^{o \to w}$, obtained from this study is comparable to the 22.8 cal mol⁻¹ Å⁻² obtained from a linear fit performed by Wimley et al. [18] while restricting the guest residue X only to amino acids with non-polar side chains. Another similar study by Fauchère and Pliška [41] using Ac-X-amides also obtained a similar value for the non-polar parameter, 20.9 cal mol⁻¹ Å⁻². Unlike these two experimental studies, our analysis included all the 20 types of guest amino acids.

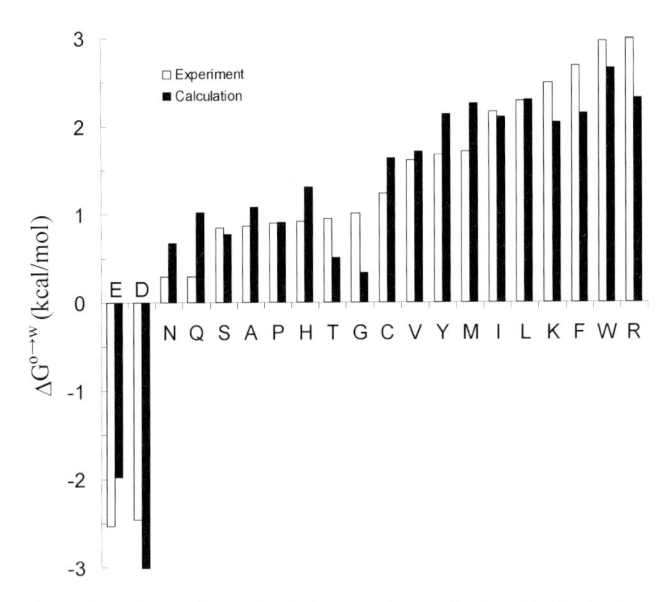

Figure 8.1 Comparison of calculated and experimental results for the octanol-to-water transfer free energy of the 20 pentapeptides AcWL-X-LL, where X is the guest residue displayed in the horizontal axis (ordered by the energy determined from experiment). Reprinted from Ref. [17] with permission from Elsevier.

8.3.2
Crystalline Phase as a Dielectric Medium

Traditionally, the idea of implicit solvent is to replace the details of the water molecules surrounding the solute in a solution with a mean-field effect of a high-dielectric medium. In the same spirit, we model the crystalline phase of a protein as a dielectric medium that has a dielectric constant between that of the protein and that of water. The expected value of the dielectric constant of the crystalline phase is less than that of water because the crystalline phase contains less water and more protein molecules arranged in some crystal arrays as compared to the solution phase.

To demonstrate the idea of an "implicit crystalline phase," we performed a comparison of two calculation methods on the solvation free energy of ribonuclease (RNase) Sa (Protein Data Bank (PDB) entry 1rgg; see Figure 8.2a) in the crystalline phase (see Ref. [17] for details). In the first series of calculations, we modeled the crystalline phase of RNase Sa as a periodic array of monomers (Figure 8.3a). The electrostatic solvation free energy of a single monomer, G_{solv}^c, was calculated in the background of the crystalline replicas. The single monomer and all its replicas were assigned a dielectric constant of 4, and the rest of space was assigned the dielectric constant of water (that is, 78.5). The crystalline array was grown by expanding the unit cell, and the corresponding solvation

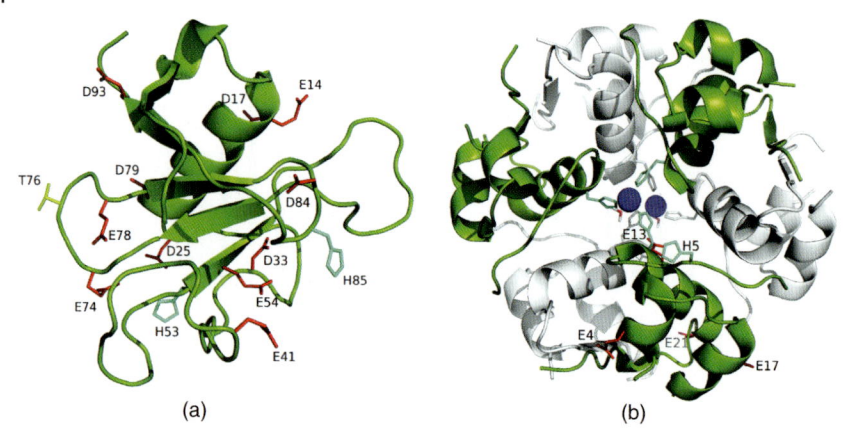

Figure 8.2 Cartoon representation of proteins in solubility predictions. (a) RNase Sa (PDB entry: 1rgg). Titratable groups are shown: 11 carboxyl side chains, two histidines, and residue T76 are labeled and displayed as red, cyan, and yellow sticks, respectively. The N-terminal residue happens to be an aspartate; the side chain of this residue was not titrated in the constant-pH simulations and is not displayed here. (b) Zinc-insulin hexamer (PDB entry: 4ins). The two trimers are displayed in the foreground and background in green and gray, respectively. The two zinc ions and the coordinating histidines and water molecules are shown. In addition, the five titrated side chains of one monomer are labeled and displayed as sticks. Reprinted from Ref. [17] with permission from Elsevier.

energy was calculated. Figure 8.4a shows that G_{solv}^c has reached its plateau value when ~100 replicas around the single monomer were included. The plateau value can be taken to be the value of G_{solv}^c for a single monomer in an infinite crystalline array. In the second series of calculations, G_{solv}^c, the electrostatic solvation energy of a single RNase Sa, was calculated without any replicas surrounding it, and a dielectric constant was applied for the exterior region of the protein (Figure 8.3b); a range of solvent dielectric constants was studied in order to reproduce the plateau of G_{solv}^c obtained from the first calculation series. The plot in Figure 8.4b was obtained when a range (from 50 to 60) of protein exterior dielectric constants was assigned while keeping that of the protein interior fixed to 4.

Based on a comparison of the two plots in Figure 8.4, we can infer that the infinite array of surrounding crystalline replicas embedded in water can be replaced by a uniform dielectric medium, with an effective dielectric constant of $\varepsilon_c \sim 53.5$. Note that the dielectric constant of 53.5 was obtained from the ideal case of a protein crystalline phase in a solution. In practice, solubility measurements are often carried out by solubilizing the protein not against a crystalline phase, but against an amorphous condensed phase [16]. However, the value 53.5 serves as a good starting point for modeling the condensed phase in a particular solubility measurement.

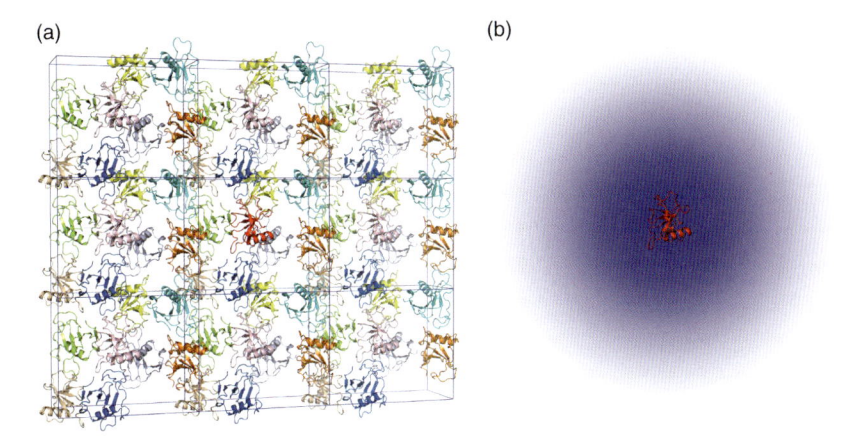

Figure 8.3 (a) A representation of crystalline RNase Sa, built according to the transformations directed by its PDB file. (b) These infinite replicas of the center protein can be replaced by an infinite dielectric medium. In the implicit model for the crystalline phase, the inhomogeneous environment of the red monomer at the center is approximated by a uniform dielectric medium (as depicted by the blue "cloud"), with a dielectric constant intermediate between a high value for water and a low value expected for protein molecules. The red monomer is the solute, which is always assigned a low dielectric constant (that is, 4). Reprinted from Ref. [17] with permission from Elsevier.

8.3.3
Protein Solubility: pH Dependence

In this section, we demonstrate the use of the implicit solvent model in modeling the pH dependence of protein solubility. Motivated by the availability of experimental data [12, 13, 15, 42, 43], we took RNase Sa and insulin (Figure 8.2) as test systems. To account for the pH effects on the protonation states, a series of pH values is set while performing molecular dynamics simulations for each protein. The constant-pH simulations were carried out with a generalized Born (GB) implicit solvent model implemented in the AMBER package [34, 44]. In the simulation program, the protonation states of titratable groups are assigned by a Monte Carlo procedure. Further details of the simulations can be found in Reference [17].

At each pH one hundred conformations were randomly sampled for further calculations. Then, the electrostatic component, $\Delta G_{el}^{c \to s}$ of the transfer free energy from the condensed phase to the solution phase was calculated as the average over the selected conformations. By choosing a reference pH, pH_{ref}, and employing the relation in Equation 8.2, one can obtain the following expression for the pH dependence of the protein solubility

$$k_B T \ln[S(pH)/S(pH_{ref})] = -[\Delta G_{el}^{c \to s}(pH) - \Delta G_{el}^{c \to s}(pH_{ref})] \tag{8.10}$$

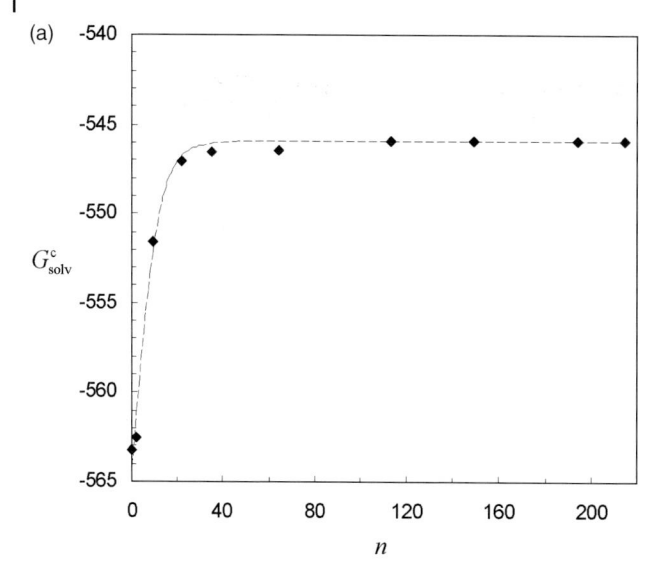

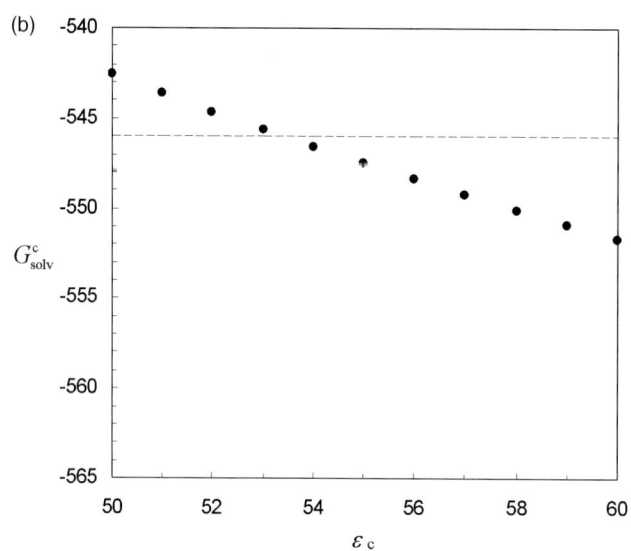

Figure 8.4 Calculation results of the electrostatic solvation energy on the explicit crystal array and an implicit model. (a) Plot of G^c_{solv} of a single RNase Sa monomer in the presence of a crystalline array of replicas. The replicas serve to change the dielectric environment of the single monomer. The infinite array is approached by including more and more distant replicas (n = total number of replicas included). The dashed curve through the data points is to guide the eye. (b) Plot of G^c_{solv} when the inhomogeneous dielectric environment is replaced by a uniform dielectric medium. The effective dielectric constant, ε_c, of the uniform dielectric medium is varied to match G^c_{solv}, which is indicated by the horizontal line. Adapted from Ref. [17] with permission from Elsevier.

where $S(\text{pH})$ and $\Delta G_{el}^{c \to s}$ (pH), respectively, are the solubility and the electrostatic component of $\Delta G^{c \to s}$ at a given pH. In our calculations, each conformation with a particular set of protonation states was used for both the solution phase and the condensed phase. Therefore, the Coulomb contributions of each conformation in the two phases were identical, and only the difference in solvation energy contributed to $\Delta G_{el}^{c \to s}$. It is assumed that the non-electrostatic component, $\Delta G_{ne}^{c \to s}$, is not affected by pH.

The protein solubility expressed in Equation 8.10 depends on the energy difference in solvation energy, resulting in $\Delta G_{el}^{c \to s}$ (pH), which is determined by only the dielectric constants in the two phases. In solving the PB equation, the two phases are characterized only by the dielectric constant of the protein exterior, that is, 78.5 for water in the solution phase and ~55 for the crystalline phase. It is interesting to note that the crystalline-phase dielectric constant of 53.5 obtained in the previous section is virtually the same as the value, 55, required for an optimal fit with experimental data on RNase Sa. We then applied the same protocol to insulin as well without further adjustment.

The reasonably good agreement of our calculations with experimental data for the two proteins is shown in Figure 8.5. The minima of the calculated solubility were reached around the pH suggested by experimental data. In general, the electrostatic solvation free energy of a protein in the solution phase increases in magnitude with increasing magnitude of the net charge on the protein. A similar trend is expected in the crystalline phase, except that the magnitude of the electrostatic solvation energy is reduced relative to the counterpart in the solution phase (under the assumption that the solvent dielectric constant in the crystalline phase is lower than in the solution phase). Taken together, $\Delta G_{el}^{c \to s}$ is expected to increase in magnitude with increasing magnitude of the net charge on the protein. The net charge can be changed by varying the pH. At the pI the net charge on the

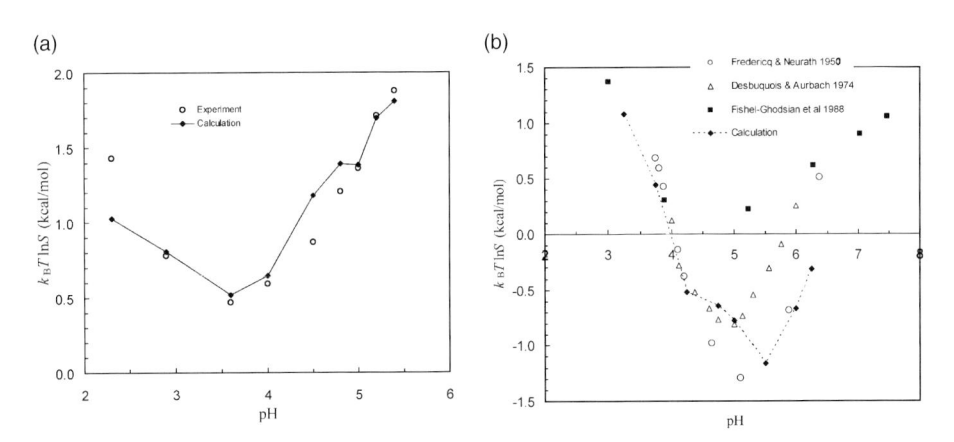

Figure 8.5 Comparison of calculated and experimental results for the pH dependence of the solubility of (a) RNase Sa, and (b) insulin. Adapted from Ref. [17] with permission from Elsevier.

protein is zero; when the pH moves away from the pI in either direction, the net charge increases in magnitude. Therefore, the pH dependence of $\Delta G_{el}^{c \rightarrow s}$ is expected to have a bell shape, with the maximum occurring around the pI. Correspondingly, the pH dependence of ln S should follow an inverted bell shape, as usually observed [3, 13, 15, 42, 43].

In comparing the experimental data and the predictions for RNase Sa (Figure 8.5a), it is clear that the calculation underestimates ln S at pH 2.3, which is the lowest pH value of the experimental study. A possible reason for the discrepancy is the lack of protonation at that pH for the side chain of the N-terminal aspartate residue and the C-terminal carboxyl in our simulation as a result of a limitation in the current implementation of the AMBER software used for the constant-pH MD simulations. In MD simulations, we allowed titration for the 11 (instead of 12) carboxyl side chains and two histidines that are shown in Figure 8.2a. As for the insulin, all the 30 titratable groups (five residues from a representative monomer are shown in Figure 8.2b) were able to be protonated and/or deprotonated by the Monte Carlo method in the constant-pH MD simulations, so that problems like that in RNase Sa under low pH were not expected to appear.

In short, our calculations are in good agreement with the results from three experimental studies [13, 42, 43] for insulin below pH 5 (Figure 8.5b), while there are no reliable experimental results to compare against above pH 5. However, increases in solubility are seen in experimental and calculated results as the pH is increased above the pI. Experimentally, the pI of zinc insulin is determined to be between 5 and 5.5 [13, 42, 43]. From the constant-pH MD simulations, the pI of our modeled insulin structure is around 6. This value is reasonable, as the net charge from each monomer, which shares one-third of a bound zinc ion, is expected to be neutral when the histidine (H5 in Figure 8.2b) is still protonated, while the four glutamates (E4, E13, E17, and E21) have nearly completed deprotonated in the presence of one arginine and one lysine (R22 and K29, both on chain B). The discrepancy of ~0.75 in the pI between the experiments and our simulations is perhaps caused by anion binding to insulin [42], which is not modeled explicitly in our study. To compensate, the plot for calculated results in Figure 8.5b has been shifted downward by 0.75 pH units.

For both proteins, the solvent dielectric constant was set to 55 for the crystalline phase and 78.5 for the solution phase. For RNase Sa, we also carried out a similar calculation with a dielectric constant of 60 assigned to the crystalline phase. The results were not significantly different from those shown in Figure 8.5a (with root-mean-square deviation between calculation and experiment changing from 0.18 to 0.22 kcal mol^{-1}). Our calculations thus suggest that the dielectric property of the condensed phase can be modeled with a solvent dielectric constant in the range of 55 to 60. Note that our calculations can only predict relative values of protein solubility against pH, namely, the $k_B T \ln S$ in the calculation plot has been shifted by a constant for each protein to be best compared with the respective experimental data.

8.3.4
Protein Solubility: Effects of Mutations

In a similar fashion as for the pH dependence of protein solubility, it is straightforward to model the effects of point mutations on protein solubility. Suppose that a native residue T in a protein undergoes a mutation to amino acid X. Then the change of transfer free energy from a crystalline phase to a solution phase by this mutation can be written as

$$\Delta\Delta G^{c \to s}(T \to X) = \Delta\Delta G^{c \to s}_{el}(T \to X) + \Delta\Delta G^{c \to s}_{ne}(T \to X) \quad (8.11)$$

Both the electrostatic and the non-electrostatic components are included. The mutational effect on the electrostatic component was obtained by calculating the change in electrostatic solvation energy from the condensed phase to the solution phase twice, once for the wild-type protein and once for the mutant, and then taking the difference:

$$\Delta\Delta G^{c \to s}_{el}(T \to X) = \Delta G^{c \to s}_{el}(X) - \Delta G^{c \to s}_{el}(T) \quad (8.12)$$

The mutational effect on the non-electrostatic component was assumed to have the form

$$\Delta\Delta G^{c \to s}_{ne}(T \to X) = \Delta\sigma^{c \to s}_{np}[A_{np}(X) - A_{np}(T)] + \Delta\sigma^{c \to s}_{p}[A_{p}(X) - A_{p}(T)] \quad (8.13)$$

where $\Delta\sigma^{c \to s}_{np}$ and $\Delta\sigma^{c \to s}_{p}$, respectively, are non-polar and polar solvation parameters for the transfer of a solute from the condensed phase to the solution phase; and $A_{np}(X)$ and $A_{p}(X)$ are the areas of the non-polar and polar portions, respectively, of the solvent-accessible surface calculated over the side chain of residue X, whose values are set to be the same as those used in Equation 8.9.

The relative solubility, $S(X)/S(T)$, of the mutant in reference to the wild-type protein can be estimated by constructing an expression similar to Equation 8.10,

$$k_B T \ln[S(X)/S(T)] = -\Delta\Delta G^{c \to s}(T \to X) \quad (8.14)$$

or the solubility of a mutant X is

$$S(X) = S(T) \exp\left(-\frac{\Delta\Delta G^{c \to s}_{el}(T \to X) + \Delta\Delta G^{c \to s}_{ne}}{k_B T}\right) \quad (8.15)$$

The two free parameters of Equation 8.15, $\Delta\sigma^{c \to s}_{np}$ and $\Delta\sigma^{c \to s}_{p}$, can be obtained by a multi-linear regression of the solubility data of the 19.

Changes in solubility of RNase Sa at pH 4.25 after mutating residue T76 into the 19 other amino acids have been measured by Trevino et al. [12]. We used these data to further test our approach for modeling the crystalline phase. We collected the RNase Sa conformations from MD simulations in implicit solvent run at pH 4.25, and then mutated each of those 100 samples into 19 mutants. A mutant was created by replacing residue T76 with any of the other 19 amino acids and then energy-minimized. PB calculations were done to both the mutant and the wild type for each conformation in the crystalline phase and solution phase, and Equation 8.12 was used to obtain the electrostatic term. Note that all values of dielectric constants used in the earlier applications were also used here without further adjustment.

A multi-linear regression of the experimental data resulted in a correlation R^2 of 0.83 and the coefficient of the electrostatic component, $\Delta\Delta G_{el}^{c \to s}(T \to X)$, was found to be very close to unity. This finding gives strong validation of our treatment of the crystalline phase as an effective dielectric medium in calculating the electrostatic component of the transfer free energy. Equation 8.15 was found to reproduce well the experimental data as shown in Figure 8.6. The optimized values of $\Delta\sigma_{np}^{c \to s}$ and $\Delta\sigma_{p}^{c \to s}$ were 8.6 and 1.1 cal mol^{-1} Å^{-2}, respectively. The calculation results correctly rank the D and E variants as having the highest solubility levels, and the F and W variants as having the lowest solubility levels. In addition, variants with measured solubility levels either higher or lower than the T variant are all correctly ranked as such by the results of the calculation.

Additional comparisons of our calculated results to experimental data were made for the changes in solubility of RNase Sa at pH 7 when residue T76 was mutated into the five other types of amino acids: D, S, A, K, and R. The procedure of generating the wild-type and mutant conformations with MD simulations for pH 7 was the same as that for pH 4.25 described above. Without further adjustment of any parameters, we calculated the solubility levels of the mutants on the basis of the solubility level of 41 g l^{-1} for the wild-type protein, resulting in values of 196, 61, 45, 23, and 28 g l^{-1} for the D, S, A, K, and R mutants, respectively. These

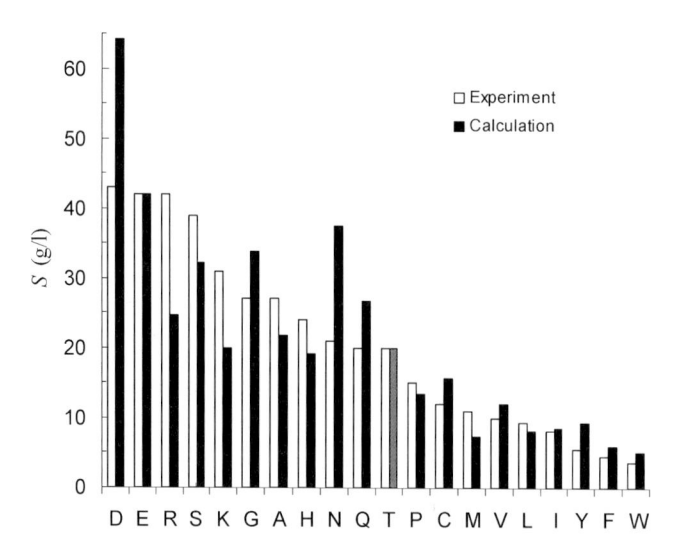

Figure 8.6 RNase Sa solubility of the 20 variants with point mutations at residue 76. The calculated result for T76 (gray) was set to be the same as the experimental result. Adapted from Ref. [17] with permission from Elsevier.

are to be compared with the experimental results of 160, 130, 92, 60, and 50 g l^{-1}. Despite some inaccuracy of the magnitudes, our calculations correctly predict that the D and S mutants are more soluble than the A mutant, whereas the K and R mutants are less soluble.

8.4 Summary and Outlook

We have demonstrated promising results of implicit solvent modeling on relative protein solubility. The key is to model the crystalline phase of a protein as a uniform dielectric medium. As a result, calculation of solubility becomes equivalent to calculation of transfer free energy from one type of solvent to another. The present study has dealt only with the solubility of protein dissolved in water, where the crystalline phase is represented by a dielectric constant intermediate between those of water and the protein. One may ask what the representative dielectric constant would be for the condensed phase of a protein dissolved in some non-aqueous solvent. This question may be irrelevant under physiological conditions, but may have an impact in biotechnology dealing with non-aqueous enzymology and implications in basic protein science. The value is expected to be between the dielectric constants of protein (say 4) and the solvent, based on our study of the protein crystalline arrays in Section 8.3.2. Repeating the quick calculations described in Section 8.3.2 for a solvent dielectric of 50 and 100 (note that for water it is 78.5), the condensed phase effective dielectric constants are always around 70% of the solvent dielectric constant like in the aqueous solution case. It is still an open question, however, why that percentage is maintained for an effective condensed phase dielectric constant.

Our methods involve structural modeling, MD simulations in the AMBER force field, the GB model for treating constant pH, and the PB solvation parameters. It appears that the current force fields and parameters are adequate to deal with protein solubility modeling. Obviously, there is still plenty of room to improve the treatment of the electrostatic and non-electrostatic contributions. For example, allowing protein terminals to be titratable in the constant-pH MD simulations will be desirable for properly modeling charge distributions at very high/low pH. Another direction that could improve our modeling is the use of a polarizable force field. In our electrostatic calculations, the solute charges are fixed upon transferring from one phase to another or from one solvent to another. In reality, electronic polarization of the solute can be different in different solvent environments. Neglect of solute electronic polarization may in part explain why an effective dielectric constant of 15 was obtained to model the octanol solvent, which is known to have a dielectric constant of 10.3 from bulk measurement [45]. In the non-electrostatic treatment, more advanced formulations have been proposed [30, 32, 33]. Along with experimental data of transfer free energies, it is promising that protein solubility data coupled with improved treatments of electrostatic and non-electrostatic contributions will lead to better solvation models.

The success of predicting the effects of pH on solubility opens a new way to search more efficiently for an optimal solvent condition for protein crystallization. Likewise, the ability of physics-based computations to predict mutational effects on solubility has a significant impact on the selection of protein targets in structural genomics projects and on the design of therapies against diseases associated with protein aggregation. The computational speed for predicting solubility is limited by the constant-pH MD simulations and PB calculations. As for the PB calculations, the speed can be tremendously increased by using alternative electrostatic solvation models, such as the GB model [46–57]. However, simply replacing the PB model with the GB model may not be advisable, since the accuracy of the current GB methods is very low for transfer free energy [58]. A promising solution is to combine PB and GB which preserves the accuracy of PB and reaches the speed of GB [58]. With this solution, fast scanning for amino acid replacement or for solvent conditions in solving solubility problems is feasible computationally. Finally, the successful modeling of the protein condensed phase as a simple implicit solvent may inspire modeling of other complex environments, such as the crowded milieu inside cells.

References

1 Arakawa, T. and Timasheff, S.N. (1985) Theory of protein solubility. *Methods Enzymol.*, **114**, 49–77.

2 Schein, C.H. (1990) Solubility as a function of protein structure and solvent components. *Nat. Biotechnol.*, **8** (4), 308–317.

3 Schmittschmitt, J.P. and Scholtz, J.M. (2003) The role of protein stability, solubility, and net charge in amyloid fibril formation. *Protein Sci.*, **12** (10), 2374–2378.

4 Howe, P.W.A. (2004) A straightforward method of optimising protein solubility for NMR. *J. Biomol. NMR*, **30** (3), 283–286.

5 Jancarik, J., Pufan, R., Hong, C., Kim, S.-H., and Kim, R. (2004) Optimum solubility (OS) screening: an efficient method to optimize buffer conditions for homogeneity and crystallization of proteins. *Acta Crystallogr. D*, **60** (9), 1670–1673.

6 Machida, S., Yu, Y., Singh, S.P., Kim, J.-D., Hayashi, K., and Kawata, Y. (1998) Overproduction of β-glucosidase in active form by an *Escherichia coli* system coexpressing the chaperonin GroEL/ES. *FEMS Microbiol. Lett.*, **159** (1), 41–46.

7 Zhou, P., Lugovskoy, A.A., and Wagner, G. (2001) A solubility-enhancement tag (SET) for NMR studies of poorly behaving proteins. *J. Biomol. NMR*, **20**, 11–14.

8 Waugh, D.S. (2005) Making the most of affinity tags. *Trends Biotechnol.*, **23** (6), 316–320.

9 Kato, A., Maki, K., Ebina, T., Kuwajima, K., Soda, K., and Kuroda, Y. (2007) Mutational analysis of protein solubility enhancement using short peptide tags. *Biopolymers*, **85** (1), 12–18.

10 Bianchi, E., Venturini, S., Pessi, A., Tramontano, A., and Sollazzo, M. (1994) High level expression and rational mutagenesis of a designed protein, the minibody: from an insoluble to a soluble molecule. *J. Mol. Biol.*, **236** (2), 649–659.

11 Mosavi, L.K. and Peng, Z.-Y. (2003) Structure-based substitutions for increased solubility of a designed protein. *Protein Eng.*, **16** (10), 739–745.

12 Trevino, S.R., Scholtz, J.M., and Pace, C.N. (2007) Amino acid contribution to protein solubility: Asp, Glu, and Ser contribute more favorably than the other hydrophilic amino acids in RNase Sa. *J. Mol. Biol.*, **366** (2), 449–460.

13 Fischel-Ghodsian, F., Brown, L., Mathiowitz, E., Brandenburg, D., and Langer, R. (1988) Enzymatically controlled drug delivery. *Proc. Natl. Acad. Sci. U.S.A.*, **85** (7), 2403–2406.

14 Tan, P.H., Chu, V., Stray, J.E., Hamlin, D.K., Pettit, D., Wilbur, D.S., Vessella, R.L., and Stayton, P.S. (1998) Engineering the isoelectric point of a renal cell carcinoma targeting antibody greatly enhances scFv solubility. *Immunotechnology*, **4** (2), 107–114.

15 Shaw, K.L., Grimsley, G.R., Yakovlev, G.I., Makarov, A.A., and Pace, C.N. (2001) The effect of net charge on the solubility, activity, and stability of ribonuclease Sa. *Protein Sci.*, **10** (6), 1206–1215.

16 Trevino, S.R., Scholtz, J.M., and Pace, C.N. (2008) Measuring and increasing protein solubility. *J. Pharm. Sci.*, **97** (10), 4155–4166.

17 Tjong, H. and Zhou, H.-X. (2008) Prediction of protein solubility from calculation of transfer free energy. *Biophys. J.*, **95** (6), 2601–2609.

18 Wimley, W.C., Creamer, T.P., and White, S.H. (1996) Solvation energies of amino acid side chains and backbone in a family of host–guest pentapeptides. *Biochemistry*, **35** (16), 5109–5124.

19 Madura, J.D., Briggs, J.M., Wade, R., Davis, M.E., Luty, B.A., Ilin, A., Antosiewicz, J., Gilson, M.K., Bagheri, B., Scott, L.R., and McCammon, J.A. (1995) Electrostatic and diffusion of molecules in solution: simulations with the University of Houston Brownian Dynamics program. *Comput. Phys. Commun.*, **91**, 57–95.

20 Gilson, M.K. and Honig, B. (1988) Calculation of the total electrostatic energy of a macromolecular system: solvation energies, binding energies, and conformational analysis. *Proteins*, **4** (1), 7–18.

21 Sitkoff, D., Sharp, K.A., and Honig, B. (1994) Accurate calculation of hydration free energies using macroscopic solvent models. *J. Phys. Chem.*, **98**, 1978–1988.

22 Roux, B. and Simonson, T. (1999) Implicit solvent models. *Biophys. Chem.*, **78** (1–2), 1–20.

23 Vijayakumar, M. and Zhou, H.-X. (2001) Salt bridges stabilize the folded structure of barnase. *J. Phys. Chem. B*, **105**, 7334–7340.

24 Baker, N.A., Sept, D., Joseph, S., Holst, M.J., and McCammon, J.A. (2001) Electrostatics of nanosystems: application to microtubules and the ribosome. *Proc. Natl. Acad. Sci. U.S.A.*, **98** (18), 10037–10041.

25 Luo, R., David, L., and Gilson, M.K. (2002) Accelerated Poisson–Boltzmann calculations for static and dynamic systems. *J. Comput. Chem.*, **23** (13), 1244–1253.

26 Bordner, A.J., Cavasotto, C.N., and Abagyan, R.A. (2002) Accurate transferable model for water, n-octanol, and n-hexadecane solvation free energies. *J. Phys. Chem. B*, **106**, 11009–11015.

27 Dong, F., Vijayakumar, M., and Zhou, H.-X. (2003) Comparison of calculation and experiment implicates significant electrostatic contributions to the binding stability of barnase and barstar. *Biophys. J.*, **85** (1), 49–60.

28 Feig, M. and Brooks, C.L. III (2004) Recent advances in the development and application of implicit solvent models in biomolecule simulations. *Curr. Opin. Struct. Biol.*, **14** (2), 217–224.

29 Baker, N.A. (2005) Improving implicit solvent simulations: a Poisson-centric view. *Curr. Opin. Struct. Biol.*, **15**, 137–143.

30 Levy, R.M., Zhang, L.Y., Gallicchio, E., and Felts, A.K. (2003) On the nonpolar hydration free energy of proteins: surface area and continuum solvent models for the solute–solvent interaction energy. *J. Am. Chem. Soc.*, **125**, 9523–9530.

31 Dzubiella, J., Swanson, J.M., and McCammon, J.A. (2006) Coupling nonpolar and polar solvation free energies in implicit solvent models. *J. Chem. Phys.*, **124**, 084905.

32 Wagoner, J.A. and Baker, N.A. (2006) Assessing implicit models for nonpolar mean solvation forces: the importance of dispersion and volume terms. *Proc. Natl. Acad. Sci. U.S.A.*, **103** (22), 8331–8336.

33 Tan, C., Tan, Y.H., and Luo, R. (2007) Implicit nonpolar solvent models. *J. Phys. Chem. B*, **111**, 12263–12274.

34 Case, D.A., Darden, T.A., Cheatham, T.E. III, Simmerling, C.L., Wang, J., Duke,

R.E., Luo, R., Merz, K.M., Wang, B., Pearlman, D.A., Crowley, M., Brozell, S., Tsui, V., Gohlke, H., Mogan, J., Hornak, V., Cui, G., Beroza, P., Schafmeister, C., Caldwell, J.W., Ross, W.S., and Kollman, P.A. (2004) AMBER 8, University of California, San Francisco.

35 Dong, F. and Zhou, H.-X. (2006) Electrostatic contribution to the binding stability of protein–protein complexes. *Proteins*, **65** (1), 87–102.

36 Qin, S. and Zhou, H.-X. (2007) Do electrostatic interactions destabilize protein–nucleic acid binding? *Biopolymers*, **86** (2), 112–118.

37 Alsallaq, R. and Zhou, H.-X. (2008) Electrostatic rate enhancement and transient complex of protein–protein association. *Proteins*, **71**, 320–335.

38 Li, S., Yang, W., Maniccia, A.W., Barrow, D. Jr, Tjong, H., Zhou, H.-X., and Yang, J.J. (2008) Rational design of a conformation-switchable Ca^{2+}- and Tb^{3+}-binding protein without the use of multiple coupled metal-binding sites. *FEBS J.*, **275** (20), 5048–5061.

39 Tjong, H. and Zhou, H.-X. (2008) On the dielectric boundary in Poisson–Boltzmann calculations. *J. Chem. Theory Comput.*, **4**, 507–514.

40 Yi, M., Tjong, H., and Zhou, H.-X. (2008) Spontaneous conformational change and toxin binding in α7 acetylcholine receptor: insight into channel activation and inhibition. *Proc. Natl. Acad. Sci. U.S.A.*, **105** (24), 8280–8285.

41 Fauchère, J.-L. and Pliška, V. (1983) Hydrophobic parameters π of amino acid side chains from the partitioning of N-acetyl-amino acid amides. *Eur. J. Med. Chem.–Chim. Ther.*, **18**, 369–375.

42 Fredericq, E. and Neurath, H. (1950) The interaction of insulin with thiocyanate and other anions. The minimum molecular weight of insulin. *J. Am. Chem. Soc.*, **72**, 2684–2691.

43 Desbuquois, B. and Aurbach, G.D. (1974) Effects of iodination on the distribution of peptide hormones in aqueous two-phase polymer systems. *Biochem. J.*, **143** (1), 83–91.

44 Mongan, J., Case, D.A., and McCammon, J.A. (2004) Constant pH molecular dynamics in generalized Born implicit solvent. *J. Comput. Chem.*, **25** (16), 2038–2048.

45 Lide, D.R. (ed.) (2009) *CRC Handbook of Chemistry and Physics*, 89th edn, CRC Press/Taylor and Francis, Boca Raton, FL.

46 Still, W.C., Tempczyk, A., Hawley, R.C., and Hendrickson, T. (1990) Semianalytical treatment of solvation for molecular mechanics and dynamics. *J. Am. Chem. Soc.*, **112** (16), 6127–6129.

47 Bashford, D. and Case, D.A. (2000) Generalized Born models of macromolecular solvation effects. *Annu. Rev. Phys. Chem.*, **51**, 129–152.

48 Lee, M.S., Salsbury, F.R. Jr, and Brooks, C.L. III (2002) Novel generalized Born methods. *J. Chem. Phys.*, **116** (24), 10606–10614.

49 Im, W., Lee, M.S., and Brooks, C.L. III (2003) Generalized Born model with a simple smoothing function. *J. Comput. Chem.*, **24** (14), 1691–1702.

50 Lee, M.S., Feig, M., Salsbury, F.R. Jr, and Brooks, C.L. III (2003) New analytic approximation to the standard molecular volume definition and its application to generalized Born calculations. *J. Comput. Chem.*, **24** (11), 1348–1356.

51 Feig, M. and Brooks, C.L. III (2004) Recent advances in the development and application of implicit solvent models in biomolecule simulations. *Curr. Opin. Struct. Biol.*, **14** (2), 217–224.

52 Feig, M., Onufriev, A., Lee, M.S., Im, W., Case, D.A., and Brooks, C.L. III (2004) Performance comparison of generalized Born and Poisson methods in the calculation of electrostatic solvation energies for protein structures. *J. Comput. Chem.*, **25** (2), 265–284.

53 Gallicchio, E. and Levy, R.M. (2004) AGBNP: an analytic implicit solvent model suitable for molecular dynamics simulations and high-resolution modeling. *J. Comput. Chem.*, **25** (4), 479–499.

54 Mongan, J., Simmerling, C., McCammon, J.A., Case, D.A., and Onufriev, A. (2007) Generalized Born model with a simple, robust molecular volume correction. *J. Chem. Theory Comput.*, **3** (1), 156–169.

55 Mongan, J., Svrcek-Seiler, W.A., and Onufriev, A. (2007) Analysis of integral

expressions for effective Born radii. *J. Chem. Phys.*, **127** (18), 185101.

56 Tjong, H. and Zhou, H.-X. (2007) GBr^6: a parameterization-free, accurate, analytical generalized Born method. *J. Phys. Chem. B*, **111**, 3055–3061.

57 Paul, L. (2008) The generalized Born/volume integral implicit solvent model: estimation of the free energy of hydration using London dispersion instead of atomic surface area. *J. Comput. Chem.*, **29** (10), 1693–1698.

58 Tjong, H. and Zhou, H.-X. (2008) Accurate calculations of binding, folding, and transfer free energies by a scaled generalized Born method. *J. Chem. Theory Comput.*, **4** (10), 1733–1744.

9
Fast Analytical Continuum Treatments of Solvation
François Marchand and Amedeo Caflisch

9.1
Introduction

Successful applications of molecular dynamics (MD) to study the structure and function of a biomolecule depend on the quality of the underlying force field and the sampling efficiency of the simulation protocol. In particular, an accurate representation of the aqueous solvent environment is important to reproduce the structural, functional, and dynamic behavior of soluble biomolecules. The most realistic and physically rigorous way to treat solvation effects is to include explicitly the solvent molecules in the simulation system, at the price of high computational cost. In fact, the solvent molecules greatly increase the number of degrees of freedom and interaction centers. Even with today's computational infrastructure, simulations of single-domain proteins (about 100 residues) cannot sample more than 0.1–1 µs. Such a short time scale prohibits the study of long-time processes like protein folding, large-scale structural transitions, multimeric assembly processes like complex formation and protein aggregation, as well as the derivation of accurate thermodynamic quantities. This computational drawback has motivated the development of fast implicit solvent models [1–3], where the mean influence of solvent molecules around the solute is described by a potential of mean force that depends only on the atom coordinates of the solute [2, 4]. An implicit solvent model not only considerably reduces the system size, but also avoids the need to average over the extremely large number of solvent configurations, and reduces the viscosity of the solvent environment by eliminating the friction from the solvent molecules, thus accelerating molecular motions [5]. Furthermore, such a model directly yields the so-called effective energy, which is the sum of the solute potential energy *in vacuo* and the solvation free energy. In contrast, explicit water simulations have to be post-processed, for example by finite-difference Poisson–Boltzmann calculations, to obtain the effective energy.

The overall free energy cost of solvating a solute molecule (ΔG_{solv}) is decomposed into a non-polar component and a polar component in most implicit solvent models [2]: $\Delta G_{solv} = \Delta G_{pol} + \Delta G_{nonpol}$. The term ΔG_{pol} is the free-energy change in

Modeling Solvent Environments. Applications to Simulations of Biomolecules. Edited by Michael Feig
Copyright © 2010 WILEY-VCH Verlag GmbH & Co. KGaA, Weinheim
ISBN: 978-3-527-32421-7

the system resulting from the electrostatic interactions. The reorientation and polarization of the individual molecules in the medium cause the solvent to act on the polar contribution in two ways: first, it interacts directly with the individual charges, giving rise to the so-called self-energy contribution to the total free energy of the system; and second, it screens the strength of the Coulomb interactions between charges in the macromolecule. The term ΔG_{nonpol} is the free energy of introducing the solute into the solvent when the electrostatic interactions between the solute and solvent are turned off. It can be further decomposed into a cavity formation term (ΔG_{cav}) and a solute–solvent van der Waals dispersion term (ΔG_{VDW}) [6].

Implicit solvent models can be classified into three main families: surface area models [7–9] (which are the simplest and were developed first), Gaussian solvent-exclusion models [10, 11], and dielectric continuum electrostatics models. The latter can be further classified into finite-difference Poisson–Boltzmann (PB) [12] and generalized Born (GB) [13, 14] models. While PB models are more accurate than GB models, they suffer from high computational cost and difficulties in the derivation of forces. The GB model is related to the PB model but contains several approximations that increase the speed of calculation. There exist models that combine the different approximations, like "generalized Born surface area" (GBSA) [15–17], where the polar part is treated through the GB formalism and the non-polar part through a surface area term.

Here, two fully analytical implicit solvent models are reviewed, the SASA ("solvent-accessible surface area") model [18], a surface area model, and the FACTS ("fast analytical continuum treatment of solvation") model [19], a recently introduced GBSA method. SASA and FACTS are both very efficient (only about 1.5 and 4 times slower, respectively, than *in vacuo*) and have been implemented in CHARMM [20]. Because they are fully analytical energy functions, analytical force vectors and the Hessian matrix of second derivatives, which is used in techniques like normal-mode analysis (NMA) [21, 22], can be derived.

9.2
The SASA Implicit Solvent Model: A Fast Surface Area Model

It is assumed that the main contributions to the solvation energy are proportional to the solvent-accessible surface area (SASA) [7] or solvent-accessible volume [10]. Several parameterizations have been proposed in the past [7–9, 23]. The SASA model implemented in CHARMM makes use of a very efficient analytical evaluation of the SASA [24] and was parameterized for the polar hydrogen force field (param19). In SASA, electrostatic screening effects are approximated by a distance-dependent dielectric function and ionic groups are neutralized [11]. The surface area approximation is used for the direct solvation (both polar and non-polar) as introduced by Eisenberg and McLachlan [7]. Because an exact analytical or numerical computation of the SASA is too slow to compete with simulations in explicit solvent, an approximate analytical expression [24] was used. This drastically

9.2 The SASA Implicit Solvent Model: A Fast Surface Area Model

reduces the computational cost with respect to an explicit solvent simulation. The model discussed here is based on the assumptions that most of the solvation energy arises from the first water shell around the protein [7], and that two atomic solvation parameters are sufficient to describe the solvation of polar groups (negative, that is, favorable, surface tension-like parameter) and non-polar groups (positive surface tension-like parameter).

Here a description of the SASA implicit solvent model and its calibration is given. This is followed by a discussion on the limitations of the model and a review of its application in studies of conformational transitions of structured peptide, aggregation of peptidic systems, and ligand–receptor interactions.

9.2.1
Description of the Model

In most empirical force fields, the Hamiltonian of the solute–solvent system is additive and consists of the sum of solute–solute, solute–solvent, and solvent–solvent terms. After integration over the solvent coordinates, the potential of mean force $W(r)$, or effective energy, can be divided into two contributions,

$$W(r) = E_{\text{solute}}(r) + G_{\text{solv}}(r) \tag{9.1}$$

for a solute having N atoms with Cartesian coordinates $r = (r_i, ..., r_N)$. The term "effective energy" for $W(r)$ is used here as in Ref. [11]; it is the sum of intra-solute and mean solvation terms. In the present study, we assume that the mean solvation energy is linearly related to the SASA of the solute:

$$G_{\text{solv}}(r) = \sum_{i=1}^{N} \sigma_i S_i(r) \tag{9.2}$$

where σ_i and $S_i(r)$ are the atomic solvation parameter and SASA of atom i, respectively. The SASA $S_i(r)$ is computed by an approximate analytical expression [24]:

$$S_i(r) = A_i \prod_{j \neq i}^{N} [1 - p_i p_{ij} b_{ij}(r_{ij})/A_i] \tag{9.3}$$

where A_i denotes the SASA of an isolated atom i of radius R_i,

$$A_i = 4\pi(R_i + R_{\text{probe}})^2 \tag{9.4}$$

and R_{probe} is the radius of the solvent probe. In Equation 9.3, $b_{ij}(r_{ij})$ represents the SASA removed from A_i due to the overlap between atoms i and j separated by a distance $r_{ij} = |r_i - r_j|$ and is given by

$$b_{ij}(r_{ij}) = \begin{cases} 0 & \text{if } r_{ij} > R_i + R_j + 2R_{\text{probe}} \\ \pi(R_i + R_{\text{probe}})(R_i + R_j + 2R_{\text{probe}} - r_{ij})[1 + (R_j - R_i)r_{ij}^{-1}] \\ \text{otherwise} \end{cases} \tag{9.5}$$

Using 270 small molecules, the atom type parameters p_i and connectivity parameters p_{ij} have been optimized to reproduce the exact SASA with $R_{\text{probe}} = 1.4\,\text{Å}$

[24]. The complete list of parameters can be found in the original publication [18].

The SASA model includes the free-energy cost of burying a charged residue in the interior of a protein. However, it does not take into account the solvent screening on the interactions between solute charges. This effect is approximated here using a distance-dependent dielectric function, $\varepsilon(r) = 2r$ and was chosen instead of $\varepsilon(r) = r$ mainly to reduce the strength of the hydrogen bonds. Larger values of the dielectric constant are expected to lead to partial unfolding of proteins in simulations at room temperature. A cutoff for long-range interactions is used (see below), so that a linear distance-dependent dielectric function does not differ significantly from a more sophisticated one, such as a sigmoidal function [25–27], because the deviation from linearity is negligible for distances smaller than 10 Å [27]. A distance-dependent dielectric function is a very simplified way of accounting for the solvent polarization effects on the solute. In particular, the screening of the electrostatic interactions between charged groups is insufficient, as shown by the formation of too stable salt bridges in MD simulations of the RGDW (Arg–Gly–Asp–Trp) peptide [28]. The limitations of this approximation can be partly overcome by using a set of partial charges with a zero total charge for every residue. In the current SASA implementation, the ionizable amino acids are neutralized [11].

The solvation model has been implemented in CHARMM and is used with a polar hydrogen CHARMM force field (param19), where the only modified parameters are the partial charges of the ionic side chains [11]. The CHARMM param19 default cutoff for long-range interactions is used (that is, a shift function [20] is used with a cutoff at 7.5 Å for both the electrostatic and van der Waals terms). This cutoff length was chosen to be consistent with the original parameterization of CHARMM param19 [20]. Even though the SASA solvation term is calculated at every dynamics step, the CPU time required for simulations with SASA is only about 50% larger than that for a simulation *in vacuo* with the same cutoff (7.5 Å).

As in a previous work [9], only two σ parameters are considered in SASA: one for carbon and sulfur atoms ($\sigma_{C,S} > 0$), and one for nitrogen and oxygen atoms ($\sigma_{N,O} < 0$). The solvation parameter of the hydrogen atoms is set to zero. The two σ parameters were optimized from 1 ns MD simulations of six small proteins at 300 K by a trial-and-error approach. The two resulting σ values that gave the minimal C_α root-mean-square deviation (RMSD) from the native state are 0.012 kcal mol^{-1} Å^{-2} for carbon and sulfur atoms, and –0.060 kcal mol^{-1} Å^{-2} for nitrogen and oxygen atoms, and correspond to those determined previously [9]. With this parameter set and the CHARMM param19 force field, SASA seems to correctly model the strength of hydrogen bonds; the MD simulations of the six small proteins showed that SASA closely reproduces the number of hydrogen bonds present in the respective X-ray structure, while other solvent models or electrostatic treatment (often) overestimate it [18, 29]. A correct treatment of the strength of hydrogen bonds is important to obtain meaningful energetics in folding–unfolding studies.

9.2.2
Applications of the SASA Implicit Solvent Model

9.2.2.1 Reversible Folding of Structured Peptides

The combination of the fast implicit solvent SASA with the united-atom force field param19, which results in a 40% reduction of the number of atoms compared to all-atom force fields and makes use of short cutoffs, allows the fast and extensive sampling of the conformational space of small to medium-sized systems. One system that was thoroughly studied is β3s, a designed 20-residue peptide whose solution conformation has been investigated by nuclear magnetic resonance (NMR) spectroscopy [30]. The NMR data indicate that β3s in aqueous solution forms a monomeric triple-stranded antiparallel β-sheet, in equilibrium with the denatured state. In Rao and Caflisch [31] the conformations sampled during long equilibrium folding–unfolding MD simulations (>10 µs in total) were mapped onto a network, with nodes representing clusters of similar conformations and links representing the observed transitions between nodes. With this representation, free-energy minima and their connectivity emerge without requiring projections onto arbitrarily chosen reaction coordinates (Figure 9.1). As previously observed for a variety of networks as diverse as the Internet and the protein interactions within a cell, the conformational space network of polypeptide chains is a scale-free network, that is, the distribution of the number of possible connections of a conformation follows a power law. Interestingly, a correlation was found between the statistical weight (size of the node) and connectivity (number of links to a node) – the most connected nodes are also low-lying minima on the free-energy landscape.

Another observation was that the native basin of the structured peptide shows a hierarchical organization of conformations. Such an organization was not observed for a random heteropolymer that lacks a native state (that is, a predominant free-energy minimum) [31]. The network projection allows the representation of the complexity of the denatured state ensemble, which is very heterogeneous and includes high-entropy, high-enthalpy conformations as well as low-entropy, low-enthalpy traps. Furthermore, the network properties were used to identify transition-state conformations and two main average folding pathways. Such a complexity of the conformational space and kinetic pathways disappears in conventional projections onto one or two progress variables [32]. Other applications of SASA include the investigation of the folding mechanism of structured peptides [33–36] and small proteins [37], as well as the reversible mechanical unfolding of a helical peptide [38]. MD simulations with SASA were used to interpret the kinetic behavior of a photo-switchable cross-linked α-helical peptide [39] and to test a new method to compute the density of states of proteins [40].

9.2.2.2 Peptide Aggregation

SASA was also successfully applied in aggregation studies of amyloidogenic peptides. Simulations of the early steps of aggregation of amyloid-forming peptides

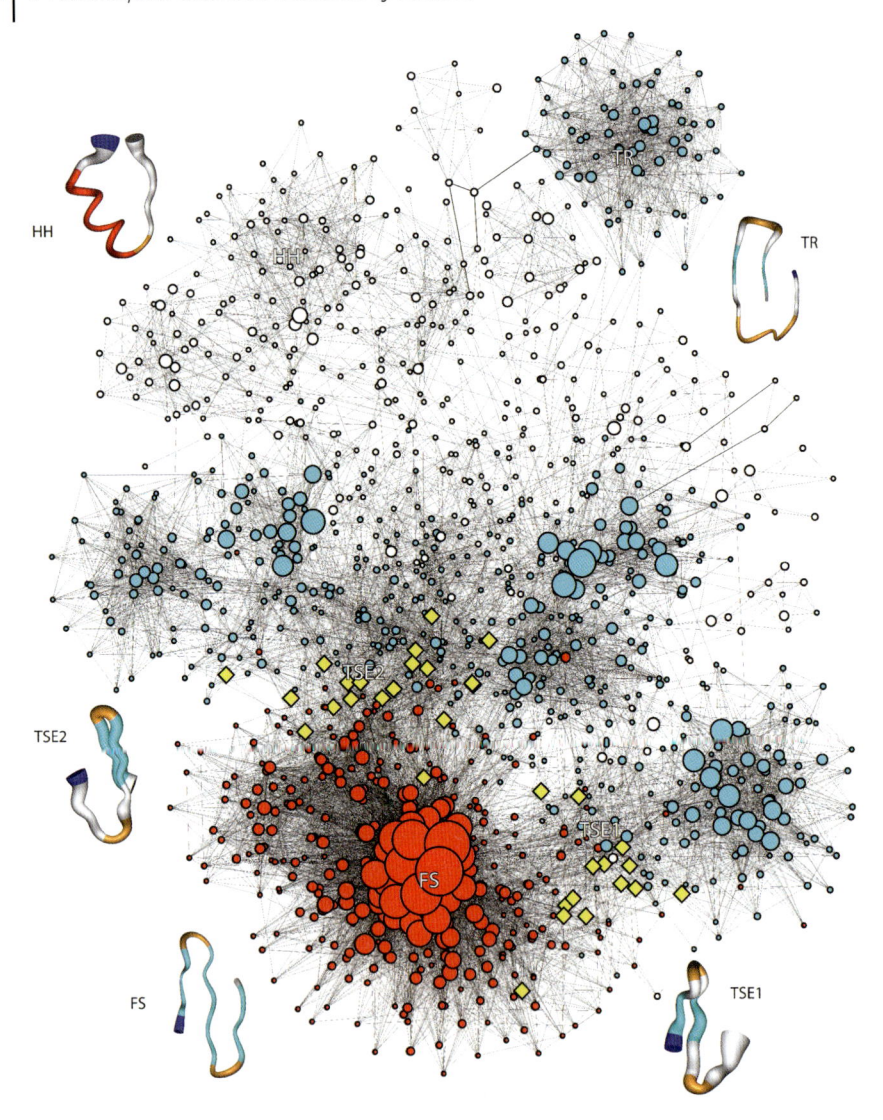

Figure 9.1 Conformational space network of the designed three-stranded antiparallel β-sheet peptide β3s. Nodes represent conformations and links represent transitions between them, as sampled during 10 μs implicit solvent molecular dynamics simulations at the melting temperature of 330 K. The size and color of the nodes reflect the statistical weight and average neighbor connectivity, respectively [31]. Representative conformations are shown by a pipe colored according to secondary structure: white for coil, red for α-helix, orange for turn or bend, cyan for β-strand, and blue for the N-terminus. The variable radius of the pipe reflects the structural variability of the snapshots within a node. The yellow diamonds are folding transition state conformations. HH, TR, TSE, and FS are the helical, trap, transition state ensemble, and folded states, respectively. Reproduced from [31] with permission from Elsevier.

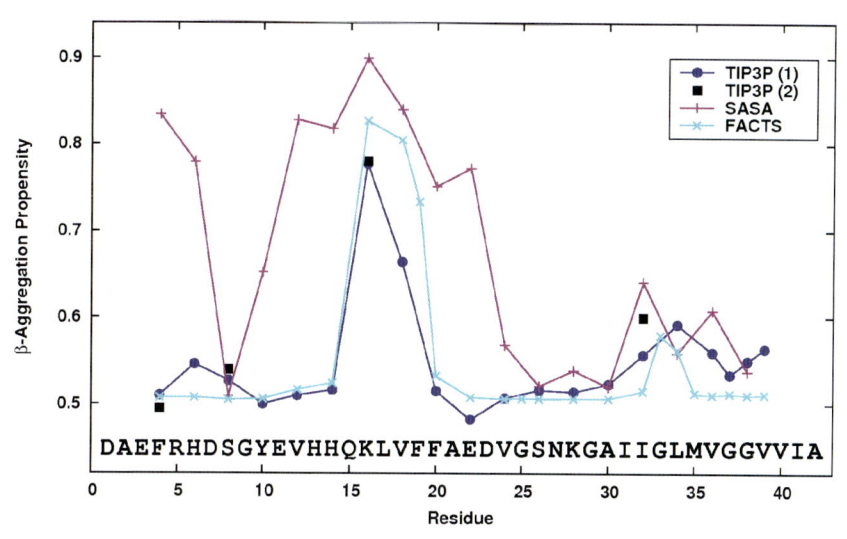

Figure 9.2 The β-aggregation propensity profile of the Alzheimer's amyloid-β peptide (Aβ$_{42}$). The peptide was decomposed into a set of overlapping heptapeptide segments, each shifted by two residues along the sequence, and three copies of each segment were simulated at 310 K with param22 and TIP3P explicit water (blue), SASA (magenta), and FACTS param19 with $\varepsilon_m = 1$ and a surface tension-like parameter $\gamma = 0.015$ kcal mol^{-1} Å^{-2} (cyan). Data from previous explicit water simulations were taken from Cecchini et al. [43] (black squares). The simulations were started from a parallel in-register arrangement in the explicit water simulations and from a conformation in which all peptides were isolated from each other in the implicit solvent simulations. The β-aggregation propensity was calculated as the average of the nematic order parameter $\overline{P}_2$ [44, 45], which was previously shown to describe the orientational order of the system and discriminate between ordered and disordered conformations [46]. All simulations identify the central hydrophobic cluster H$_{13}$HQKLVFFA$_{21}$ as a strong aggregation-promoting sequence.

using the SASA model have provided evidence of the importance of side-chain interactions [41, 42]. Cecchini et al. devised a strategy where an amyloidogenic sequence is decomposed into overlapping short stretches, and then long MD simulations of multiple copies of each stretch are run in order to sample their tendency to build ordered parallel aggregates [43]. The resulting amyloidogenicity profiles highlight so-called aggregation "hot-spots", short stretches that promote aggregation of full-length sequences (Figure 9.2). For one such system, the yeast prion Ure2p, this method was used to predict a double-point mutant with lower β-aggregation propensity that was later confirmed by an experimental test [43].

9.2.2.3 Other Applications

The SASA implicit solvent model was also used in the characterization of the unbinding mechanism of odorant molecules from the odorant binding protein

(OBP) [47]. MD simulations with SASA allowed the identification of a consensus pathway for thymol unbinding from a rat OBP and the description of the associated conformational changes in the receptor. The binding affinity calculated from potentials of mean force (PMFs) was comparable to the value measured by isothermal titration calorimetry.

9.2.3
Limitations of the SASA Implicit Solvent Model

The SASA model is not expected to describe correctly the stability of large proteins: first, because the screening between partial charges does not depend on the local environment; and second, because it is unaffected by atoms that are near the surface but remain completely inaccessible to solvent. SASA should work best for small systems where most or all atoms are at least partially exposed. Limitations in modeling correctly the behavior of large systems is exemplified in Ferrara et al. [18], where MD simulations of three proteins—barnase (1a2p, 110 residues), hen egg-white lysozyme (1hel, 129 residues), and cutinase (1cus, 197 residues)—result in fast unfolding even at 300 K, with a C_α RMSD from their respective native conformation above 3.5 Å. The fact that SASA is not appropriate for large systems is also reflected in its evaluation as a scoring function for the CASP4 protein structure prediction competition, where its performance was lower than that of more sophisticated approaches like Poisson–Boltzmann or generalized Born methods [48].

Limitations for small systems were also apparent in the simulation of two designed mini-protein motifs BBA5 and $\alpha t \alpha$. BBA5 is a 23-residue peptide with a $\beta\beta\alpha$ architecture [49], whereas $\alpha t \alpha$ is a 38-residue peptide designed to adopt a helical hairpin conformation in aqueous solution [50]. During simulations of the two systems at 280 K starting from their respective native conformation, the C_α RMSD relative to the native state rapidly increased above 4 Å. Detailed analysis of the structural deviation showed that, while the distinct secondary structure elements were mostly preserved (the β-hairpin and the helix in BBA5, and helix 1 and helix 2 in $\alpha t \alpha$), most of the deviation arises from the loss of their respective native tertiary structures (the packing of the hairpin and helix in BBA5, and the packing of helix 1 and helix 2 in $\alpha t \alpha$). Another source of deviation was an increased content of π-helicity. It is likely that the largest error in implicit solvent models originates from the treatment of the charged groups. The use of a distance-dependent dielectric function and the EEF1 modifications of the CHARMM param19 force field, where ionic side chains are neutralized, lead to a rather crude approximation of the electrostatic contribution. This can be a major source of error for BBA5 and $\alpha t \alpha$, since both have a high charge density (7 and 15, respectively). On the other hand, it was experimentally observed that, at pH 10.5, $\alpha t \alpha$, which has four lysine residues, is largely disordered even at a temperature of 5 °C [51]. Finally, BBA5 was not stable in a 300 K MD simulation with explicit water and the AMBER force field with an 8 Å cutoff, whereas it was stable with a 10 Å cutoff [52].

9.3
The FACTS Implicit Solvent Model: A Fast Generalized Born Approach

The limitations of the SASA model motivated the development of the FACTS implicit model [19]. FACTS implements a more rigorous treatment of electrostatic solvation, which does not require the neutralization of ionic groups, and also takes into account the degree of solvent exposure for the calculation of screening effect.

Despite the significant variability of the dielectric constant in the interior of a protein molecule [53, 54], several implicit solvent models are based on the assumption that the protein is a uniform, low-dielectric region. The essential approximation in such continuum electrostatics models is to represent the solvent as a featureless high-dielectric medium, and the macromolecule as a region with a low-dielectric constant and a spatial charge distribution [1–4, 55–62]. In this way, the solvent degrees of freedom and interaction centers are not taken into account explicitly. The Poisson equation would provide an exact description of such a solute–solvent system, and its numerical solution, obtained either by a finite-difference algorithm [12, 63–65] or by a boundary-element algorithm [66–69], is more efficient than the explicit treatment of the solvent, but still not fast enough for effective utilization in computer simulations of macromolecules. Based on the Born model for ionic solvation [70], the generalized Born (GB) model extends this formalism to treat solutes containing multiple charged particles and an arbitrarily shaped molecular surface. The GB formalism has become an efficient method for the evaluation of continuum electrostatic energies [14] and approximates the PB electrostatic solvation energy as an efficient pairwise summation that allows analytical force calculations:

$$\Delta G^{el,GB} = -\frac{1}{2}\tau \sum_{i,j=1}^{N} \frac{q_i q_j}{\sqrt{r_{ij}^2 + R_i R_j \exp(-r_{ij}^2/\kappa R_i R_j)}} \tag{9.6}$$

where r_{ij} is the distance between charges q_i and q_j, $r_{ii} = 0$, the constant κ is usually set to 4 or 8, $\tau = 1/\varepsilon_m - 1/\varepsilon_s$, and N is the number of atoms in the solute. The volume occupied by the solute is assigned a low dielectric constant ε_m (typically 1, 2, or 4) and the charge distribution is defined by the partial charges of the solute atoms. The solvent is replaced by a uniform medium with a high dielectric constant ε_s (typically 78.5 or 80 in the case of water).

The effective Born radius, R_i, is a key quantity in the GB formalism. It measures the degree of burial of individual solute charges and corresponds to the distance between a particular atom and its hypothetical spherical dielectric boundary, chosen such that the self (or atomic) electrostatic solvation energy, ΔG_i^{el}, satisfies the Born equation [70]:

$$\Delta G_i^{el} = -\frac{\tau q_i^2}{2R_i} \tag{9.7}$$

In principle, the "exact" effective Born radii can be calculated from Equation 9.7 using the self electrostatic solvation energy obtained through the PB theory, but this would bring no computational advantage. One key observation was that Equation 9.6 yields very accurate results if ΔG_i^{el} (or equivalently R_i) is a good approximation of the value obtained by solving the PB equation [71].

The first generation of GB models used the Coulomb field approximation for the evaluation of ΔG_i^{el}, where the electric displacement $\boldsymbol{D}_i$ for each atom i is calculated by supposing that the hypothetical dielectric boundary is spherical and that atom i lies at the center of this sphere. A large variety of procedures for calculating effective Born radii within the Coulomb field approximation have been presented. These include numerical surface or volume integrations [14, 72–75], analytical integral expressions [15], and pairwise summation approximations [76–78]. After recognizing that the Coulomb field approximation was a major source of deviation from PB values [3, 79], corrections to the Coulomb field approximation have been suggested and shown to greatly increase the accuracy of the effective Born radii [16, 74, 75, 80]. Current accurate GB implementations are between 10 and 20 times slower than simulations *in vacuo* [81]. Moreover, for proteins of about 100 residues, the computational cost per MD time step is about the same for accurate GB models and explicit water simulations with periodic boundary conditions [75].

In FACTS, the self electrostatic solvation energy and SASA of individual atoms are calculated using intuitive geometric properties of the solute whose evaluation requires only solute interatomic vectors. For each solute atom, the volume and spatial symmetry of its neighboring atoms (or, equivalently, of the solvent displaced by the neighboring atoms) are approximated (Figure 9.3). A combination of these two measures is used as variable of a sigmoidal function (see below). The parameters of the sigmoidal function, together with those of the linear combination with cross-term, are derived by fitting to atomic electrostatic solvation energy values calculated by numerical solution of the PB equation. The GB formula (Equation 9.6) is used to obtain the electrostatic solvation free energy of the macromolecule. The FACTS model does not assume the Coulomb field approximation and does not require the definition of a dielectric discontinuity surface (such dielectric boundary is only required to calculate the PB reference data to which the parameters of the FACTS model are fitted). The same two measures of solvent displacement are combined and used in another sigmoidal function to estimate the SASA of individual atoms. The parameters of the sigmoidal function are derived by fitting to SASA values calculated by an exact analytical method [82]. Finally, the non-polar contribution to the solvation free energy is assumed to be proportional to the sum of the atomic SASA values [7, 8].

Both electrostatic solvation energy and SASA are determined using the same geometrical properties and analytical framework, which makes FACTS a comprehensive and efficient implicit solvation model. When compared with one of the most accurate GB methods [81], like "generalized Born using molecular volume" (GBMV) [75], it is shown that solvation energies computed by FACTS are of similar accuracy to the best available GB implementations, and MD simulations with FACTS are only four times slower than using the *in vacuo* energy.

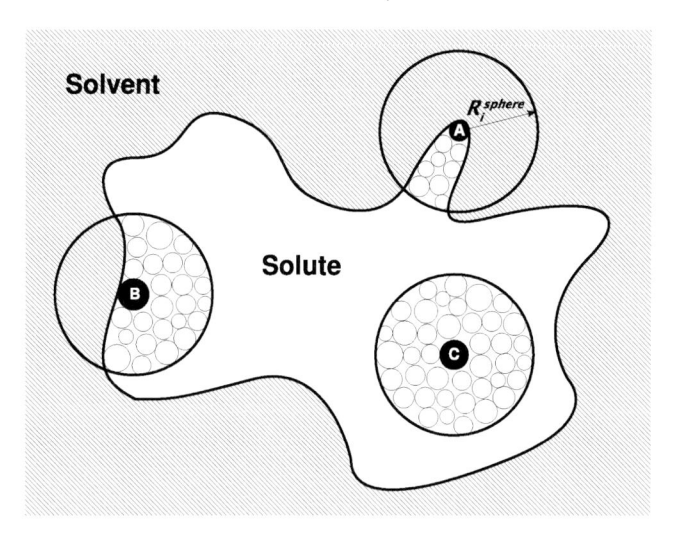

Figure 9.3 Schematic illustration of the measure of degree of burial in FACTS. The solvent exposure of an atom depends on the location of neighboring atoms. The large circle represents the sphere of radius R_i^{sphere} used in FACTS to quantify the atomic solvation energy. FACTS computes the volumes occupied by the neighboring atoms inside the sphere and the symmetry of their spatial distribution. For example, for atom A (which may be part of a protruding side chain), the volume occupied by neighboring atoms is small and the symmetry is low, so that its atomic solvation energy ΔG_A^{el} is large. The surface atom B has an intermediate measure of volume and a low symmetry value, which result in an intermediate ΔG_B^{el}. Atom C is completely buried, its volume and symmetry values are maximal, and ΔG_C^{el} vanishes to zero.

9.3.1
Description of the Model

9.3.1.1 Atomic (or Self) Electrostatic Solvation Energy

The essential idea in FACTS is that the electrostatic solvation free energy of atom i, ΔG_i^{el}, is evaluated by considering a sphere of radius R_i^{sphere} around atom i [19]. The radius is large enough so that the atom distribution outside the sphere has only a negligible effect on ΔG_i^{el}. If only atom i of the macromolecule were present within the sphere of radius R_i^{sphere}, solving the Poisson equation would result in $\Delta G_i^{\text{el}} \cong -\tau q_i^2 / 2 r_i^{\text{VDW}}$. As more and more atoms are gradually added (as for atoms A and B in Figure 9.3), ΔG_i^{el} becomes less favorable depending in a complex way on the spatial distribution of the additional atoms. When all the solvent has finally been flushed out from within the sphere (as for atom C in Figure 9.3), solving the Poisson equation would result in $\Delta G_i^{\text{el}} \cong 0$.

The solvent exposure of a given atom i in a macromolecule depends obviously on the volume occupied by neighboring atoms, a second factor being the distance between the atom i and a neighboring atom j; atom i will be more desolvated by a close atom j than by a distant one. Then, for a given number of neighboring atoms inside the sphere of radius R_i^{sphere} with fixed distances from atom i, numerous different atomic distributions can be obtained by rotations of the individual neighboring atoms (especially in three dimensions), and each of these configurations may result in a different electrostatic solvation free energy of atom i. To account for this effect, FACTS introduces a measure of symmetry. In Figure 9.3 atom B lies at the surface of a protein, directly contacting the dielectric discontinuity surface, and is therefore well exposed to solvent. In this atomic configuration, most neighboring atoms are packed in one side of the sphere R_i^{sphere} and the atomic distribution is thus highly asymmetric. One can imagine different configurations with more symmetric atom distributions, where atom i is no longer in direct contact with the solvent and is thus more shielded from it, resulting in a less favorable solvation free energy ΔG_i^{el}. It should be noted that the boundary between the macromolecular and solvent environments in the PB evaluation is defined by the molecular surface [83] and, contrary to the van der Waals surface, it avoids the presence of solvent inside interstitial volumes and microcavities.

To cast the above ideas into mathematical form, the abbreviations $x_{ij} = x_i - x_j$, $r_{ij} = |x_{ij}|$, and $\hat{x}_{ij} = x_{ij}/r_{ij}$ are introduced. The measure of *volume* occupied by the solute around atom i is defined by

$$A_i = \sum_{j=1, j \neq i}^{N} V_j \Theta_{ij} \tag{9.8}$$

and the measure of *symmetry* by

$$B_i = \frac{\left| \sum_{j=1, j \neq i}^{N} (V_j/r_{ij}) \Theta_{ij} \hat{x}_{ij} \right|}{1 + \sum_{j=1, j \neq i}^{N} (V_j/r_{ij}) \Theta_{ij}} \tag{9.9}$$

where

$$\Theta_{ij} := \begin{cases} \left[1 - (r_{ij}/R_i^{\text{sphere}})^2\right]^2 & r_{ij} \leq R_i^{\text{sphere}} \\ 0 & r_{ij} > R_i^{\text{sphere}} \end{cases} \tag{9.10}$$

The measure of volume A_i is simply the sum of the van der Waals volumes V_j of the atoms surrounding atom i within the sphere, weighted by Θ_{ij}. Typically A_i ranges between 100 and 2000 Å^3 in a sphere of radius $R_i^{\text{sphere}} \cong 10$ Å.

The measure of symmetry B_i is a weighted Euclidean norm of the sum of the unit vectors pointing from the central atom i to the neighboring atoms. Thereby each unit vector is weighted by Θ_{ij}, and additionally by the volume of the neighboring atom V_j to which it points, divided by its distance r_{ij} from atom i. There is no other reason for the additional weighting factor V_j/r_{ij} except for the fact that it was found to improve the correlation between the values of B_i and atomic solvation energies calculated by PB. The value of B_i is normalized to range between 0 and 1. For a fully symmetric distribution, B_i equals 0; whereas for a totally asymmetric distribution (for example, only one neighboring atom), B_i is close to 1. The additive constant of 1 in the denominator of Equation 9.9 prevents the denominator becoming zero for a completely isolated ion.

The purpose of the function Θ_{ij} is two-fold: weighting and smoothing. The function Θ_{ij} is equal to 1 for $r_{ij} = 0$ and drops continuously until Θ_{ij} equals 0 at $r_{ij} = R_i^{sphere}$. Thus, on the one hand, Θ_{ij} accounts for the fact that, the further an atom is placed from atom i, the less it influences its solvation energy. On the other hand, Θ_{ij} ensures the existence of continuous (first and second) derivatives.

When the PB-derived atomic solvation energies $\Delta G_i^{el,PB}$ for unit charges are plotted against A_i and B_i, a sigmoidal distribution of data is observed (Figure 9.4). Therefore, the measures of volume and symmetry are combined linearly and by a cross-term into a single measure of solvent displacement

$$C_i = A_i + b_1 B_i + b_2 A_i B_i \qquad (9.11)$$

and a sigmoidal shaped function of C_i is used to calculate the electrostatic solvation energy $\Delta G_i^{el,FACTS}$ of atom i for a unit charge:

$$\Delta G_i^{el,FACTS} = a_0 + \frac{a_1}{1+e^{-a_2(C_i-a_3)}} \qquad (9.12)$$

The parameters a_0 and a_1 are determined using the limiting cases of a fully buried and fully exposed atom. In the case of a fully buried atom (that is, $C_i \to +\infty$), the value of ΔG_i^{el} should vanish, which implies that $a_0 = -a_1$ and $a_2 > 0$. For a fully exposed atom (that is, $C_i \to 0$), the Born formula applies, so that $a_0 = -(\tau/2r_i^{VDW})(1+e^{-a_2 a_3})$ for a unit charge. Hence, for each van der Waals radius, the five parameters b_1, b_2, a_2, a_3, and R^{sphere} have to be determined by an optimization procedure. The sigmoidal function (Equation 9.12) gives an accurate fit to $\Delta G_i^{el,PB}$ (Figure 9.4). Intuitively, C_i measures the solvent displacement around atom i, and the solvation energy of atom i is a sigmoidal function of this measure.

9.3.1.2 Total Electrostatic Solvation Energy

The total electrostatic solvation energy in the FACTS model is the sum of the atomic self-energies and the GB interaction term:

$$\Delta G^{el,FACTS} = \sum_{i=1}^{N} \Delta G_i^{el,FACTS} - \tau \sum_{1 \le i < j \le N} \frac{q_i q_j}{\sqrt{r_{ij}^2 + R_i^{FACTS} R_j^{FACTS} \exp\left(-r_{ij}^2/\kappa R_i^{FACTS} R_j^{FACTS}\right)}} \qquad (9.13)$$

Here $\Delta G_i^{el,FACTS}$ is calculated according to Equation 9.12, and $R_i^{FACTS} = -\tau q_i^2/2\Delta G_i^{el,FACTS}$, with N being the number of atoms in the macromolecule. Equivalently,

$$\Delta G^{el,FACTS} = -\frac{1}{2}\tau \sum_{i,j=1}^{N} \frac{q_i q_j}{\sqrt{r_{ij}^2 + R_i^{FACTS} R_j^{FACTS} \exp\left(-r_{ij}^2/\kappa R_i^{FACTS} R_j^{FACTS}\right)}} \qquad (9.14)$$

with $r_{ii} = 0$. Note that the second sum in Equation 9.13 implies an infinite cutoff, while a truncation scheme (shifting [20]) is used in the FACTS implementation. Also, a multiplicative factor of 332.0716 is used in front of τ to obtain energy values in kilocalories per mole (kcal mol^{-1}) with interatomic distances in Ångström (Å) and partial charges in electronic units (e).

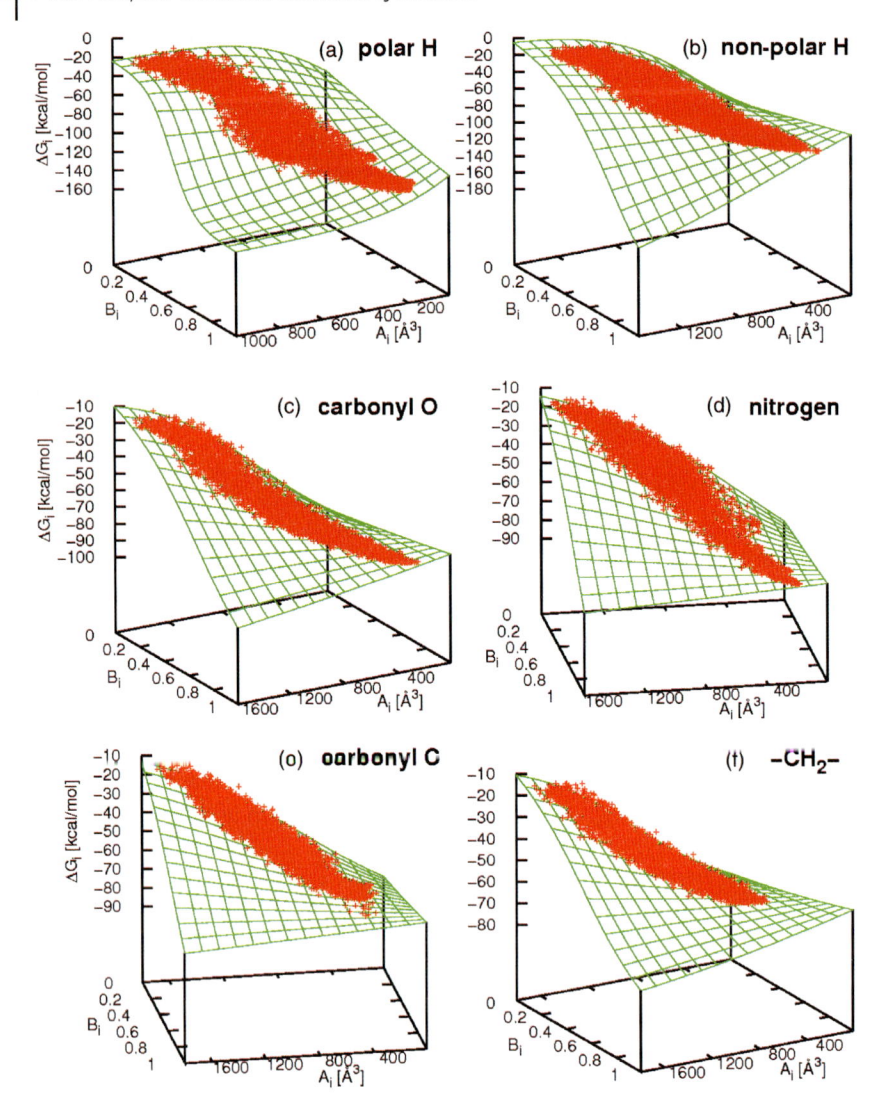

Figure 9.4 The green surface represents Equation 9.12, that is, the FACTS atomic electrostatic solvation energy as a function of A_i and B_i for param22. The red data points are atomic solvation energy values calculated by PB using unit charges and $\varepsilon_m = 1$. The dependence on the symmetry is more pronounced for the polar hydrogen atoms (a) than for the aliphatic carbon atoms (f) because the latter are almost always buried, whereas the former have similar probabilities to be buried or exposed. Note that a fully symmetric distribution yields $B_i = 0$.

9.3.1.3 Atomic Solvent-Accessible Surface Area

Estimating the amount and symmetry of the solvent that is displaced around a given atom provides information on how much the atom is accessible to solvent. Therefore, the geometric concepts described above for approximating the atomic electrostatic solvation energy offer a straightforward way to approximate the SASA of atom i, S_i, by taking into account the relative positions of the surrounding atoms. Analogously to Equations 9.11 and 9.12 one can define

$$D_i = A_i + d_1 B_i + d_2 A_i B_i \tag{9.15}$$

and

$$S_i^{FACTS} = c_0 + \frac{c_1}{1 + e^{-c_2(D_i - c_3)}} \tag{9.16}$$

for the SASA of atom i. The parameters c_0 and c_1 are determined using the limiting cases of a fully buried and fully exposed atom. In the case of a fully buried atom (that is, $D_i \to +\infty$), the value of S_i should vanish, which implies that $c_0 = -c_1$ and $c_2 > 0$. For a fully exposed atom (that is, $D_i \to 0$), the analytical formula applies, so that $c_0 = 4\pi(r_i^{VDW} + 1.4)^2(1 + e^{-c_2 c_3})$ using a probe sphere of 1.4 Å radius. The parameters d_1, d_2, c_2, and c_3 are derived by fitting to exact values of the SASA [82].

9.3.1.4 Total Solvation Free Energy in the FACTS Model

The solvation free energy of a macromolecule is written as the sum of a polar and a non-polar term,

$$\Delta G^{FACTS} = \Delta G^{el,FACTS} + \gamma \sum_{i=1}^{N} S_i^{FACTS} \tag{9.17}$$

where γ denotes the empirical surface tension parameter. Values of γ between 0.015 and 0.025 kcal mol^{-1} Å^{-2} have been used to run MD simulations [19].

9.3.2 Parameterization of FACTS

The parameterization was done separately for the polar hydrogen parameter set param19 and the all-atom parameter set param22. Briefly, two training sets, of 81 (param19) and 72 (param22) conformations, from 29 peptides and proteins containing native, molten globule-like and extended conformations and spanning a wide spectrum of secondary structures and irregular shapes (cavities, open loops, and so on), were used for the parameterization. Atomic solvation energies $\Delta G_i^{el,PB}$ were calculated by numerical solution of the PB equation with the PBEQ module [84] in CHARMM. Reference atomic SASA values (S_i^{exact}) were computed with an analysis module in CHARMM. For each van der Waals radius, two sets of parameters had to be optimized separately: the five parameters b_1, b_2, a_2, a_3, and R^{sphere} for the atomic solvation energies, and the four parameters d_1, d_2, c_2, and c_3 for the atomic SASA. An upper bound of 10 Å was imposed for the optimization of R^{sphere}. Furthermore, R^{sphere} was optimized only for electrostatic solvation energies. For atomic SASA values the R^{sphere} parameters determined for the electrostatic solvation were used to increase the efficiency in MD simulations, as the same list of atom pairs can be used for the evaluation of electrostatic solvation energy and SASA. Optimal parameters were obtained by minimizing the deviations of $\Delta G_i^{\text{el,FACTS}}$ from $\Delta G_i^{\text{el,PB}}$ and of S_i^{FACTS} from S_i^{exact}. A particle swarm algorithm [85] was used for parameter optimization.

9.3.3
Validation and Applications of FACTS

The FACTS model was compared with GBMV [81], one of the most accurate GB methods available (we refer the reader to the original FACTS publication [19] for details). Quantities like atomic solvation energies, screened interaction energies and SASA values are slightly better approximated by GBMV, but the correlation between FACTS and reference values is still very high, while FACTS is about 10 times faster than GBMV. The biggest deviations of FACTS from "exact" values occur with the param19 parameter set (a force field with short cutoffs and less hydrogenatoms than param22), and especially in the evaluation of atomic SASA values, mainly because, as mentioned above, the sphere radii were not optimized *ad hoc* for the atomic SASA evaluation, but set equal to those of the electrostatic atomic solvation energy for computational efficiency.

9.3.3.1 Potential of Mean Forces of Side-Chain Dimers
One important test of the quality of an implicit solvent model is the comparison of potentials of mean forces (PMFs) between pairs of ionic or polar side-chain analogs or backbone fragments computed in implicit solvent with PMFs obtained in explicit water. Six PMFs calculated in explicit water and the implicit models GBMV and FACTS, are shown in Figure 9.5. A general trend is that, compared to TIP3P (transferable intermolecular potential, three-point [86]) explicit water, GBMV underestimates the interaction energies whereas FACTS overestimates them. Close agreement of FACTS with the explicit water profile is seen in the backbone–Arg and Ser–Asn systems, and a large deviation is observed in the Lys–Glu system. In most cases FACTS does not reproduce the desolvation barrier, while GBMV tends to overestimate it. It has been observed that the reproduction of the desolvation barrier, which only influences the kinetics, is less important than the correct estimation of the minimum of the interaction energy, which influences the overall thermodynamics of a system [29].

9.3.3.2 Atomic Fluctuations
MD simulations *in vacuo* suffer from too small atomic fluctuations. The RMSD fluctuations of the C_α atoms of ubiquitin (Protein Data Bank (PDB) ID: 1UBQ) and chymotrypsin inhibitor 2 (PDB: 2CI2) computed from param19 300 K MD simulations with either EEF1, SASA or FACTS as implicit solvent model are shown in Figure 9.6. In the case of ubiquitin (Figure 9.6a), the fluctuations in FACTS and SASA trajectories are slightly higher than in the EEF1 trajectory, and

9.3 The FACTS Implicit Solvent Model: A Fast Generalized Born Approach

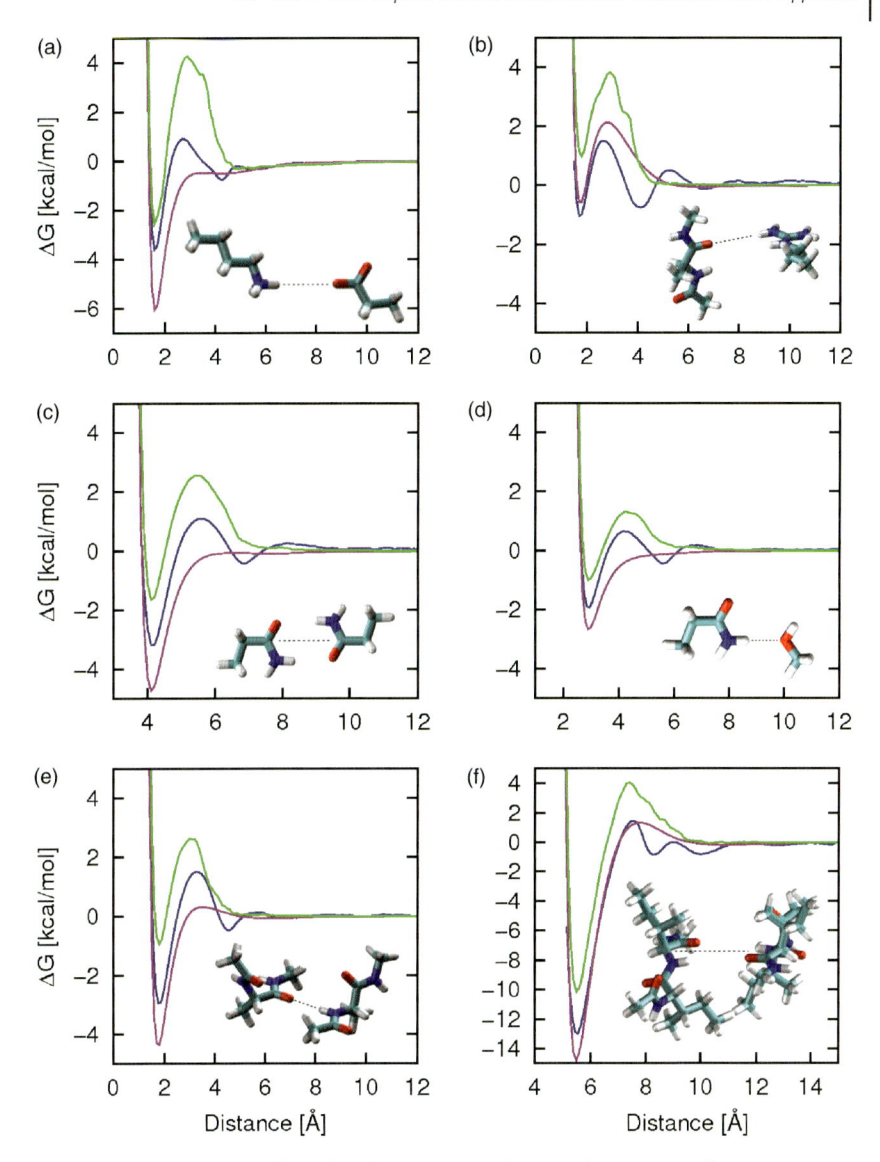

Figure 9.5 Free-energy profiles of six dimers in TIP3P water (blue), GBMV implicit solvent (green), and FACTS param22 with $\varepsilon_m = 1$ (magenta). The dimer configurations are shown in the inserts. The reaction coordinates plotted along the x-axis are (a) r(Lys HZ1–Glu OE1), (b) r(Ala O–Arg HH22), (c) r(Gln CD–Gln CD), (d) r(Ser OG–Gln NE2), (e) r(Ala HN–Ala O), and (f) r(Ile-Ile CA–Ile-Ile CA). The molecules were constrained so that their only degree of freedom was translation along the dashed line.

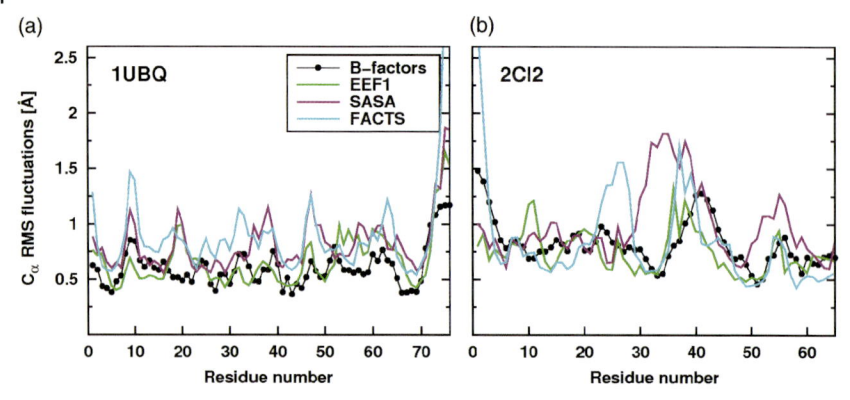

Figure 9.6 RMS fluctuations (Å) of the C_α atoms of (a) ubiquitin and (b) chymotrypsin inhibitor 2, extracted from 300 K simulations started from their respective native structures. Simulations were performed with param19 and either with EEF1 (green), SASA (magenta), and FACTS with $\varepsilon_m = 1$ and a surface tension-like parameter $\gamma = 0.025\,\text{kcal mol}^{-1}\,\text{Å}^{-2}$ (cyan). The plotted values are averages of fluctuations of five consecutive 2 ns trajectory segments. The bold line with circles represents the fluctuations derived from the crystallographic B-factors [87] using the formula RMS fluctuation $= [3B/(8\pi^2)]^{0.5}$.

slightly overestimate the crystallographic B-factors [87]. A small increase of the fluctuation may be expected for proteins in solution due to the lack of crystal packing. More importantly, FACTS and SASA reproduce the peaks (that is, the regions of highest mobility) as can be seen in the case of 2CI2 (Figure 9.6b), where the N-terminal segment and the loop (residues 38–44) are the most flexible regions according to both MD simulations and X-ray data.

9.3.3.3 Peptide Aggregation

Following the work of Cecchini *et al.* on the β-aggregation propensity profile of the Alzheimer's amyloid-β peptide (Aβ$_{42}$) [43], simulations were repeated with either TIP3P explicit water or FACTS as solvent model. As shown in Figure 9.2, the profile calculated in explicit water is better reproduced by FACTS. The high charge density in the N-terminal region of Aβ$_{42}$ is responsible for the strong deviation between the β-aggregation propensities calculated by the SASA and explicit water simulations.

9.3.3.4 Scalar and Parallel Performance

The main advantage of FACTS is its speed; it is only four times slower than *in vacuo* and about 10 times faster than GBMV. The CPU time scales linearly with system size, as shown in Figure 9.7a. Furthermore, FACTS has been parallelized and scales well on up to eight CPUs. A scaling factor of 5.6 is obtained in simulations of the 389-residue protein β-secretase (PDB: 1SGZ) on eight Xeon 2.33 GHz cores (70% of ideal scaling; Figure 9.7b). Remarkably, simulations of a much smaller system, a 56-residue designed protein (PDB: 2JWS), scale almost as well as the large 1SGZ system (a scaling factor of 5.3 on eight cores).

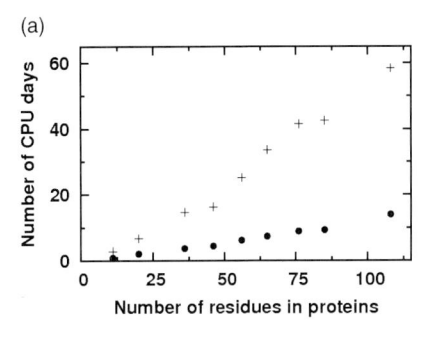

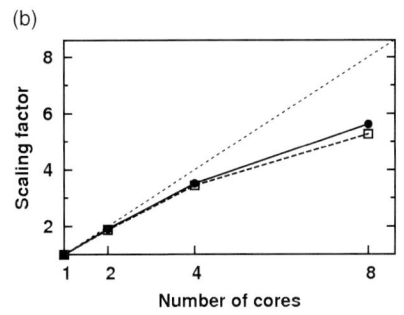

Figure 9.7 FACTS timing analysis. (a) System size scaling of CPU time required for 100 ns MD simulations with FACTS. Circles and plus symbols correspond to simulations with param19 and param22, respectively. (b) Scaling factors as a function of number of cores used in parallel MD simulations of the 389-residue protein β-secretase (PDB: 1SGZ; solid line with circles) and of a 56-residue designed protein (PDB: 2JWS; bold dashed line with squares). The thin dashed line represents an ideal scaling behavior. All simulations were performed on a dual motherboard with Xeon E5410 2.33 GHz quad-core processors.

9.4 Conclusions

The SASA implicit solvent model combines a fast approximation of the surface area with two atomic solvation parameters, and uses a linear distance-dependent dielectric function and neutralized ionic side chains to approximate the electrostatic screening effects. It reasonably describes solvation effects for peptides and small proteins, where most of the charges are exposed to solvent. The typical problems that arise in *in vacuo* MD simulations, that is, large deviations from the native conformation and an excessive number of intra-solute hydrogen bonds, are reduced when including the mean solvation term. The low computational burden of SASA allowed the study of slow processes like aggregation of amyloidogenic peptides [41–43, 46], and reversible folding of structured peptides [33–35, 52], which in turn helped to develop network-based analysis methods of the conformational space [31, 88, 89]. SASA shows its limitations in MD simulations of proteins or highly charged molecules, where the distance-dependent dielectric function and the neutralization of ionic side chains fails to correctly describe the electrostatics in aqueous solution [18].

Attempts to alleviate these limitations led to the development of FACTS [19], an implicit solvent model based on the generalized Born treatment of electrostatics. Notably, FACTS is very efficient because it uses simple measures of solvent displacement and requires only distances between solute atoms that are close in three-dimensional space and are therefore included in standard non-bonding lists. With FACTS the structural integrity of globular proteins in long MD simulations is preserved, and the atomic fluctuations correlate well with values derived from crystallographic *B*-factors. FACTS can be further improved, as the deviations in the PMFs from explicit water curves demonstrate. A source of deviation may be in the use of van der Waals radii for the definition of the solute–solvent dielectric boundary in the derivation of PB atomic free energy. The van der Waals radii are not optimized to reproduce solvation energy, and a different set of radii may lead to improved agreement between explicit water free energies and PB values [90], which in turn will improve the overall dynamical behavior of biomolecular systems. A more realistic description of solvation effects can also be obtained with an improved treatment of the non-polar component beyond the linear approximation of the solvent-accessible surface area, as it is increasingly recognized that different hydrophobic solvation regimes exist, depending on the length scale of the molecular system [91–93].

Acknowledgments

We thank Jianhan Chen and Charles L. Brooks III for generously sharing their data on potential of mean forces (PMFs). We thank Stefanie Muff and Andrea Prunotto for helpful discussions and comments. This work was supported in part by the Swiss National Competence Center for Research in Neural Plasticity and Repair.

References

1 Feig, M. and Brooks, C.L. III (2004) Recent advances in the development and applications of implicit solvent models in biomolecule simulations. *Curr. Opin. Struct. Biol.*, **14**, 217–224.

2 Roux, B. and Simonson, T. (1999) Implicit solvent models. *Biophys. Chem.*, **78**, 1–20.

3 Bashford, D. and Case, D.A. (2000) Generalized Born models of macromolecular solvation effects. *Annu. Rev. Phys. Chem.*, **51**, 129–152.

4 Cramer, C.J. and Truhlar, D.G. (1999) Implicit solvation models: equilibria, structure, spectra, and dynamics. *Chem. Rev.*, **99**, 2161–2200.

5 Zagrovic, B. and Pande, V. (2003) Solvent viscosity dependence of the folding rate of a small protein: distributed computing study. *J. Comput. Chem.*, **24**, 1432–1436.

6 Levy, R.M., Zhang, L.Y., Gallicchio, E., and Felts, A.K. (2003) On the nonpolar hydration free energy of proteins: surface area and continuum solvent models for the solute–solvent interaction energy. *J. Am. Chem. Soc.*, **125**, 9523–9530.

7 Eisenberg, D. and McLachlan, A.D. (1986) Solvation energy in protein folding and binding. *Nature*, **319**, 199–203.

8 Ooi, T., Oobatake, M., Némethy, M., and Scheraga, H.A. (1987) Accessible surface areas as a measure of the thermodynamic parameters of hydration of peptides. *Proc. Natl. Acad. Sci. U.S.A.*, **84**, 3086–3090.

9 Fraternali, F. and van Gunsteren, W.F. (1996) An efficient mean solvation force model for use molecular dynamics simulations of proteins in aqueous solution. *J. Mol. Biol.*, **256**, 939–948.

10 Stouten, P.F.W., Froemmel, C., Nakamura, H., and Sander, C. (1993) An effective solvation term based on atomic occupancies for use in protein simulations. *Mol. Simul.*, **10**, 97–120.

11 Lazaridis, T. and Karplus, M. (1999) Effective energy function for proteins in solution. *Proteins*, **35**, 133–152.

12 Warwicker, J. and Watson, H.C. (1982) Calculation of the electric potential in the active site cleft due to α-helix dipoles. *J. Mol. Biol.*, **157**, 671–679.

13 Constanciel, R. and Contreras, R. (1984) *Theor. Chim. Acta*, **65**, 1.
14 Still, W.C., Tempczyk, A., Hawley, R.C., and Hendrickson, T. (1990) Semianalytical treatment of solvation for molecular mechanics and dynamics. *J. Am. Chem. Soc.*, **112**, 6127–6129.
15 Schaefer, M. and Karplus, M. (1996) A comprehensive analytical treatment of continuum electrostatics. *J. Phys. Chem.*, **100**, 1578–1599.
16 Im, W., Lee, M.S., and Brooks, C.L. III (2003) Generalized Born model with a simple smoothing function. *J. Comput. Chem.*, **24**, 1691–1702.
17 Mongan, J., Simmerling, C., McCammon, J.A., Case, D.A., and Onufriev, A. (2007) Generalized born model with a simple, robust molecular volume correction. *J. Chem. Theory Comput.*, **3**, 156–169.
18 Ferrara, P., Apostolakis, J., and Caflisch, A. (2002) Evaluation of a fast implicit solvent model for molecular dynamics simulations. *Proteins*, **46**, 24–33.
19 Haberthür, U. and Caflisch, A. (2008) Fast analytical continuum treatment of solvation. *J. Comput. Chem.*, **29**, 701–715.
20 Brooks, B.R., Bruccoleri, R.E., Olafson, B.D., States, D.J., Swaminathan, S., and Karplus, M. (1983) CHARMM: a program for macromolecular energy, minimization, and dynamics calculations. *J. Comput. Chem.*, **4**, 187–217.
21 Tama, F. and Sanejouand, Y.H. (2001) Conformational change of proteins arising from normal mode calculations. *Protein Eng. Des. Sel.*, **14**, 1–6.
22 Cecchini, M., Houdusse, A., and Karplus, M. (2008) Allosteric communication in myosin V: from small conformational changes to large directed movements. *PLoS Comput. Biol.*, **4**, e1000129.
23 Wesson, L. and Eisenberg, D. (1992) Atomic solvation parameters applied to molecular dynamics of proteins in solution. *Protein Sci.*, **1**, 227–235.
24 Hasel, W., Hendrickson, T.F., and Still, W.C. (1988) A rapid approximation to the solvent accessible surface areas of atoms. *Tetrahedron Comput. Methodol.*, **1**, 103–116.
25 Mehler, E.L. and Eichele, G. (1984) Electrostatic effects in water-accessible regions of proteins. *Biochemistry*, **23**, 3887–3891.
26 Ramstein, J. and Lavery, R. (1988) Energetic coupling between DNA bending and base pair opening. *Proc. Natl. Acad. Sci. U.S.A.*, **85**, 7231–7235.
27 Mehler, E.L. (1990) Comparison of dielectric response models for simulating electrostatic effects in proteins. *Protein Eng.*, **3**, 415–417.
28 Bartels, C., Stote, R., and Karplus, M. (1998) Characterization of flexible molecules in solution: the RGDW peptide. *J. Mol. Biol.*, **284**, 1641–1660.
29 Chen, J.H., Im, W.P., and Brooks, C.L. III (2006) Balancing solvation and intramolecular interactions: toward a consistent generalized Born force field. *J. Am. Chem. Soc.*, **128**, 3728–3736.
30 De Alba, E., Santoro, J., Rico, M., and Jiménez, M.A. (1999) *De novo* design of a monomeric three-stranded antiparallel β-sheet. *Protein Sci.*, **8**, 854–865.
31 Rao, F. and Caflisch, A. (2004) The protein folding network. *J. Mol. Biol.*, **342**, 299–306.
32 Caflisch, A. (2006) Network and graph analyses of folding free energy surfaces. *Curr. Opin. Struct. Biol.*, **16**, 71–78.
33 Cavalli, A., Haberthür, U., Paci, E., and Caflisch, A. (2003) Fast protein folding on downhill energy landscape. *Protein Sci.*, **12**, 1801–1803.
34 Ferrara, P., Apostolakis, J., and Caflisch, A. (2000) Targeted molecular dynamics simulations of protein unfolding. *J. Phys. Chem.*, **104**, 4511–4518.
35 Paci, E., Cavalli, A., Vendruscolo, M., and Caflisch, A. (2003) Analysis of the distributed computing approach applied to the folding of a small beta-peptide. *Proc. Natl. Acad. Sci. U.S.A.*, **100**, 8217–8222.
36 Settanni, G., Rao, F., and Caflisch, A. (2005) Φ-value analysis by molecular dynamics simulations of reversible folding. *Proc. Natl. Acad. Sci. U.S.A.*, **102**, 628–633.
37 Gsponer, J. and Caflisch, A. (2001) Role of native topology investigated by multiple unfolding simulations of four SH3 domains. *J. Mol. Biol.*, **309**, 285–298.
38 Rathore, N., Yan, Q.L., and de Pablo, J.J. (2004) Molecular simulation of the

reversible mechanical unfolding of proteins. *J. Chem. Phys.*, **120**, 5781–5788.

39 Ihalainen, J.A., Bredenbeck, J., Pfister, R., Helbing, J., Woolley, G.A., and Hamm, P. (2007) Folding and unfolding of a photoswitchable peptide from picoseconds to microseconds. *Proc. Natl. Acad. Sci. U.S.A.*, **104**, 5383–5388.

40 Rathore, N., Knotts, T.A., and de Pablo, J.J. (2003) Configurational temperature density of states simulations of proteins. *Biophys. J.*, **85**, 3963–3968.

41 Gsponer, J., Haberthür, U., and Caflisch, A. (2003) The role of sidechain interactions in the early steps of aggregation: molecular dynamics simulations of an amyloid forming peptide from the yeast prion Sup35. *Proc. Natl. Acad. Sci. U.S.A.*, **100**, 5154–5159.

42 Paci, E., Gsponer, J., Salvatella, X., and Vendruscolo, M. (2004) Molecular dynamics studies of the process of amyloid aggregation of peptide fragments of transthyretin. *J. Mol. Biol.*, **340**, 555–569.

43 Cecchini, M., Curcio, R., Pappalardo, M., Melki, R., and Caflisch, A. (2006) A molecular dynamics approach to the structural characterization of amyloid aggregation. *J. Mol. Biol.*, **357**, 1306–1321.

44 Chandrasekhar, S. (1992) *Liquid Crystals*, Cambridge University Press, Cambridge.

45 Zannoni, C. (2001) Molecular design and computer simulations of novel mesophases. *J. Mater. Chem.*, **11**, 2637–2646.

46 Cecchini, M., Rao, F., Seeber, M., and Caflisch, A. (2004) Replica exchange molecular dynamics simulations of amyloid peptide aggregation. *J. Chem. Phys.*, **121**, 10748–10756.

47 Hajjar, E., Perahia, D., Debat, H., Nespoulous, C., and Robert, C.H. (2006) Odorant binding and conformational dynamics in the odorant binding protein. *J. Biol. Chem.*, **281**, 29929–29937.

48 Feig, M. and Brooks, C.L. III (2002) Evaluating casp4 predictions with physical energy functions. *Proteins*, **49**, 232–245.

49 Struthers, M.D., Cheng, R.P., and Imperiali, B. (1996) Design of a monomeric 23-residue polypeptide with defined tertiary structure. *Science*, **271**, 342–345.

50 Fezoui, Y., Connolly, P.J., and Osterhout, J.J. (1997) Solution structure of αtα, a helical hairpin peptide of *de novo* design. *Protein Sci.*, **6**, 1869–1877.

51 Fezoui, Y., Weaver, D.L., and Osterhout, J.J. (1994) De novo design and structural characterization of an α-helical hairpin peptide: a model system for the study of protein folding intermediates. *Proc. Natl. Acad. Sci. U.S.A.*, **91**, 3675–3697.

52 Ferrara, P., Apostolakis, J., and Caflisch, A. (2000) Thermodynamics and kinetics of folding of two model peptides investigated by molecular dynamics simulations. *J. Phys. Chem. B*, **104**, 5000–5010.

53 Warshel, A. and Russell, S.T. (1984) Calculations of electrostatic interactions in biological systems and in solutions. *Q. Rev. Biophys.*, **17**, 283–422.

54 Warshel, A. and Papazyan, A. (1998) Electrostatic effects in macromolecules: fundamental concepts and practical modeling. *Curr. Opin. Struct. Biol.*, **8**, 211–217.

55 Tomasi, J. and Persico, M. (1994) Molecular interactions in solution: an overview of methods based on continuous distributions of the solvent. *Chem. Rev.*, **94**, 2027–2094.

56 Gilson, M.K. (1995) Theory of electrostatic interactions in macromolecules. *Curr. Opin. Struct. Biol.*, **5**, 216–223.

57 Orozco, M. and Luque, F.J. (2000) Theoretical methods for the description of the solvent effect in biomolecular systems. *Chem. Rev.*, **100**, 4187–4225.

58 Simonson, T. (2001) Macromolecular electrostatics: continuum models and their growing pains. *Curr. Opin. Struct. Biol.*, **11**, 243–252.

59 Werner, P. and Caflisch, A. (2003) A sphere-based model for the electrostatics of globular proteins. *J. Am. Chem. Soc.*, **125**, 4600–4608.

60 Baker, N.A. (2005) Improving implicit solvent simulations: a Poisson-centric view. *Curr. Opin. Struct. Biol.*, **15**, 137–143.

61 Im, W., Chen, J., and Brooks, C.L. III (2006) Peptide and protein folding and conformational equilibria: theoretical treatment of electrostatics and hydrogen bonding with implicit solvent models. *Adv. Protein Chem.*, **72**, 171–195.

62 Feig, M., Chocholousova, J., and Tanizaki, S. (2006) Extending the horizon: towards the efficient modeling of large biomolecular complexes in atomic detail. *Theor. Chem. Accts.*, **116**, 194–205.

63 Gilson, M.K. and Honig, B.H. (1988) Calculation of the total electrostatic energy of a macromolecular system: solvation energies, binding energies, and conformational analysis. *Proteins*, **4**, 7–18.

64 Bashford, D. and Karplus, M. (1990) pK_as of ionizable groups in proteins: atomic detail from a continuum electrostatic model. *Biochemistry*, **29**, 10219–10225.

65 Davis, M.E., Madura, J.D., Luty, B.A., and McCammon, J.A. (1991) Electrostatics and diffusion of molecules in solution: simulations with the University of Houston Brownian Dynamics program. *Comput. Phys. Commun.*, **62**, 187–197.

66 Zauhar, R.J. and Morgan, R.S. (1985) A new method for computing the macromolecular electric potential. *J. Mol. Biol.*, **186**, 815–820.

67 Cortis, C.M. and Friesner, R.A. (1997) An automatic three-dimensional finite element mesh generation system for the Poisson–Boltzmann equation. *J. Comput. Chem.*, **18**, 1570–1590.

68 Cortis, C.M. and Friesner, R.A. (1997) Numerical solution of the Poisson–Boltzmann equation using tetrahedral finite-element meshes. *J. Comput. Chem.*, **18**, 1591–1608.

69 Vorobjev, Y.N. and Scheraga, H.A. (1997) A fast adaptive multigrid boundary element method for macromolecular electrostatic computations in a solvent. *J. Comput. Chem.*, **18**, 569–583.

70 Born, M. (1920) Volumen und Hydratationswärme der Ionen. *Z. Phys.*, **1**, 45–48.

71 Onufriev, A., Bashford, D., and Case, D.A. (2002) Effective Born radii in the generalized Born approximation: the importance of being perfect. *J. Comput. Chem.*, **23**, 1297–1304.

72 Scarsi, M., Apostolakis, J., and Caflisch, A. (1997) Continuum electrostatic energies of macromolecules in aqueous solutions. *J. Phys. Chem. A*, **101**, 8098–8106.

73 Ghosh, A., Rapp, C.S., and Friesner, R.A. (1998) Generalized Born model based on a surface integral formulation. *J. Phys. Chem. B*, **102**, 10983–10990.

74 Lee, M.S., Salsbury, F.R., and Brooks, C.L. III (2002) Novel generalized Born methods. *J. Chem. Phys.*, **116**, 10606–10614.

75 Lee, M.S., Feig, M., Salsbury, F.R., and Brooks, C.L. III (2003) New analytic approximation to the standard molecular volume definition and its application to generalized Born calculations. *J. Comput. Chem.*, **24**, 1348–1356.

76 Hawkins, G.D., Cramer, C.J., and Truhlar, D.G. (1996) Parametrized models of aqueous free energies of solvation based on pairwise descreening of solute atomic charges from a dielectric medium. *J. Phys. Chem.*, **100**, 19824–19839.

77 Qiu, D., Shenkin, P.S., Hollinger, F.P., and Still, W.C. (1997) The GB/SA continuum model for solvation. A fast analytical method for the calculation of approximate Born radii. *J. Phys. Chem. A*, **101**, 3005–3014.

78 Dominy, B.N. and Brooks, C.L. III (1999) Development of a generalized Born model parametrization for proteins and nucleic acids. *J. Phys. Chem. B*, **103**, 3765–3773.

79 Schaefer, M. and Froemmel, C. (1990) A precise analytical method for calculating the electrostatic energy of macromolecules in aqueous solution. *J. Mol. Biol.*, **216**, 1045–1066.

80 Grycuk, T. (2003) Deficiency of the Coulomb-field approximation in the generalized Born model: an improved formula for Born radii evaluation. *J. Chem. Phys.*, **119**, 4817–4826.

81 Feig, M., Onufriev, A., Lee, M.S., Im, W., Case, D.A., and Brooks, C.L. III (2003) Performance comparison of generalized Born and Poisson methods in the calculation of electrostatic solvation energies for protein structures. *J. Comput. Chem.*, **25**, 264–284.

82 Lee, B. and Richards, F.M. (1971) The interpretation of protein structures: estimation of static accessibility. *J. Mol. Biol.*, **55**, 379–400.

83 Richards, F.M. (1977) Areas, volumes, packing, and protein structure. *Annu. Rev. Biophys. Bioeng.*, **6**, 151–176.

84 Im, W., Beglov, D., and Roux, B. (1998) Continuum solvation model: computation of electrostatic forces from numerical solutions to the Poisson–Boltzmann equation. *Comput. Phys. Commun.*, **111**, 59–75.

85 Kennedy, J. and Eberhart, R.C. (2001) *Swarm Intelligence*, Morgan Kaufmann, San Francisco.

86 Jorgensen, W.L., Chandrasekhar, J., Madura, J., Impey, R.W., and Klein, M.L. (1983) Comparison of simple potential functions for simulating liquid water. *J. Chem. Phys.*, **79**, 926–935.

87 McPhalen, C.A. and James, M.N.G. (1987) Crystal and molecular structure of the serine proteinase inhibitor CI-2 from barley-seeds. *Biochemistry*, **26**, 261–269.

88 Krivov, S.V., Muff, S., Caflisch, A., and Karplus, M. (2008) One-dimensional barrier preserving free-energy projections of a beta-sheet miniprotein: new insights into the folding process. *J. Phys. Chem. B*, **112**, 8701–8714.

89 Muff, S. and Caflisch, A. (2008) Kinetic analysis of molecular dynamics simulations reveals changes in the denatured state and switch of folding pathways upon single-point mutation of a β-sheet miniprotein. *Proteins*, **70**, 1185–1195.

90 Nina, M., Beglov, D., and Roux, B. (1997) Atomic radii for continuum electrostatics calculations based on molecular dynamics free energy simulations. *J. Phys. Chem. B*, **101** (26), 5239–5248.

91 Chandler, D. (2005) Interfaces and the driving force of hydrophobic assembly. *Nature*, **437**, 640–647.

92 Rajamani, S., Truskett, T.M., and Garde, S. (2005) Hydrophobic hydration from small to large lengthscales: understanding and manipulating the crossover. *Proc. Natl. Acad. Sci. U.S.A.*, **102**, 9475–9480.

93 Graziano, G. (2007) Cavity thermodynamics and surface tension of water. *Chem. Phys. Lett.*, **442**, 307–310.

10
On the Development of State-Specific Coarse-Grained Potentials of Water

Hyung Min Cho and Jhih-Wei Chu

10.1
Introduction

Water is a ubiquitous solvent for living matter and is indispensable to human life. With a simple molecular structure, water can serve as the medium for complicated chemical and biochemical reactions that not only sustain life on Earth but also enable the realization of modern technologies. The small size of the water molecule plays a part in producing the unique patterns of the hydrogen bonding networks in different phases, which contribute to the anomalous equation of state, phase transition behavior, and dynamic properties of water as compared to other liquids [1, 2]. The unique properties of water also lead to the hydrophobic effects that cause the immiscibility between oil and water. Hydrophobic effects are the essential driving forces for protein folding and self-assembly of complex materials [3, 4] (see Chapter 1).

A fundamental way of elucidating the mechanisms by which water achieves its multifaceted functionality is to develop physical models of water to reproduce and predict the behavior of water via computer simulation. Despite the tremendous progress and success made in this area over the past several decades, quantitative reproduction of all known water properties via computer simulation still has not been achieved. In this work, we focus on water models based on molecular mechanics (MM), that is, the breaking and forming of chemical bonds will not be considered. Many structural and thermodynamic properties of water can be captured with quantitative accuracy using MM force fields with atomic resolution [5–8].

Unfortunately, all-atom simulations of water become exceedingly expensive when a large periodic cell is needed, such as for simulating biomolecules and their assemblies in explicit water. In most cases, 80–90% of the computational effort in all-atom molecular dynamics (MD) simulation of biomolecules is spent on water. To extend the accessible time and length scales of molecular modeling and simulation, a plausible strategy is to develop a simpler representation of the water molecule, for example, using a single site instead of three atoms. This coarse-grained (CG) procedure can significantly reduce the number of degrees of freedom used

Modeling Solvent Environments. Applications to Simulations of Biomolecules. Edited by Michael Feig
Copyright © 2010 WILEY-VCH Verlag GmbH & Co. KGaA, Weinheim
ISBN: 978-3-527-32421-7

for modeling a molecular system and avoid the calculation of long-range electrostatic interactions. The interaction potential at the CG scale is also expected to be smoother after removing the detailed features of a molecule. As a result, a larger time step can be used for CG MD simulation as compared to that of all-atom MD simulations. On the other hand, with reduced degrees of freedom in a CG model, it is inevitable that certain properties of water cannot be retained. Examining the behaviors of a CG model as compared to an all-atom one thus provides a novel way of characterizing the molecular origin of specific properties of water.

There are numerous ways of developing CG force fields [9–21]. We adopt a philosophy of conceiving a CG force field as the potential of mean force of a fully detailed reference system with the non-CG degrees of freedom integrated out according to the equilibrium distribution function of the fully detailed system. Based on such theory, a variational principle can be developed to uniquely determine a CG force field by minimizing the difference in mean forces predicted by a CG force field and those given by the reference system with full details. This variational principle is referred to as the multi-scale coarse graining (MS-CG) method [22–26]. The resulting force matching (FM) procedure [27, 28] of the MS-CG method corresponds to solving of the Yvon–Born–Green (YBG) equation [29–31] that correlates three-body and two-body correlation functions through mean forces. Detailed derivation of theory and numerical procedures of the MS-CG method can be found elsewhere [24, 25].

Recently we showed that the MS-CG framework can also be used to invert radial distribution functions (RDFs) to the underlying pair forces [32], in other words, to determine a molecular model from target data. Target RDFs can be obtained from all-atom MD simulations or experiments. The FM equation, or the YBG equation, is solved self-consistently until convergence is achieved to reproduce target RDFs. This method is thus called the iterative-YBG method. There are two major advantages of the iterative-YBG method over the commonly used reverse Monte Carlo approach [33, 34], which also aims to find a pair potential to reproduce a target RDF. First, the update of the force field is exactly described by the YBG equation – no ad hoc assumption is needed. Second, the initial CG force field for self-consistent iteration is determined by applying the FM method – specifying an initial condition is not necessary. Based on the uniqueness theorem of Henderson, all pair potentials that give rise to a given RDF are unique up to a trivial constant, independent of the method used for finding them [35].

In this chapter we examine CG force fields obtained by applying the iterative-YBG method using all-atom MD simulations based on a series of all-atom water models: "transferable intermolecular potential, three-point" (TIP3P) [36], "simple point charge" (SPC) [37], "transferable intermolecular potential, four-point" (TIP4P) [36], and "simple water model with four sites and Drude polarizability" (SWM4-DP) [38]). In particular, the SWM4-DP model takes the effects of polarizability into account via Drude oscillators [39]. The goal is to illustrate how molecular interactions at a CG scale affect the resulting structural and thermodynamic properties of liquid water. Since the iterative-YBG method gives a CG force field that corresponds to a potential of mean force associated with CG degrees of freedom, the resulting force field is specific to the thermodynamic state and is referred to as a state-specific coarse-grained force field.

In the following, we first briefly describe the theory and procedure of the iterative-YBG method. We then present the results of applying this method to retrieve CG force fields from all-atom MD simulations using different water potentials. The resulting differences in structural and thermodynamic properties of different CG water models will be compared based on the differences in pair forces. Finally, conclusions are given.

10.2
Methods of Computing Coarse-Grained Potentials of Liquid Water

In this section, the iterative-YBG procedure is briefly summarized. Based on the results of all-atom MD simulation of water, the iterative-YBG method extracts an effective CG force field that reproduces the RDF observed in the reference all-atom MD. The resulting force field corresponds to a potential of mean force with non-CG degrees of freedom integrated out and is thus state specific.

10.2.1
Multi-Scale Coarse Graining (MS-CG) Method with Force Matching (FM)

A general derivation of the MS-CG method with force matching can be found elsewhere [24, 25] and we only provide a short summary for the case of coarse graining an all-atom representation of water into a single-site CG model. Extension to multi-component systems is straightforward [24]. The basic idea of the MS-CG method is that configurations and the force vector of each configuration sampled in all-atom MD can be employed to variationally determine a CG potential. The variational principle is to minimize the mean-square difference, χ_{AA}^2, between the instantaneous forces exerted on CG sites observed in the all-atom reference system, F_I, that is, computed by using an all-atom force field and adding together the forces of atoms that constitute a CG site, and those determined by a CG force field, F_I^{CG}, at the same CG configuration:

$$\chi_{AA}^2 = \left\langle \frac{1}{3N^{CG}} \sum_{I=1}^{N^{CG}} |F_I - F_I^{CG}|^2 \right\rangle_{AA} \tag{10.1}$$

In Equation 10.1, the summation runs over N^{CG} CG sites in the system, and the angular brackets with subscript "AA" indicate an ensemble average over the equilibrium distribution of the reference all-atom system. In this work, we employ the assumptions of pair additive and isotropic forces for the CG model, and the force on a CG site I given a configuration is expressed as

$$F_I^{CG} = \sum_{J \neq I} F(R_{IJ}) \hat{u}_{IJ} \tag{10.2}$$

where R_{IJ} is the distance between CG sites I and J, and the unit vector $\hat{\mathbf{u}}_{IJ}$ is defined as pointing to site I from site J.

The interactions between a given pair of CG sites are thus described by a scalar function $F(R)$. To provide a general way of representing the emergent interactions at a CG scale, we use a tabulated form of $F(R)$ on a set of evenly spaced grid points:

$$F(R) = \sum_{d=1}^{N_d} F_d \delta(R - R_d) \tag{10.3}$$

where $\delta(R - R_d)$ is one when $|R - R_d| < \Delta R/2$ and zero elsewhere; and ΔR is the grid size. With the definition of pair interaction in the form of Equation 10.3, the minimization of the mean square difference in forces, χ^2_{AA} in Equation 10.1, becomes a linear least-squares problem and the solution satisfies the following set of algebraic equations:

$$\sum_{d'} H^{AA}_{dd'} F_{d'} = G^{AA}_d \tag{10.4}$$

The simplicity of solving linear equations is another motivation for using a tabulated representation of pair interactions. In Equation 10.4, $H^{AA}_{dd'}$ and G^{AA}_d are defined as

$$H^{AA}_{dd'} = \left\langle \sum_I \sum_{J \neq I} \sum_{K \neq I} \hat{\mathbf{u}}_{IJ} \cdot \hat{\mathbf{u}}_{JK} \delta(R_{IJ} - R_d) \delta(R_{JK} - R_{d'}) \right\rangle_{AA} \tag{10.5}$$

and

$$G^{AA}_d = \left\langle \sum_I \sum_{J \neq I} (\mathbf{F}_I \cdot \hat{\mathbf{u}}_{IJ}) \delta(R_{IJ} - R_d) \right\rangle_{AA} \tag{10.6}$$

The MS-CG method thus determines a CG force field (F_d values) by solving Equation 10.4. The force matching (FM) equation (Equation 10.4) is essentially a discrete representation of the YBG equation of classical liquid-state theory and provides an exact relationship between two- and three-body correlation functions through pairwise and additive forces [24]. The correspondence between the MS-CG method and the YBG equation highlights the significance of many-body interactions among CG sites in determining the underlying interactions. On the right-hand side of Equation 10.4, G^{AA}_d can be reformulated using the definition of the derivative of the radial distribution function, $g(R)$ [24]:

$$\left. \frac{dg^{AA}(R)}{dR} \right|_{R=R_d} = \frac{1}{k_B T} \sum_{d'} \left[\left(\frac{1}{4\pi R_d^2 \rho^2 V_0} \right) H^{AA}_{dd'} F_{d'} \right] \tag{10.7}$$

where V_0 is the total volume and ρ denotes the number density of the CG sites, which is assumed to be constant at given temperature T.

10.2.2
The Iterative-YBG Method

The effective CG force field obtained by solving the FM equation in the previous section often gives a radial distribution function different from the target profile obtained from all-atom MD simulation [26]. Here, we first rationalize the reason for this discrepancy and then propose a method to resolve this issue [32]. With a pair additive CG force field, the RDF of a CG MD simulation also satisfies the YBG equation:

$$\sum_{d'} H_{dd'}^{CG} F_{d'}^{CG} = G_d^{CG} \tag{10.8}$$

where $H_{dd'}^{CG}$ and G_d^{CG} are defined in the same way as in Equations 10.5 and 10.6 but the ensemble averages are taken via the equilibrium distribution function according to the CG potential instead of that of the all-atom model. Therefore, $H_{dd'}^{CG}$ and $H_{dd'}^{AA}$ are also different from each other in addition to G_d^{AA} and G_d^{CG}. Since the CG force field is obtained by solving Equation 10.4, it becomes clear that the difference between $H_{dd'}^{CG}$ and $H_{dd'}^{AA}$ leads to inconsistency in the RDFs, that is, G_d^{CG} differs from G_d^{AA}. In other words, the coarse graining process affects spatial correlation functions at different hierarchies to different extents in a manner according to Equation 10.4. Therefore, a computational scheme can be designed to find a CG force field that reproduces the target RDF by solving the YBG equation self-consistently such that the following equation is satisfied [32]:

$$\sum_{d'} (H_{dd'}^{CG})^* F_{d'}^* = (G_d^{CG})^* = G_d^{AA} \tag{10.9}$$

Therefore, we propose to solve Equation 10.8 iteratively with the right-hand side set at G_d^{AA} according to the target RDF. Once convergence is achieved, the resulting CG force reproduces the target RDF. This iterative method is thus called the iterative-YBG method. The overall procedure is summarized as follows:

1) Perform an all-atom MD simulation and calculate the spatial correlation functions G_d^{AA} and $H_{dd'}^{AA}$; then solve the FM equation to obtain an initial CG force field for self-consistent iteration.
2) Perform a CG MD simulation with the computed CG force field from step 1 and calculate G_d^{CG} and $H_{dd'}^{CG}$.
3) If G_d^{CG} differs from G_d^{AA}, solve Equation 10.8 again with G_d^{AA} on the right-hand side to update the CG force field.
4) Repeat steps 2 and 3 until self-consistency in the CG force field is accomplished.

10.2.3
Numerical Issues of the Iterative-YBG Method

Although the overall procedure outlined above for the iterative-YBG method is straightforward, several key numerical issues need to be resolved for actual applications. A major difficulty is the statistical uncertainty in computing the spatial correlation functions G_d and $H_{dd'}$ from MD simulations [25]. Statistical noise would result in a jagged profile of the CG force table (for example, see Figure 10.1a) and eventually cause instability of the CG MD simulations. To resolve this issue, we employ a smoothing procedure based on the Savitzky–Golay (SG) algorithm [40, 41] to remove the jagged features of the force profile at each step of the self-consistent iteration.

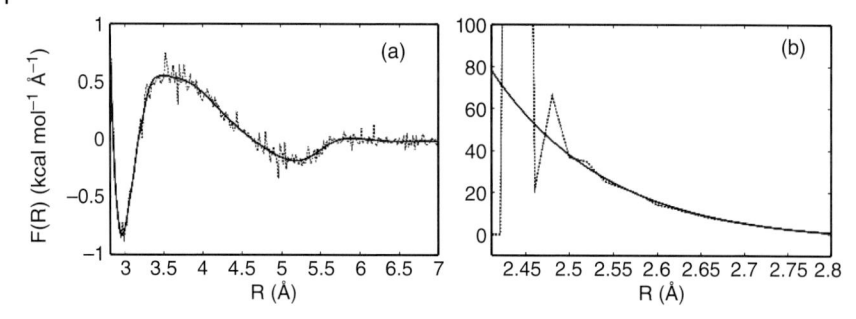

Figure 10.1 Pairwise force between CG water sites as a function of inter-site distance R. Raw data with statistical noise (dotted line) are compared with the smoothed force profile (solid line) obtained by (a) applying the Savitzky–Golay (SG) scheme for $R > R_c$ and (b) using an analytic function for $R < R_c$. See text for the selection rule for the cutoff distance R_c.

Another issue is that the repulsive core of the pair interactions results in poor sampling in the region of small pair distances. Since configurations with short distances between CG sites are rarely sampled, the statistics for short-range force coefficients are particularly poor (see Figure 10.1b). Even worse, statistical uncertainty for short-range forces can propagate to other elements of the force table during iteration through $H_{dd'}$. To resolve this issue, we employ an analytic function to describe the repulsive part of the CG force field near the core but use the tabulated representation for the rest. For the core potential between two CG sites, a simple analytic function is used:

$$\frac{A}{R^{12}} + \frac{B}{R^8} + \frac{C}{R^6} \qquad (10.10)$$

The parameters A, B, and C in Equation 10.10 are adjusted to obtain a smooth connection to the tabulated region that ensures continuity up to the second derivative of force at the patching radius, R_c. Typically, the value of R_c for a given target RDF is chosen to be the distance at which the RDF reaches the value of one from the core. We found that the results for force coefficients are not sensitive to the choice of R_c nor to the functional form used to represent the core interaction.

All of the all-atom and CG MD simulations are performed using the CHARMM software [42], based on which we implement the functionality of using tabulated pair forces to compute the interactions between CG sites.

10.3
Structural Properties and the "Representability" Problem of Coarse-Grained Liquid Water Models

In this work, we develop pair potentials for CG water with a single site per molecule that corresponds to the center of mass. This choice makes the molecular

10.3 Structural Properties and the "Representability" Problem

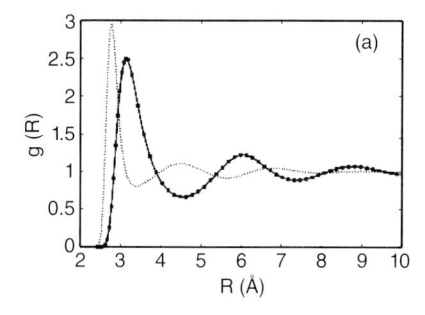

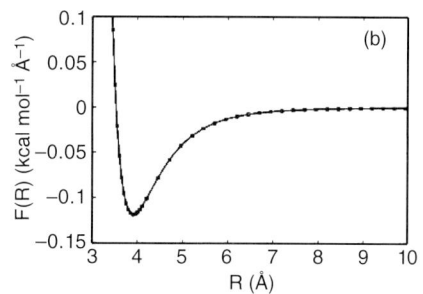

Figure 10.2 (a) Radial distribution function and (b) effective force field of an LJ fluid obtained from an MD simulation using the LJ parameters of the oxygen atom in the SPC water model (filled squares) compared with those computed by using the iterative-YBG method (solid line). The RDF for the SPC water model (dotted line) is also shown for comparison.

system of CG water similar to monatomic fluids except for differences in pair potential. Since the Lennard-Jones (LJ) potential is often used for simulating simple monatomic fluids, for which the inter-particle forces are due to induced-dipole/induced-dipole interactions, we first apply the iterative-YBG method to an LJ fluid to see if we can retrieve the LJ potential used in an MD simulation. Since no coarse graining operation is applied, the FM equation only needs to be solved once and the resulting force table should be identical to that used in the simulation. By using the LJ parameters of the oxygen atom in the simple point charge (SPC) water model [37], we perform the MD simulation with 1000 LJ particles at 300 K using the Nosé–Hoover thermostat [43, 44]. The volume of the system was fixed to give the same number density as for all-atom SPC water at 1 atm and 300 K. The resulting RDFs from MD simulations using LJ potential or force tables are indistinguishable, as shown in Figure 10.2a; the center-of-mass RDF of all-atom water is also shown for comparison. The force table recovers the original LJ pair potential with quantitative accuracy, as shown in Figure 10.2b. This result indicates that applying the smoothing procedure mentioned above to remove jagged features in the force table does not cause noticeable changes in the RDF.

In the following, we apply the iterative-YBG method to compute the CG force fields of water from MD simulations using different all-atom water models to examine how molecular interactions at the CG scale affect the structural and thermodynamic properties of water. Furthermore, we also quantify the emergent effects of charge polarizability on mean forces at the CG scale.

10.3.1
Coarse-Grained Water Model Computed from All-Atom Water Potentials

A set of all-atom water models is chosen for developing the CG water model. Two simple three-site models (TIP3P [36] and SPC [37]), a four-site model (TIP4P [36]), and a polarizable water model using Drude oscillator (SWM4-DP [38]) are adopted. In the CG representation, a water molecule is mapped onto a single CG site. From an all-atom configuration of water, the corresponding CG configuration is obtained by computing the center of mass of water. The interaction forces between CG particles are tabulated on numerical grids and are determined by the iterative-YBG procedure described in the previous section.

For each of the aforementioned atomistic models, an MD simulation is first performed to determine the average density of the liquid water at a temperature of 300 K under a pressure of 1 atm. The Nosé–Hoover thermostat [43, 44] and Langevin piston barostat [45] are used to control the target pressure and temperature. For each simulation, a system of 1000 all-atom water molecules is prepared in a cubic simulation box with periodic boundary conditions. The long-range electrostatic interactions are calculated by the particle mesh Ewald (PME) method [46, 47]. The short-range non-bonded interactions are shifted to zero at a cutoff radius of 10 Å. The intramolecular vibrations are suppressed by keeping bond lengths and angles at their equilibrium values using the SHAKE algorithm [48], and a time step of 2 fs is used for stable integration of the equations of motion. After determining the averaged volumes by *NPT* simulations, MD simulations in a constant *NVT* ensemble are then performed at the averaged volumes for 1 ns, with coordinates and forces saved every 1 ps to calculate the CG force field.

For the tabulated representation of non-bonded interactions between CG sites, 159 force parameters are assigned on evenly spaced grid points over an inter-site distance ranging from 2.1 to 10.0 Å. To conduct force matching iterations in computing CG force fields, a CG MD simulation is performed at each step of the iteration. The Langevin equations of motion are solved using the tabulated force table at a given iteration step for 7.5 ns using a time step of 5 fs. A CG configuration is saved every 300 steps to update the force table at each step of the iterative-YBG iteration. Along with reduced degrees of freedom, a larger time step of integration is also used for the CG simulation because the detailed features of the potential energy surface at the atomic scale are averaged out, thus resulting in a smoother profile at the CG scale. Another important factor contributing to reduction in computational cost is that our CG model does not carry charges by construction, since interactions between CG sites are decay rapidly after integrating out the internal degrees of freedom of water.

The overall computational cost for this CG simulation is significantly lower than that for simulation using all-atom force fields. In the cases of the three-site all-atom models (TIP3P and SPC), it is observed that a CG simulation has about 20 times reduction of computational cost for a given length of trajectory. Furthermore, the convergence in the CG force field is accomplished in only four to six iterations, with most of the correction achieved after the very first iteration. This result confirms that the iterative-YBG method is an efficient computational tool for deducing underlying CG force fields from distribution functions of all-atom configurations at a fairly low computational cost. Quick convergence is a consequence of enforcing the correct relationship between the two- and three-body correlation functions among the CG sites through the YBG equation. In other words, a trial CG force field is steered toward the target pair correlation function by the three-body correlation function assembled during the previous CG simulation. If the initial CG force

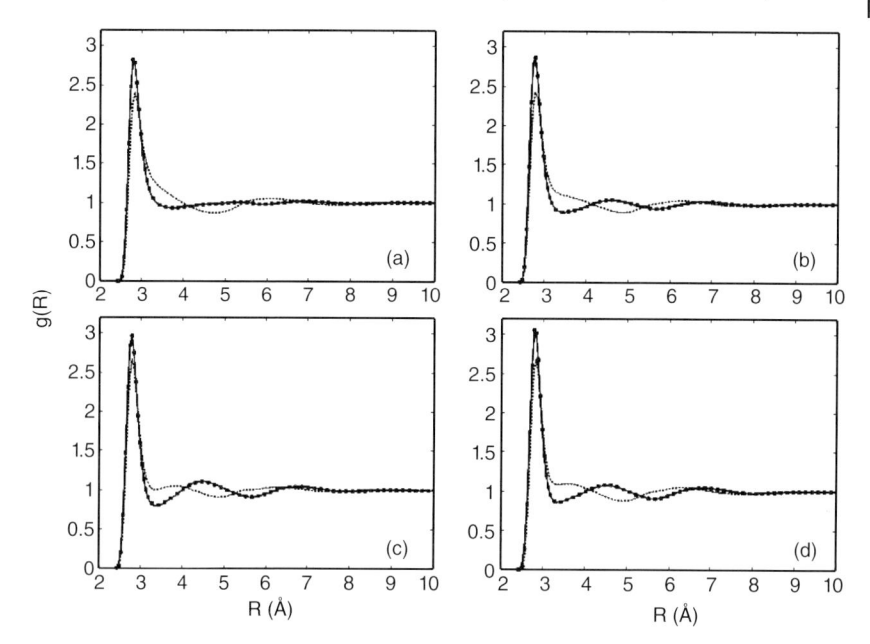

10.3 Structural Properties and the "Representability" Problem

Figure 10.3 Center-of-mass radial distribution functions (RDFs) between water molecules obtained from all-atom MD simulations and those from CG MD simulations using the iterative-YBG force field are compared for a selected set of liquid water models at a temperature of 300 K: (a) TIP3P, (b) SPC, (c) TIP4P, and (d) SWM4-DP. Filled squares: RDFs computed from all-atom MD simulation. Solid line: RDFs computed from CG MD simulation using the iterative-YBG force field. Dashed line: RDFs computed from CG MD simulation using the CG force field from force matching without self-consistent iteration.

field is close to the final self-consistent solution, convergence can usually be achieved within a small number of iterations. We found that the CG force fields obtained by force matching serve as good initial guesses even for much more complex systems to reach convergence in four to ten steps.

The center-of-mass RDFs between water molecules obtained from the all-atom simulations of each water model are shown in Figure 10.3. The RDFs computed by the CG force fields from force matching without self-consistent iteration are also shown in Figure 10.3 for comparison. Considerable deviations in RDFs are clearly observed by comparing the results obtained from using the force-matched CG force field (dotted lines) and those from the all-atom simulations (filled squares). Although the location of the first peak is well reproduced, the peak height is significantly underestimated by using the CG force field without self-consistent iteration.

During the course of self-consistent iteration starting from the force-matched CG force field, the CG force table readily leads to the converged force profile. The converged force table does reproduce the all-atom profile quantitatively, as illustrated in Figure 10.3 (solid lines). Converged force tables obtained by applying the

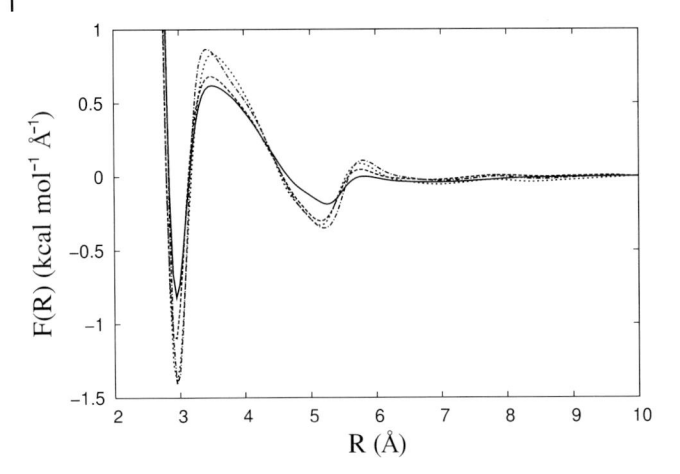

Figure 10.4 Effective pairwise forces between CG water molecules used to generate the RDFs in Figure 10.3 for the same set of liquid water models: TIP3P (full curve), SPC (dashed curve), TIP4P (dotted curve), and SWM4-DP (dashed-dotted curve).

iterative-YBG method are shown in Figure 10.4. The iterative-YBG force tables obtained from different all-atom models show similar qualitative features: a deep attractive well near 3 Å, a repulsive barrier around 3.5 Å, and a second shallower attractive well that gives the second peak of RDF around 5 Å. It is instructive to see how differences in pair forces affect the resulting RDFs. Figure 10.4 indicates that the first attractive well in the CG force profiles of TIP4P and SWM4-DP water are slightly deeper than that of the TIP3P and SPC models. A similar trend is also seen in the region of repulsive interactions after the first well. These differences result in a higher and narrower first peak in the RDF for TIP4P and SWM4-DP, as shown in Figure 10.3. Another observation from Figure 10.4 is that the second attractive well of the force table of TIP3P is much less prominent than that of the other models. This observation is in line with the less pronounced second peak in the RDF of TIP3P water, indicating how pairwise forces may modulate the second peak in the RDF. The iterative-YBG method can thus be used to qualify how liquid structures are determined by pair interactions at the CG scale.

10.3.2
Anisotropic Structural Property of Liquid Water

One of the unique properties of liquid water is its tendency to form a tetrahedral hydrogen bonding network [49, 50]. Hydrogen bonds are highly directional and are responsible for much of the anomalous behavior of water [2, 51]. In this section we aim to analyze how much a purely isotropic CG model can capture the higher-order structural properties of water, that is, order parameters other than RDF. In Figure 10.5, we compare the tetrahedral structure of two water models using TIP3P and SWM4-DP potentials. The order parameter q introduced by Errington and Debenedetti is employed to measure the tetrahedral structure of water [52]:

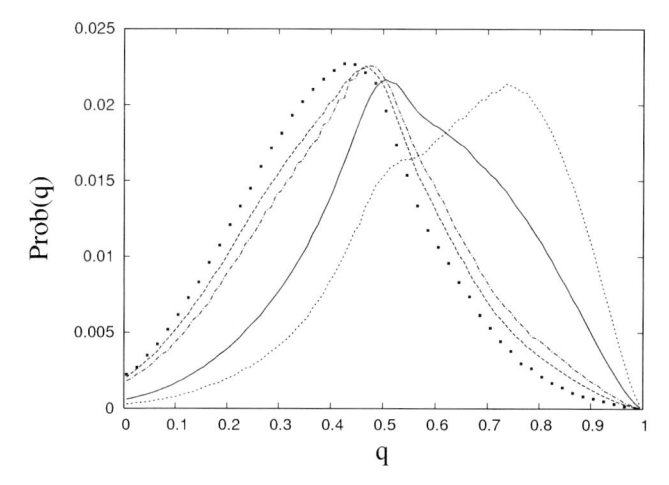

Figure 10.5 Probability distributions of tetrahedral order parameter q from CG simulation using the iterative-YBG force field for TIP3P and SWM4-DP, compared with those obtained from all-atom (AA) MD simulations with full-atomic details: AA TIP3P, full line; CG TIP3P, dashed line; AA SWM4-DP, dotted line; and CG SWM4-DP, dashed-dotted line. The probability distribution of q for an LJ fluid (filled squares) is also shown for comparison.

$$q = 1 - \frac{3}{8}\sum_{J=1}^{3}\sum_{K>J}^{4}\left(\cos\theta_{JK} + \frac{1}{3}\right)^2 \tag{10.11}$$

where θ_{JK} is the angle between two vectors from a given CG site to a pair of its neighbors (J and K) but the summation only runs over the four nearest neighbors of a CG water. The average value of q for a perfectly tetrahedral arrangement is one, and for a random distribution without tetrahedrality it is zero. Figure 10.5 indicates that the SWM4-DP all-atom model gives much more pronounced tetrahedrality than the TIP3P model due to the inclusion of charge polarizability.

Does self-consistent iteration enhance the tetrahedrality of CG water by optimizing the force table to reproduce the all-atom RDF? The distributions of q for CG TIP3P and CG SWM-DP indicate that the deeper well and higher peak in the force table of CG SWM4-DP (Figure 10.4) do lead to higher tetrahedrality than those of CG TIP3P, but the difference is much less than that between the all-atom TIP3P and SWM4-DP distributions of q, for which a qualitative difference can be seen. To further analyze this result, the probability distribution of q for an LJ fluid is also shown in Figure 10.5. The LJ fluid is found to shift to a distribution of q with less tetrahedral structure but with similar qualitative behavior to those of CG TIP3P and CG SWM4-DP. Therefore, analysis of using q as the metric of tetrahedrality indicates that single-site CG models of water are not able to recover the qualitative behavior of tetrahedrality of all-atom models, although the magnitude of the tetrahedrality as measured by q can still be increased by pair forces, as indicated by the comparison between LJ, CG TIP3P, and CG SWD4-DP simulations. This result reveals a limitation of using a CG model in capturing certain properties of the fully detailed case. The iterative-YBG method, on the other hand,

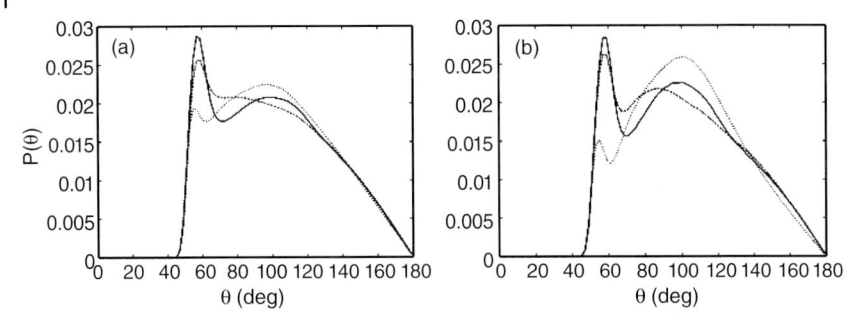

Figure 10.6 Angle distributions $P(\theta)$ defined in Equation 10.12 obtained from CG MD simulations using the iterative-YBG force field (solid line) and using the CG force field from force matching without self-consistent iteration (dashed line) are compared with those from all-atom MD simulations (dotted line) for (a) TIP3P and (b) SWM4-DP models.

can be used to obtain an optimal force field given a choice of CG model to systematically and quantitatively characterize the outcomes of adopting a CG instead of an all-atom model.

To further analyze how self-consistent iteration affects the tetrahedral structure of CG water, we measure the distribution of the angle between any three nearest neighbors of water [53]:

$$P(\theta) = Z_\rho \int_0^{R_s} \int_0^{R_s} g^{(3)}(R_{12}, R_{13}, \theta)\sin\theta R_{12}^2 R_{13}^2 dR_{12} dR_{13} \qquad (10.12)$$

Here, $g^{(3)}$ is a three-body correlation function of the CG model and Z_ρ is the normalization constant at system density ρ. The integration is conducted with an upper limit of radius R_s, which is chosen to be 3.4 Å. In Figure 10.6, we compare this angle distribution computed from all-atom MD simulations with those computed from CG simulations. The probability profiles computed from all-atom simulations show a clear feature at the tetrahedral angle around 110°, while those from CG simulations have a higher peak around 60°, which is due to molecular packing of particles with isotropic attractive interactions [54]. By comparing $P(\theta)$ obtained by CG force fields deduced from FM and iterative-YBG methods, we aim to illustrate the effects of self-consistent iteration on water tetrahedrality from a different perspective. It is clear from Figure 10.6 that, by modulating pair forces, the peak of $P(\theta)$ that corresponds to a tetrahedral structure can be enhanced. However, it is achieved at the cost of enhancing the peak due to molecular packing as well, since the attractive interaction between CG sites is increased by self-consistent iteration (see Figure 10.4).

10.3.3
Properties of CG Water by Using an iterative-YBG Force Field and Pressure Constraint

The next question is how thermodynamic properties obtained by using the iterative-YBG force field of a CG model compare with those of the all-atom reference system. In other words, what is the representability of other properties when the RDF is used as the objective function for force field parameterization? Since a significant portion of the details in an all-atom representation are removed by coarse graining, a CG force field is not expected to accurately represent all the properties of an all-atom model. Therefore, an important question in CG modeling revolves around what the properties of the original system are that one would like to reproduce. The answer to this question determines the procedure and strategy of developing a CG force field.

The iterative-YBG method determines a CG force field that reproduces the RDF of liquid water at a given thermodynamic state. Other than the RDF, we do not expect that the iterative-YBG CG force field will recover all the other properties of liquid water, such as pressure, density, thermal compressibility, or self-diffusion coefficient, although it would be desirable to be able to reproduce as many other properties as possible in addition to the RDF. We demonstrate that the requirement of reproducing a given property can be achieved by implementing this requirement as a constraint during self-consistent iteration. For example, if the pressure of an NVT simulation using an iterative-YBG CG force field differs from that of an all-atom simulation, one can impose a pressure constraint during self-consistent iteration. In doing so, we found that the target RDF can be reproduced with a force table that also gives the target pressure in an NVT simulation. This approach modifies the FM objective function to include the virial pressure constraint through the use of a Lagrange multiplier:

$$\chi^2 = \left\langle \frac{1}{3N^{CG}} \sum_{I=1}^{N^{CG}} |F_I - F_I^{CG}|^2 \right\rangle + \lambda \left[P_{AA} - \left(\rho k_B T + \frac{1}{3V} \left\langle \sum_I \sum_{J>I} R_{IJ} \cdot (F(R_{IJ}) \hat{u}_{IJ}) \right\rangle \right) \right]$$

(10.13)

where P_{AA} is the target pressure of the all-atom reference system and λ is a Lagrange multiplier. The RDF computed from using the resulting pressure-constraint iterative-YBG force field for TIP3P water (dashed line) is shown in Figure 10.7a and is compared with that without the pressure constraint (solid line). The two RDFs are identical. The converged force tables obtained from self-consistent iteration with and without the pressure constraint are shown in Figure 10.7b. It can be seen that the difference between the two force tables is nearly negligible. In other words, the pressure constraint can be satisfied with very minor changes in the force profile. Similar trends are observed in all of the all-atom reference models of water investigated in the present study. It is also important to notice that a slight modification of the CG force field may lead to a considerable change in the thermodynamic properties of a liquid. Without using a pressure constraint during self-consistent iteration, the resulting iterative-YBG force field gives a pressure on the order of thousands of atmospheres; similar results have been observed in other studies as well [55]. Another observation is that the core region of CG force fields is not affected by employing a pressure constraint; longer-range interactions are affected more. This result is consistent with the common notion that the structure of a liquid is mostly determined by short-range repulsive

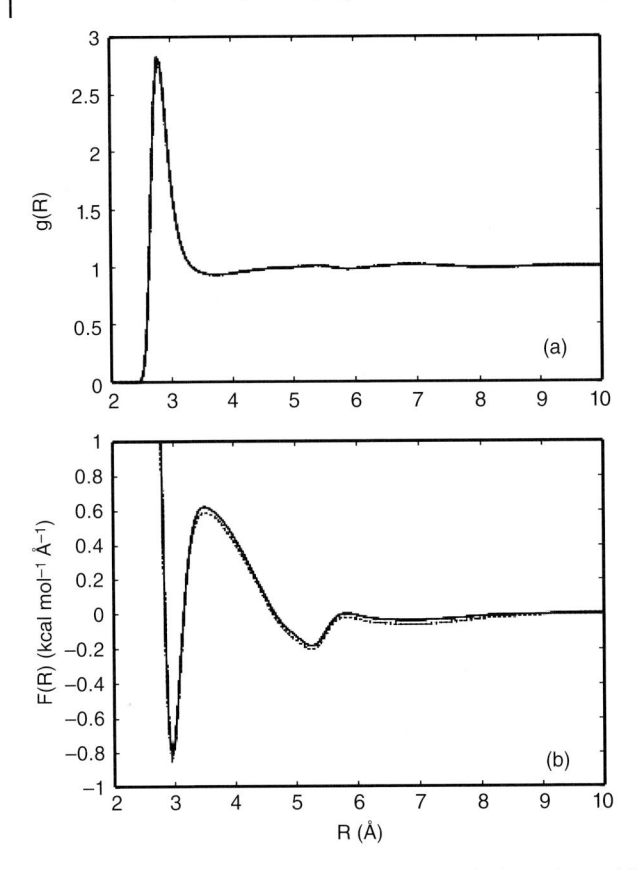

Figure 10.7 (a) The radial distribution function and (b) effective pairwise force between CG water molecules obtained from the iterative-YBG force field calculated with a pressure constraint (dotted line) compared with those obtained from the iterative-YBG force field calculated without a pressure constraint (solid line) for TIP3P liquid water model.

interactions, while thermodynamic properties and energetics are sensitive to the longer-range portion of the potential.

In Table 10.1, several properties of liquid water computed by using the iterative-YBG CG force field with pressure constraint are listed. Values of the TIP3P all-atom reference system and experiments are also shown for comparison. Different from previous simulations, the results shown in Table 10.1 are obtained by conducting simulations in the *NPT* ensemble at 1 atm and 300 K. The average density of CG water is $1.02 \, g \, cm^{-3}$, close to the results of all-atom MD and experiments. The heat of vaporization is significantly underestimated by the CG force field since the interaction energies are not consistent with that of the all-atom model. It is in fact possible to develop an iterative-YBG CG force field that reproduces the heat of vaporization value corresponding to the all-atom system by employing an energy constraint. However, we found that the pressure and energy constraints of a CG

Table 10.1 Properties of liquid water at 300 K and 1 atm using a one-site CG model with pair interaction obtained by the iterative-YBG method with a pressure constraint. Results obtained by using the TIP3P all-atom model and experiments are shown for comparison. Properties compared are liquid density ρ, heat of vaporization ΔH_{vap}, thermal expansion coefficient α_P, isothermal compressibility κ_T, constant-pressure heat capacity C_p, and self-diffusion coefficient D_{diff}.

Property	Iterative-YBG	All-atom	Experiment[a]
ρ (g cm^{-3})	1.02	0.998	0.997
ΔH_{vap} (kcal mol^{-1})	3.70	10.27	10.51
α_P (10^{-5} K^{-1})	418	88.9	25.7
κ_T (10^{-5} atm^{-1})	21.1	5.9	4.58
C_p (cal mol^{-1} K^{-1})	14.5	18.7	18.0
D_{diff} (10^{-9} m^2 s^{-1})	16.4	5.8	2.3

a) Experimental data obtained from Refs [56–59].

system cannot be reproduced simultaneously with the presence of both constraints, which results in the distortion of the RDF compared to the target profile. This result reflects the limited representability of a CG model. Furthermore, the thermal expansion coefficient α_P and isothermal compressibility κ_T of CG water are considerably higher compared to the values of the all-atom model, similar to the trends observed in previous works [23, 55], but the heat capacity C_p of CG water is in reasonable agreement with that of the TIP3P model and experiment. In fact, most of the difference between the all-atom and CG heat capacities can be compensated by taking into account the rotational degrees of freedom neglected in the CG model, which would contribute to an increase in heat capacity by 2.98 cal mol^{-1} K^{-1} (at 300 K).

10.4
Conclusions

In this chapter, the coarse graining of liquid water is described by applying a variational principle and the iterative-YBG method. This variational principle is based on the viewpoint that a CG potential aims to model the potential of mean force of the fully detailed system with the non-CG degrees of freedom integrated out. The resulting CG force field obtained by the iterative-YBG method is thus state specific and reproduces the RDF observed in an all-atom MD simulation of water. If pairwise interactions are employed in the CG model, the representability of the anisotropic structural properties of liquid water is highly limited. The iterative-YBG method can improve upon the agreement in these properties, such as tetrahedral order parameters, but the qualitative discrepancy cannot be corrected.

The representability of a CG model to thermodynamic properties is also limited due to the reduction of degrees of freedom. In order to develop a CG force field, an important question faced by the modeler is thus what physical properties need to be accurately represented by a CG model for the problem of interest. The iterative-YBG method is based on a variational principle of matching the potential of mean force of the CG degrees of freedom and is hence powerful in reproducing structural properties, such as RDF, with high accuracy. Coupled with constraints on thermodynamic properties, the scope of representability of the resulting CG force field can be enhanced. We suggest that the CG force field of water developed via the iterative-YBG approach can be used for cases where an accurate representation of liquid structure is desired.

Acknowledgment

The authors would like to acknowledge University of California, Berkeley, for supporting this research.

References

1 Cho, C.H., Singh, S., and Robinson, G.W. (1996) Liquid water and biological systems: the most important problem in science that hardly anyone wants to see solved. *Faraday Discuss.*, **103**, 19.

2 Brovchenko, I. and Oleinikova, A. (2008) Multiple phases of liquid water. *ChemPhysChem*, **9**, 2660.

3 Tanford, C. (1980) *The Hydrophobic Effect: Formation of Micelles and Biological Membranes*, John Wiley & Sons, Inc., New York.

4 Chandler, D. (2005) Interfaces and the driving force of hydrophobic assembly. *Nature*, **437**, 640.

5 Wallqvist, A. and Mountain, R.D. (1999) Molecular models of water: derivation and description. *Rev. Comput. Chem.*, **13**, 183.

6 Jorgensen, W.L. and Tirado-Rives, J. (2005) Potential energy functions for atomic-level simulations of water and organic and biomolecular systems. *Proc. Natl. Acad. Sci. U.S.A.*, **102**, 6665.

7 Guillot, B. (2002) A reappraisal of what we have learnt during three decades of computer simulations on water. *J. Mol. Liq.*, **101**, 219.

8 Vega, C., Abascal, J.L.F., Conde, M.M., and Aragones, J.L. (2008) What ice can teach us about water interactions: a critical comparison of the performance of different water models. *Faraday Discuss.*, **141**, 251.

9 Nielsen, S.O., Lopez, C.F., Srinivas, G., and Klein, M.L. (2004) Coarse grain models and the computer simulation of soft materials. *J. Phys. Condens. Matter*, **16**, R481.

10 Tozzini, V. (2005) Coarse-grained models for proteins. *Curr. Opin. Struct. Biol.*, **15**, 144.

11 Ayton, G.A., Noid, W.G., and Voth, G.A. (2007) Multiscale modeling of biomolecular systems: in serial and in parallel. *Curr. Opin. Struct. Biol.*, **17**, 192.

12 Chen, L.J., Qian, H.J., Lu, Z.Y., Li, Z.S., and Sun, C.C. (2006) An automatic coarse-graining and fine-graining simulation method: application on polyethylene. *J. Phys. Chem. B*, **110**, 24093.

13 Murtola, T., Falck, E., Patra, M., Karttunen, M., and Vattulainen, I. (2004) Coarse-grained model for phospholipid/cholesterol bilayer. *J. Chem. Phys.*, **121**, 9156.

14 Qian, H.J., Carbone, P., Chen, X., Karimi-Varzaneh, H.A., Liew, C.C., and Muller-Plathe, F. (2008) Temperature-transferable coarse-grained potentials for ethylbenzene, polystyrene, and their mixtures. *Macromolecules*, **41**, 9919.

15 Han, W., Wan, C.-K., and Wu, Y.-D. (2008) Toward a coarse-grained protein model coupled with a coarse-grained solvent model: solvation free energies of

amino acid side chains. *J. Chem. Theory Comput.*, **4**, 1891.

16 Han, W. and Wu, Y.-D. (2007) Coarse-grained protein model coupled with a coarse-grained water model: molecular dynamics study of polyalanine-based peptides. *J. Chem. Theory Comput.*, **3**, 2146.

17 Marrink, S.J., de Vries, A.H., and Mark, A.E. (2004) Coarse grained model for semiquantitative lipid simulations. *J. Phys. Chem. B*, **108**, 750.

18 Karaborni, S., Esselink, K., Hilbers, P.A.J., Smit, B., Karthauser, J., Vanos, N.M., and Zana, R. (1994) Simulating the self-assembly of gemini (dimeric) surfactants. *Science*, **266**, 254.

19 Shen, V.K., Cheung, J.K., Errington, J.R., and Truskett, T.M. (2006) Coarse-grained strategy for modeling protein stability in concentrated solutions. II: Phase behavior. *Biophys. J.*, 1949.

20 Adhangale, P.S. and Gaver, D.P. III (2006) Equation of state for a coarse-grained DPPC monolayer at the air/water interface. *Mol. Phys.*, **104**, 3011.

21 Li, T. and Nies, E. (2007) Coarse-grained molecular dynamics modeling of associating fluids: thermodynamics, liquid structure, and dynamics in the limit of zero association strength. *J. Phys. Chem. B*, **111**, 2274.

22 Izvekov, S. and Voth, G.A. (2005) A multiscale coarse-graining method for biomolecular systems. *J. Phys. Chem. B*, **109**, 2469.

23 Izvekov, S. and Voth, G.A. (2005) Multiscale coarse graining of liquid-state systems. *J. Chem. Phys.*, **123**, 134105.

24 Noid, W.G., Chu, J.-W., Ayton, G.S., and Voth, G.A. (2007) Multiscale coarse-graining and structural correlations: connections to liquid-state theory. *J. Phys. Chem. B*, **111**, 4116.

25 Noid, W.G., Chu, J.-W., Ayton, G.S., Krishna, V., Izvekov, S., Voth, G.A., Das, A., and Anderson, H.C. (2008) The multiscale coarse-graining method. I. A rigorous bridge between atomistic and coarse-grained models. *J. Chem. Phys.*, **128**, 244114.

26 Noid, W.G., Liu, P., Wang, Y., Chu, J.-W., Ayton, G.S., Izvekov, S., Anderson, H.C., and Voth, G.A. (2008) The multiscale coarse-graining method. II. Numerical implementation for coarse-grained molecular models. *J. Chem. Phys.*, **128**, 244115.

27 Ercolessi, F. and Adams, J.B. (1994) Interatomic potentials from first-principles calculations: the force-matching method. *Europhys. Lett.*, **26**, 583.

28 Izvekov, S., Parrinello, M., Burnham, C.J., and Voth, G.A. (2004) Effective force fields for condensed phase systems from *ab initio* molecular dynamics simulation: a new method for force-matching. *J. Chem. Phys.*, **120**, 10896.

29 Yvon, J. (1935) *Le Théorie Statistique des Fluides et l'Equation d'Etat*, Actes Scientifique et Industrie, no. **203**, Hermann, Paris.

30 Born, M. and Green, H.S. (1964) A general kinetic theory of liquids. I. The molecular distribution functions. *Proc. R. Soc. Lond. Ser. A*, **188**, 10.

31 Hansen, J.P. and McDonald, I.R. (1990) *Theory of Simple Liquids*, 2nd edn, Academic Press, San Diego, CA.

32 Cho, H.M., and Chu, J.-W. (2009) Inversion of Radial Distribution Functions to Pair Forces by Solving the Yvon-Born-Green Equation Iteratively. *J. Chem. Phys.* **131**, 134107.

33 Lyubartsev, A.P. and Laaksonen, A. (1995) Calculation of effective interaction potentials from radial distribution functions: a reverse Monte Carlo approach. *Phys. Rev. E*, **52**, 3730.

34 McGreevy, R.L. (2001) Reverse Monte Carlo modelling. *J. Phys. Condens. Matter*, **13**, R877.

35 Henderson, R.L. (1974) A uniqueness theorem for fluid pair correlation functions. *Phys. Lett.*, **49A**, 197.

36 Jorgensen, W.L., Chandrasekhar, J., Madura, J.D., Impey, R.W., and Klein, M.L. (1983) Comparison of simple potential functions for simulating liquid water. *J. Chem. Phys.*, **79**, 926.

37 Berendsen, H.J.C., Postma, J.P.M., van Gunsteren, W.F., and Hermans, J. (1981) *Intermolecular Forces* (ed. B. Pullman), Reidel, Dordrecht.

38 Lamoureux, G., MacKerell, A.D. Jr, and Roux B. (2003) A simple polarizable model of water based on classical Drude oscillators. *J. Chem. Phys.*, **119**, 5185.

39 Hirschfelder, J.O., Curtiss, C.F., and Bird, R.B. (1964) *The Molecular Theory of Gases and Liquids*, John Wiley & Sons, Inc., New York.

40 Savitzky, A. and Golay, M.J.E. (1964) Smoothing and differentiation of data by simplified least squares procedures. *Anal. Chem.*, **36**, 1627.

41 Steinier, J., Termonia, Y., and Deltour, J. (1972) Comments on smoothing and differentiation of data by simplified least square procedure. *Anal. Chem.*, **44**, 1906.

42 Brooks, B.R., Bruccoleri, R.E., Olafson, B.D., States, D.J., Swaminathan, S., and Karplus, M. (1983) A program for macromolecular energy, minimization, and dynamics calculations. *J. Comput. Chem.*, **4**, 187.

43 Nosé, S. (1984) A unified formulation of the constant temperature molecular dynamics methods. *J. Chem. Phys.*, **81**, 511.

44 Hoover, W.G. (1985) Canonical dynamics: equilibrium phase-space distributions. *Phys. Rev. A*, **31**, 1695.

45 Feller, S.E., Zhang, Y., Pastor, R.W., and Brooks, B.R. (1995) Constant pressure molecular dynamics simulation: the Langevin piston method. *J. Chem. Phys.*, **103**, 4613.

46 Darden, T., Perera, L., Li, L., and Pedersen, L. (1999) New tricks for modelers from the crystallography toolkit: the particle mesh Ewald algorithm and its use in nucleic acid simulations. *Structure*, **7**, R55.

47 Darden, T., York, D., and Pedersen, L. (1993) Particle mesh Ewald: an $N\log(N)$ method for Ewald sums in large systems. *J. Chem. Phys.*, **98**, 10089.

48 Ryckaert, J.-P., Ciccotti, G., and Berendsen, H.J.C. (1977) Numerical integration of the cartesian equations of motion of a system with constraints: molecular dynamics of n-alkanes. *J. Comput. Phys.*, **23**, 327.

49 Smith, J.D., Cappa, C.D., Wilson, K.R., Messer, B.M., Cohen, R.C., and Saykally, R.J. (2004) Energetics of hydrogen bond network rearrangements in liquid water. *Science*, **306**, 851.

50 Head-Gordon, T. and Johnson, M.E. (2005) Tetrahedral structure or chains for liquid water. *Proc. Natl. Acad. Sci. U.S.A.*, **103**, 7973.

51 Stillinger, F.H. (1980) Water revisited. *Science*, **209**, 451.

52 Errington, J.R. and Debenedetti, P.G. (2001) Relationship between structural order and anomalies of liquid water. *Nature*, **409**, 318.

53 Johnson, M.E., Head-Gordon, T., and Louis, A. (2007) Representability problems for coarse-grained water potentials. *J. Chem. Phys.*, 144509.

54 Conway, J.H. and Sloane, N.J.A. (1998) *Sphere Packings, Lattices and Groups*, 3rd edn, Springer, New York.

55 Wang, H., Junghans, C., and Kremer, K. (2009) Comparative atomistic and coarse-grained study of water: what do we lose by coarse-graining? *Eur. Phys. J. E*, **28**, 221.

56 Kell, G.S. (1875) Density, thermal expansivity, and compressibility of liquid water from 0° to 150°C: correlations and tables for atmospheric pressure and saturation reviewed and expressed on 1968 temperature scale. *J. Chem. Eng. Data*, **20**, 97.

57 Dorsey, N.E. (1940) *Properties of Ordinary Water Substance*, Reinhold, New York.

58 Mills, R. (1973) Self-diffusion in normal and heavy water in the range 1–45°C. *J. Phys. Chem.*, **77**, 685.

59 Krynicki, K., Green, C.D., and Sawyer, D.W. (1978) Pressure and temperature dependence of self-diffusion in water. *Faraday Discuss. Chem. Soc.*, **66**, 199.

11
Molecular Dynamics Simulations of Biomolecules in a Polarizable Coarse-Grained Solvent

Tap Ha-Duong, Nathalie Basdevant, and Daniel Borgis

11.1
Introduction

Among the various physical forces that drive folding and recognition of biomolecules, it is hardly necessary to emphasize the important influence of the solvent, particularly its electrostatic and hydrophobic effects, on their tertiary and quaternary structures. From an energetic point of view, microcalorimetry experiments, which can measure solvation free energies, combined with osmotic pressure and volumetric approaches, are able to study the role of hydration and dehydration in the stability and specificity of biomolecular assemblies. These studies have shown that the binding reactions of biomolecules are generally accompanied by a dehydration of their surfaces, which increases the entropy of the solvent and then favorably decreases the solvation free energy of the complexes [1–3]. However, in many cases, some water molecules are trapped on biomolecule surfaces to stabilize their structure or complexes, yielding an unfavorable decrease in the solvation entropy, which is compensated by a favorable decrease of the solvation enthalpy [4, 5]. Many crystallographic and nuclear magnetic resonance (NMR) experiments have shown that selected water molecules participate extensively in the stabilization of the three-dimensional conformation of many macromolecules by filling their unsatisfied hydrogen bond sites [6–9], or influence the specificity of biomolecular association by forming "aqueous bridges" between two molecular surfaces that do not fit perfectly [10–12]. Hence, correlations between microscopic solvent structure and solvation free energy variations are difficult to assess, and theoretical descriptions of biomolecule hydration are necessary to relate X-ray and NMR observations to thermodynamic measures and to help to understand the contribution of water activity to biological processes.

Modeling the solvent around a biomolecule can be performed using two rather extreme strategies. The first one consists in considering explicitly the many solvent molecules around the macromolecule, using an empirical potential compatible with the solute force fields, for example, the well-known "transferable intermolecular potential, three-point" (TIP3) [13] or "simple point charge" (SPC) [14] models. Molecular dynamics (MD) simulations using explicit water molecules seem to be the most widely used method, since, in principle, such a method should provide a complete and detailed description of the biomolecule hydration properties. However, a proper treatment of biomolecule solvation generally requires the consideration of a large number of water molecules around the solute, and the computation of the solute–solvent and solvent–solvent interactions rapidly becomes a bottleneck in the calculations. Hence, in practice, such approaches prove to be quite inefficient for calculating thermodynamics properties, such as free energies, because their computation in terms of average microscopic quantities requires excessively long and costly simulations to obtain sufficient statistics. This is especially true for the dielectric properties of polar solvents, the convergence of which is well known to be very slow.

In the second class of methods, so-called implicit solvent models, the aqueous solvent is accounted for by a structureless high-dielectric continuum obeying the macroscopic laws of electrostatics. One of the most widely used methods of that kind is based on the Poisson–Boltzmann equation, which can be solved on a grid for each microscopic internal conformation of the solute. This can be done by using various efficient numerical techniques such as finite differences [15, 16], surface boundary elements [17, 18], finite elements [19], multi-dimensional optimization [20], or a recent method based on a "Car–Parrinello" minimization of the polarization density free energy functional [21]. Among other continuum models, one should also mention the numerically efficient approaches formulated in terms of effective pairwise additive interactions such as the screened Coulomb potentials model [22] or the generalized Born approach [23, 24] (described in other chapters), which allow easy calculations of atomic forces and fast molecular dynamics simulations. All those methods have the advantage of yielding the solvation contribution directly in terms of free energy rather than energy. However, they have the drawback of neglecting the molecular nature of water and cannot provide useful information about the solvent structure around solutes, which X-ray and NMR experiments or MD simulations in explicit solvent can provide.

Recently, there have been attempts to devise methods that still cope with the molecular nature of the solvent, but do not consider explicitly all its microscopic degrees of freedom. The development of such methods, with a status in between a purely macroscopic electrostatic description and all-atom microscopic simulations, can benefit from the accumulated knowledge in the statistical mechanics of liquids [25]. Among these approaches, one should mention the three-dimensional integral equation theory around complex molecules [26, 27], lattice models such as the Langevin dipole approach of Warshel et al. [28, 29] or the multipolar lattice gas [30]. Another promising approach is the density-functional theory (DFT) of molecular liquids [31, 32]. An important "computational" application of the DFT approach can be found in the work of Lowen et al. [33], who combined the DFT and Car–Parrinello type dynamical minimization for treating the counterion environment around an assembly of large polyions. This methodology was later applied to the case of the solvent electrostatic polarization field around a solute of complex shape [21].

A natural way to handle DFT equations is to define the solvent density on a grid, whose spacing should be much finer than the solvent molecule dimensions, at least in the immediate vicinity of the solute. For large biomolecules, one should then use large grid sizes, leading to problems of memory and computer time. Furthermore, for a molecular solvent, one should optimize both the positional and orientational densities, the latter being known to present strongly nonlinear behavior near an interface and problems of convergence. In the past few years, we have proposed that another route is explored, which, although inspired by DFT ideas, explicitly keeps the notion of particles for describing the solvent, and allows these difficulties to be overcome. In our mixed approach, the translational space is explored by means of particles, whereas the orientational degrees of freedom are treated at equilibrium using an orientational density functional. As for explicit solvent models, there is an internal consistency and a number of advantages in treating both the solute and solvent with particles: natural definition of excluded volume, possibility of constant-temperature or constant-pressure algorithms, atomic-like force fields, and so on.

In this chapter, we present a treatment of biomolecule hydration, named the polarizable pseudo-particle (PPP) solvent model, which was introduced in Refs. [34–36] and goes some way to fulfilling the following three expectations: (i) to provide a rapid and proper calculation of the electrostatic solvation free energies; (ii) to account for the molecular nature and structure of solvent around solutes; and (iii) to allow long molecular dynamics simulations of biological molecules at a much smaller computational cost than with explicit water models. As will be mentioned in the next section, the PPP model is easily generalizable to a coarse-grained version, in which each grain represents a chunk of matter with several embedded water molecules, instead of one, and which is fully consistent with a coarse-grained representation of biomolecules such as in Refs. [36, 37].

11.2
Theory

To arrive at our simplified model of a solvent, we present here a "top-down" approach, going from a purely macroscopic description to a "phenomenological" mesoscopic or microscopic one. Nevertheless, interested readers will find in Ref. [38] a "bottom-up" statistical approach, in which particle-based effective models similar to the phenomenological one can be rigorously derived from a microscopic Hamiltonian of a polar solvent.

11.2.1
A Non-Local Formulation of the Dielectric Continuum Theory in Terms of Polarization Density

In methods based on the Poisson–Boltzmann equation, the dielectric continuum surrounding a solute is characterized by its electrostatic potential, which has to be solved on a grid discretizing the space. As shown by Marcus [39], an alternative equivalent approach is to formulate the solvent dielectric properties in terms of a continuous polarization density field. Within this electrostatic theory, the non-equilibrium electrostatic free energy of a charge distribution $\rho_0(r)$ immersed in a dielectric medium can be written as a functional of the polarization density $P(r)$ at position r in the system:

$$F_{nl}[P(r)] = \int \frac{P^2(r)}{2\varepsilon_0 \chi(r)} \, dr - \int P(r) E_0(r) \, dr - \frac{1}{2} \iint P(r) T(r-r') P(r') \, dr \, dr' \tag{11.1}$$

Here ε_0 is the vacuum permittivity; $\chi(r)$ is the dielectric susceptibility, related to the dielectric constant by the relation $\chi(r) = \varepsilon(r) - 1$; $E_0(r)$ is the electrostatic field created in vacuum by the solute charge distribution $\rho_0(r)$; and $T(r-r')$ is the dipolar tensor. The first integral in the above equation represents the free energy necessary to polarize the solvent, and the second one accounts for the interactions between the solute charges and the solvent dipoles. The third term accounts for the mutual induction of the solvent dipoles at different positions and makes the functional F_{nl} non-local in space. If we introduce the polarization electric field created by the dielectric solvent surrounding the volume dr and defined as $E_p(r) = \int T(r-r') P(r') dr'$, then the previous functional can also be written as

$$F_{nl}[P(r)] = \int \frac{P^2(r)}{2\varepsilon_0 \chi(r)} \, dr - \int P(r) E_0(r) \, dr - \frac{1}{2} \int P(r) E_p(r) \, dr \tag{11.2}$$

One should remember here that the non-uniform polarization density creates a local density of "polarization charge" $\rho_{pol}(r) = -\nabla \cdot P(r)$ (see Refs. [40, 41]), so that the total electric field defined as $E(r) = E_0(r) + E_p(r)$ obeys the Gauss law $\varepsilon_0 \nabla \cdot E(r) = \rho_0(r) + \rho_{pol}(r)$. Introducing the dielectric displacement vector defined as $D(r) = \varepsilon_0 E(r) + P(r)$, it is then easy to recover the general Maxwell equation for a dielectric medium: $\nabla \cdot D(r) = \rho_0(r)$.

For a fixed solute charge distribution, the minimization of the functional F_{nl} with respect to $P(r)$ leads to the constitutive equation of the dielectric at equilibrium:

$$\tilde{P}(r) = \varepsilon_0 \chi(r) E(r) \tag{11.3}$$

This last equation is strictly equivalent to the known constitutive relation $D(r) = \varepsilon(r) E(r)$, which is used to determine the two vectors $D(r)$ and $E(r)$. Nevertheless, owing to the dependence of the electric field $E(r)$ on the polarization field $E_p(r)$, one should keep in mind that the determination of the non-local polarization density requires an iterative procedure that could turn out to be very CPU timeconsuming.

11.2.2
A Local Polarization Density Free-Energy Functional

An approximate polarization density functional that has the interesting property of being *local* in space can be obtained by supposing that the dielectric displacement $D(r)$ is equal to $\varepsilon_0 E_0(r)$. This approximation originates from the fact that those two vectors obey the same Maxwell–Gauss differential equation mentioned in the previous section. In principle, this property does not imply that the two fields are equal, but only that they differ by a curl. The complete equality is strictly satisfied in some special situations, such as a uniform dielectric medium, spherical or planar geometry [40, 42], but it yields a much simpler free-energy functional than Equations 11.1 or 11.2. This approximation has been used in many solvation theories, for example, to study the charge-transfer reactions in solution [43, 44]. It has also been used to obtain a generalized Born formulation of the solvation free energy of a solute with sums of self-terms and pairwise contributions [45, 46]. In particular, Schaefer and Froemmel have tested this approximation for a model spherical protein and for the superoxide dismutase and have found a remarkable agreement with finite-difference Poisson–Boltzmann calculations [45].

Under the condition that $D(r) = \varepsilon_0 E_0(r)$ and from the definition of the dielectric displacement vector, one can determine the polarization field as $E_p(r) = -P(r)/\varepsilon_0$. Injected into the non-local functional of Equation 11.2, this relation yields the following *local* polarization density functional [43]:

$$F_{\text{lol}}[P(r)] = \int \frac{\varepsilon(r) P^2(r)}{2\varepsilon_0 \chi(r)} dr - \int P(r) E_0(r) dr \tag{11.4}$$

One should emphasize here that the solvent dipole–dipole interactions accounted for in the third term of Equations 11.1 and 11.2 have mathematically vanished in the local functional, but are still implicitly captured in its first term. Minimizing the local functional with respect to $P(r)$ allows the direct determination of the polarization density field as well as the electrostatic free energy of the system at thermodynamic equilibrium:

$$\tilde{P}(r) = \varepsilon_0 \frac{\chi(r)}{\varepsilon(r)} E_0(r) \tag{11.5}$$

$$F_{\text{eq}} \equiv F_{\text{lol}}[\tilde{P}(r)] = -\int \frac{\varepsilon_0}{2} \frac{\chi(r)}{\varepsilon(r)} E_0^2(r) dr \tag{11.6}$$

Again, the three previous equations are exact and completely equivalent to the non-local formulation for systems with particular geometry, for example, a charge in a spherical cavity. In other situations, the local functional provides an approximate description of electrostatics whose validity should be evaluated. For example, in the simple case of a point dipole μ embedded in a spherical cavity of radius R and internal dielectric constant $\varepsilon_i = 1$, immersed in a solvent with a dielectric constant ε_s, the local approximation yields the solvation free energy:

$$\Delta F_{\text{eq}} = -\frac{1}{3} \frac{\varepsilon_s - 1}{\varepsilon_s} \frac{\mu^2}{R^3} \tag{11.7}$$

whereas its exact expression is the Kirkwood formula [47]:

$$\Delta F_K = -\frac{1}{2} \frac{\varepsilon_s - 1}{\varepsilon_s + \frac{1}{2}} \frac{\mu^2}{R^3} \tag{11.8}$$

Thus, for a high-dielectric solvent such as water, the local approximation does provide the correct ε_s, μ, and R dependences, but it systematically introduces a relative error of two-thirds in the solvation free energy with respect to the "true" electrostatics. Possible ways to correct this error are to define an effective radius $R_{\text{eff}} = (2/3)^{1/3} R$ or to assign an effective dipole $\mu_{\text{eff}} = (3/2)^{1/2} \mu$ for dipolar entities (see the parameterization section 11.3.1).

11.2.3
Particular Case of Solutes with Uniform Dielectric Constant

Until now, no details have been specified about the dependence of the dielectric constant $\varepsilon(r)$ on the chemical composition of the polarizable medium. In the classical all-atom models of biomolecules, such as AMBER [48] or CHARMM [49], the solute dielectric constant is often fixed to unity, so that the integral of Equation 11.6, which normally runs over the whole space, only has to be calculated over the solvent region. However, this simplification is not possible if we want to account for the polarizability of biomolecules. In this section, we want to show that the integral of the equilibrium solvation free energy (Equation 11.6) can nevertheless be calculated over only the solvent region, in the approximate case of a solute characterized by an uniform dielectric constant.

For a heterogeneous dielectric system, we can formally write

$$\frac{\chi(r)}{\varepsilon(r)} = \frac{\chi_i}{\varepsilon_i} + \left(\frac{\chi_s}{\varepsilon_s} - \frac{\chi_i}{\varepsilon_i}\right) H(r) \tag{11.9}$$

where χ_i and ε_i are the susceptibility and dielectric constant inside the solute, whereas χ_s and ε_s are characteristic of the solvent; and $H(r)$ is a "shape" function that allows one to differentiate the solute from the solvent: $H(r) = 0$ inside the solute and $H(r) = 1$ in the solvent. Inserting Equation 11.9 into Equation 11.6 yields

$$F_{\text{eq}} = -\frac{\varepsilon_0}{2} \frac{\chi_i}{\varepsilon_i} \int E_0^2(r)\, dr - \frac{\varepsilon_0}{2}\left(\frac{\chi_s}{\varepsilon_s} - \frac{\chi_i}{\varepsilon_i}\right) \int H(r) E_0^2(r)\, dr \tag{11.10}$$

In the case where $\mathbf{E}_0$ is created by point charges q_i, the first term in the previous equation has the following simple analytical form (discarding the self-energy of the charges, which acts as a constant):

$$\int E_0^2(r)\, dr = \frac{1}{4\pi\varepsilon_0} \sum_{i \neq j} \frac{q_i q_j}{R_{ij}} = \frac{2}{4\pi\varepsilon_0} \sum_{i<j} \frac{q_i q_j}{R_{ij}} \tag{11.11}$$

and the second integral has to be calculated over only the solvent volume, because the function $H(r)$ is null inside the solute: $\int H(r) E_0^2(r)\, dr = \int_{\text{solv}} E_0^2(r)\, dr$. The equilibrium electrostatic free energy for the heterogeneous dielectric system can then be written as

$$F_{\text{eq}} = -\frac{1}{4\pi\varepsilon_0} \frac{\chi_i}{\varepsilon_i} \sum_{i<j} \frac{q_i q_j}{R_{ij}} - \frac{\varepsilon_0}{2}\left(\frac{\chi_s}{\varepsilon_s} - \frac{\chi_i}{\varepsilon_i}\right) \int_{\text{solv}} E_0^2(r)\, dr \tag{11.12}$$

A nice consequence of the above expression is that, if we now consider the total electrostatic energy V_{elec} of the system defined as the sum of the direct Coulomb interactions between the solutes charges q_i and the free energy F_{eq}, then its first term combines with the direct interactions and yields an effective Coulomb potential of point charges in a uniform medium of dielectric constant ε_i:

$$V_{elec} = \frac{1}{4\pi\varepsilon_0} \sum_{i<j} \frac{q_i q_j}{R_{ij}} + F_{eq} = \frac{1}{4\pi\varepsilon_0 \varepsilon_i} \sum_{i<j} \frac{q_i q_j}{R_{ij}} - \frac{\varepsilon_0}{2}\left(\frac{\chi_s}{\varepsilon_s} - \frac{\chi_i}{\varepsilon_i}\right)\int_{solv} E_0^2(r)\, dr \quad (11.13)$$

One should emphasize here that the last integral is no longer an electrostatic solvation free energy, but represents the necessary work to transfer the solute from a medium with dielectric ε_i to the polar solvent with dielectric ε_s.

11.2.4
Electrostatics on Particles

The numerical calculation of the integral in Equations 11.12 and 11.13 can be classically performed using a regular grid outside the solute. Introducing the density of the grid mesh $\rho = 1/dr_k$, the total electrostatic energy can be computed as

$$V_{elec} = \frac{1}{4\pi\varepsilon_0 \varepsilon_i} \sum_{i<j} \frac{q_i q_j}{R_{ij}} - \frac{\varepsilon_0}{2\rho}\left(\frac{\chi_s}{\varepsilon_s} - \frac{\chi_i}{\varepsilon_i}\right)\sum_k E_0^2(r_k) \quad (11.14)$$

If each node of the solvent grid is now characterized by an effective induced dipole $\tilde{\mu}_k = \alpha_e E_0(r_k)$ reacting to the electric field created by the solute charges with effective polarizability $\alpha_e = (\varepsilon_0/\rho)(1/\varepsilon_i - 1/\varepsilon_s)$, then the above equation can be reformulated as

$$V_{elec} = \sum_{i<j} \frac{q_i q_j}{4\pi\varepsilon_0 \varepsilon_i R_{ij}} - \frac{1}{2}\sum_k \tilde{\mu}_k E_0(r_k) \quad (11.15)$$

It should be noted here that the introduced polarizability α_e is defined in terms of the homogeneous macroscopic dielectric constants of the solute and solvent, and thus includes both *electronic* and *orientational* polarizations.

At this stage of the solvation theory, the originality of our approach is to consider that the induced dipoles $\tilde{\mu}_k$ are carried by moving soft particles, instead of the fixed nodes of a regular grid. Hence, the solvent model turns out to be a pseudo-molecular non-polar polarizable fluid placed in the external electric field of the solute. In this scheme, if we now consider an explicit molecular solute, with atomic positions R_i, surrounded by solvent pseudo-particles at positions r_k, we can define a phenomenological all-particle-based Hamiltonian that will be used to model the dynamics of the system:

$$H = K(\dot{R}_i, \dot{r}_k) + V_{bond}(R_i) + V_{VDW}(R_i, r_k) + V_{elec} \quad (11.16)$$

Here K is the total kinetic energy; V_{bond} denotes the solute bonded interactions, including its stretching, bending, and torsion potentials; and V_{VDW} represents the solute–solute, solute–solvent, and solvent–solvent van der Waals energy terms. In classical all-atom force fields, these latter terms are generally modeled with 6–12 Lennard-Jones energy functions. It should be noted that these non-polar potentials allow one to avoid all the problems related to the definition of the solute–solvent boundary, since both solute and solvent particles naturally have a van der Waals excluded volume. The last term of the above Hamiltonian is the total electrostatic energy given in Equation 11.15 and which again includes the direct Coulomb interactions between the solute charges and the polarization free energy F_{eq} that replaces the classical solute–solvent and solvent–solvent bare electrostatic interactions.

It should be emphasized that, in our particle-based solvent model, the free energy expressed in Equations 11.6 and 11.12 is now only an "orientational free energy," since it is calculated on an "instantaneous disordered grid" whose nodes are the solvent particles. Because the solvent particles have translational degrees of freedom, the complete polarization free energy of the system is simply retrieved by averaging F_{eq} over a significant number of particle positions. If molecular dynamics simulations are used to explore the solvent translational phase space, relatively short trajectories reveal enough to yield meaningful statistics, since the solvent particles formally interact with each other through only short-range van der Waals potentials, due to the "local approximation" [34].

Lastly, it can be useful to specify that the electric forces acting on the solute charges and solvent-induced dipoles arise from the total electrostatic energy V_{elec} and can be formulated in terms of partial derivatives of the vacuum electric field $E_0(r_k)$ with respect to the solute and solvent positions:

$$F_k = -\frac{\partial V_{elec}}{\partial r_k} = \tilde{\boldsymbol{\mu}}_k \frac{\partial E_0(r_k)}{\partial r_k} \tag{11.17}$$

$$F_i = -\frac{\partial V_{elec}}{\partial R_i} = \frac{q_i}{\varepsilon_i} E_0(R_i) + \sum_k \tilde{\boldsymbol{\mu}}_k \frac{\partial E_0(r_k)}{\partial R_i} \tag{11.18}$$

Since the solute–solvent electrostatic forces vary as the inverse of the distance to the power 5, another noticeable advantage of the solvent model is the possibility of switching off these interactions with a weak loss of accuracy, and thus to save computational costs.

11.2.5
A Phenomenological Dipolar Saturation

As previously mentioned, the local approximation introduces a systematic error in the estimation of the solvation free energy of point dipoles embedded in spherical cavities. The same approximation also leads to an incorrect potential of mean force for an ion pair with opposite charges. Indeed, our first tests have shown that this potential appears to be too repulsive and without the two minima associated with the contact ion pair and the solvent-separated ion pair, predicted from simulations in explicit water models [38]. In order to cure this deficiency, we have added to the model another level of phenomenology. In the spirit of the Langevin dipoles of Warshel et al. [28, 50, 51], we assumed that the microscopic induced dipoles $\tilde{\boldsymbol{\mu}}_k$

saturate when the electric field becomes too strong, instead of a simple proportionality. This can be accounted for by using the Langevin function $\mathscr{L}(x) = \coth(x) - 1/x$ and an extra parameter μ_s that fixes the saturation value:

$$\tilde{\mu}_k = \mu_s \mathscr{L}[(3\alpha_e/\mu_s)E_0(r_k)]\frac{E_0(r_k)}{E_0(r_k)} \quad (11.19)$$

It should be noted that, near zero, the Langevin function is $\mathscr{L}(x) \approx x/3$, and the induced dipoles recover their linear expression $\tilde{\mu}_k \approx \alpha_e E_0(r_k)$ for low electric field. With this saturation assumption, the simple formulations of the solute–solvent electrostatic forces (Equations 11.17 and 11.18) remain valid, provided that the total electrostatic energy (Equation 11.15) is replaced by the following expression:

$$V_{elec} = \sum_{i<j}\frac{q_i q_j}{4\pi\varepsilon_0\varepsilon_i R_{ij}} - \frac{\mu_s^2}{3\alpha_e}\sum_k \ln\left\{\frac{\sinh[(3\alpha_e/\mu_s)E_0(r_k)]}{(3\alpha_e/\mu_s)E_0(r_k)}\right\} \quad (11.20)$$

The added phenomenological dipole saturation appears necessary to treat a solvated neutral polar solute correctly using the local approximation, but it can also be justified using microscopic arguments [38]. Indeed, considering the rigorous Hamiltonian of a Stockmayer fluid, composed of dipolar Lennard-Jones microscopic particles, it can be shown that, using a standard DFT scheme [32], its equilibrium "orientational free energy" and averaged dipoles are very similar to those written in Equations 11.19 and 11.20. Compared to the "top-down" phenomenological model, the "bottom-up" statistical approach introduces a slightly different Langevin function, but yields the same physical dipolar saturation property.

11.3
Applications: Solvation of All-Atom Models of Biomolecules

The solvent polarizable pseudo-particles described above were used to study the dynamic solvation properties of biomolecules, ranging from peptides and small proteins to oligonucleotides, all of which are described at the atomic scale. We show here that the PPP solvent model first allows correct simulation of the dynamic three-dimensional structures of biomolecules and secondly provides a satisfactory description of their solvation in terms of electrostatic free energy and hydration preferential sites.

11.3.1
Parameters and Simulation Conditions

We summarize here the main steps of the solvent model parameterization procedure, whose details can be found in Ref. [34]. First, we assigned intrinsic characteristics for each solvent particle, which is assumed to have the mass and size of one water molecule. Its self Lennard-Jones parameters are fixed to the values $\sigma_{LJ} = 2.88$ Å and $\varepsilon_{LJ} = -0.764$ kcal mol^{-1}. The dipolar saturation assigned value is $\mu_s = 1.5$ D, a compromise between the values 1.25 and 1.7 D, respectively, used by

Warshel et al. in their Langevin dipole model [28] and by Pollock and Alder for their Stockmayer particle permanent dipole [52]. The solvent density is chosen close to that of water: $\rho = 0.0337\,\text{Å}^{-3}$. At the simulation temperature $T = 300\,\text{K}$, those parameters correspond to the reduced variables $\sigma^* = \rho\sigma_{LJ}^3 = 0.8$, $T^* = k_B T/\varepsilon_{LJ} = 0.8$, and $\mu^* = \mu_s/(\varepsilon_{LJ}\sigma_{LJ}^3)^{1/2} = 1.3$, which meet the conditions for dipolar Lennard-Jones particles to be in the liquid phase [53] with a high dielectric constant [52]. The calculations presented here were performed with a dielectric constant $\varepsilon_i = 1$ inside the biomolecules, as is usually done in standard all-atom force fields, and $\varepsilon_s = 80$ in the solvent. It should be noted that the solvent particle self-diffusion coefficient computed with these parameters is equal to $1.8 \times 10^{-4}\,\text{cm}^2\,\text{s}^{-1}$ at 300 K, which is about one order of magnitude higher than that of water ($2.4 \times 10^{-5}\,\text{cm}^{-2}\,\text{s}^{-1}$). This means that the solvent dynamics is greatly accelerated and the efficiency of exploration of the solute phase space is probably improved [54].

In a second step, for the solute–solvent Lennard-Jones interactions, the atomic radii were rescaled by a factor that depends on their initial radii and partial charges, and which was determined in order to best fit the Born solvation free energy [55] of spherical ions with various radius and charge [34]. Lastly, as shown in the theory section 11.2.2, the local approximation leads to an electrostatic solvation energy of dipolar hard spheres that is underestimated by a factor of 2/3 with respect to the Kirkwood theory [47]. This can be corrected by renormalizing the dipole moment by a factor $f_q = (3/2)^{1/2} \approx 1.225$. In that spirit, we renormalized the dipolar moment of the biomolecule building units (backbones, side chains, riboses, bases, and so on), but with a slightly lower factor $f_q = 1.185$. This value, which probably arises from the non-spherical solute boundary softness, provides more satisfactory results, in terms of both structural stability and solvation free energy [34].

The all-atom force field used for biomolecule modeling is AMBER94 [48]. Molecular dynamics simulations were performed with the software ORAC [56], which uses a multiple time step r-RESPA algorithm [57]. Three different time steps were used: 1 fs for the bonded and short-range non-bonded interactions (lower than 5 Å); 2 fs for the medium-range (between 5 and 8 Å) interactions; and 4 fs for the long-range interactions. The non-bonded interactions were smoothly truncated with a switching function between 11 and 12 Å. All simulations were performed in the canonical NVT (NVT, constant number of particles, volume, and temperature) ensemble, using a Nosé–Hoover thermostat [58]. The solvated biomolecules were progressively heated and generally equilibrated at 300 K for several hundred picoseconds, then simulated for several nanoseconds without any constraints except those on bonds involving hydrogens, using the SHAKE algorithm [59]. Typically, for small proteins and oligonucleotides, a simulation box of about $50 \times 50 \times 70\,\text{Å}^3$ was used, with around 5200 solvent particles.

11.3.2
Stability of Small Proteins in the PPP Solvent

The capability of our solvent model to correctly mediate the intramolecular forces that determine the solute structure stability were tested on three polypeptides: the

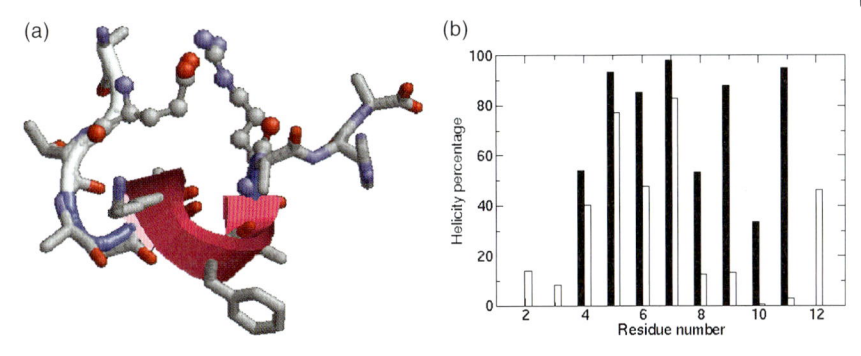

11.3 Applications: Solvation of All-Atom Models of Biomolecules | 261

Figure 11.1 (a) Snapshot of the RN24 peptide at the end of its simulation at 275 K. The RN24 residues 2 and 10 are displayed with balls. (b) RN24 helicity percentage per residue calculated from simulations at 275 K (black bars) and 355 K (white bars).

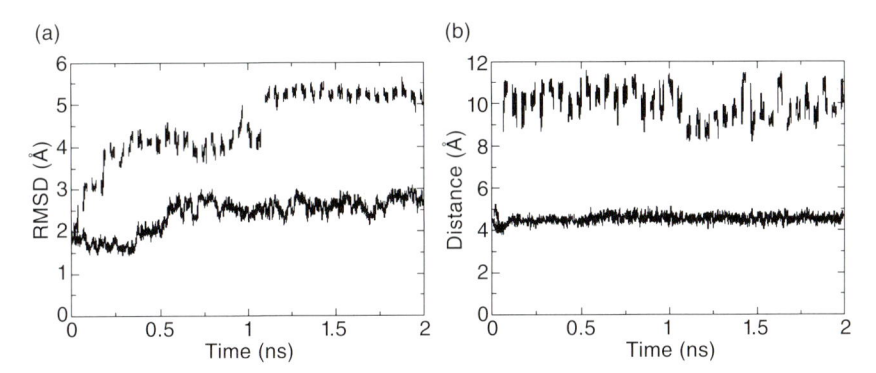

Figure 11.2 Time evolutions of (a) the RN24 peptide RMSD from its initial conformation and (b) the Glu2-Arg10 side-chain distance at 275 K (black lines) and 355 K (gray lines).

N-terminal α-helix of the ribonuclease A protein (RN24), the bovine pancreatic trypsin inhibitor (1BPI), and the B1 immunoglobulin-binding domain of streptococcal protein G (1PGB). In these three studies, no counterions were added to the simulated systems.

The RN24 sequence is Ala-Glu-Thr-Ala-Ala-Ala-Lys-Phe-Leu-Arg-Ala-His-Ala. Circular dichroism (CD) experiments show that 50–60% of its residues are in a helical conformation at low temperature [60]. NMR studies specify that its N-terminus is less helical than the remainder of the peptide and emphasize a stabilizing salt bridge between the Glu2 and Arg10 side chains (Figure 11.1) [61]. Here we compare with experimental results from two molecular dynamics simulations of the RN24 peptide, one at a low temperature of 275 K and the other one at 355 K. As shown in Figure 11.2, the peptide root-mean-square deviation (RMSD) from its initial structure is stable around 3 Å at 275 K, but it increases to a larger value around 5 Å at 355 K, indicating that the α-helix is not well conserved at high temperature, as expected from experiments.

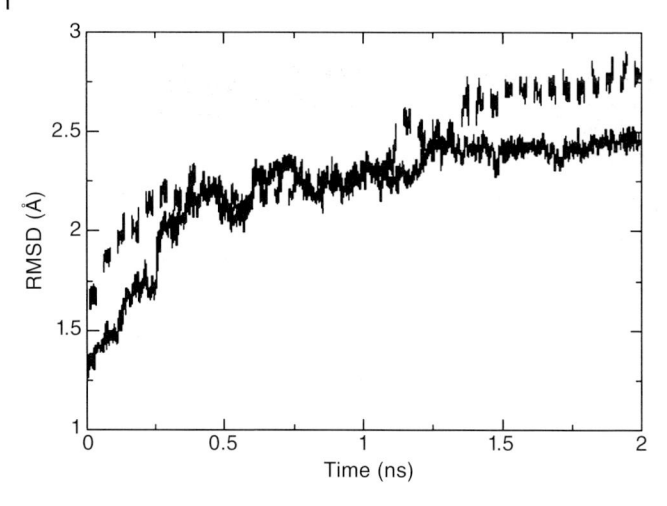

Figure 11.3 Time evolutions of the RMSD from their experimental conformation for 1PGB (black line) and 1BPI (gray line) simulations.

To be more specific, the averaged helicity ratio of the peptide is found to be around 30% at 355 K, whereas it is equal to 55% at 275 K, which is in good agreement with CD experiments [60]. When examining the time-averaged helicity ratio as a function of the residue number (Figure 11.1), it can be observed for the simulation at 275 K, but not at 355 K, that the peptide is mainly in the helical conformation from residue 4 to 11, in agreement with NMR structural studies [61]. This helical conformation is maintained as a result of the ionic interactions between the charged side chains Glu2 and Arg10, which remain in close contact at 275 K but are solvent-separated at 355 K (Figure 11.2).

The 1PGB and 1BPI proteins possess 56 and 58 residues, respectively, and both contain the two elementary α-helix and β-sheet secondary structures. They have been the focus of many experimental studies [62–65] and thus are good candidates to test the PPP solvent influence on their dynamic conformations. Figure 11.3 displays the time evolution of the RMSD from the initial conformation for both protein simulations. The 1PGB trajectory appears to be stable after about 1 ns, whereas the 1BPI simulation seems to be slower to stabilize.

The secondary structures of the two proteins were analyzed using calculations of the α-helical and β-sheet percentage of each residue (Figure 11.4). It can be observed for 1PGB that both the central helix and four strands are well conserved during the simulation. Interestingly, the loop region between the β3 and β4 strands is found to have a helical conformation, stabilized by interactions between the backbone and the side chain of Asp46, Ala48, Thr49, and Thr51. This observation is in good agreement with several NMR experiments [65, 66], which have emphasized such a conformationally restricted loop. In the 1BPI molecular dynamics simulation, the first β-strand and C-terminal helix are also clearly conserved, but its N-terminal short helix and second β-strand appear to be slightly less stable. These less structured residues, as well as the appearance of a

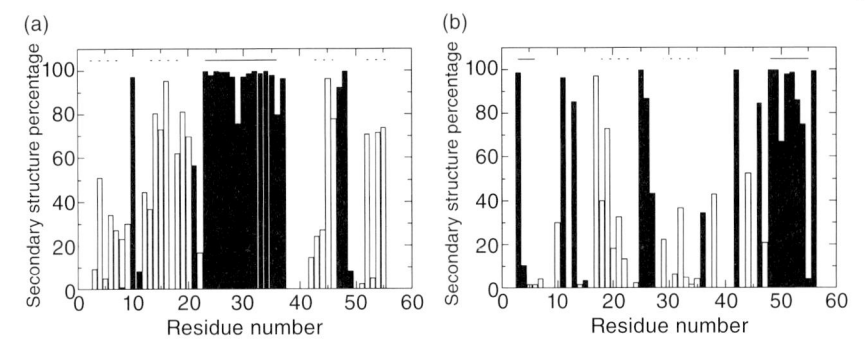

Figure 11.4 Helical (black bars) and strand (white bars) percentage per residue in (a) 1PGB and (b) 1BPI simulations. Experimental secondary structures are indicated by dark (helix) and light (strand) gray horizontal lines.

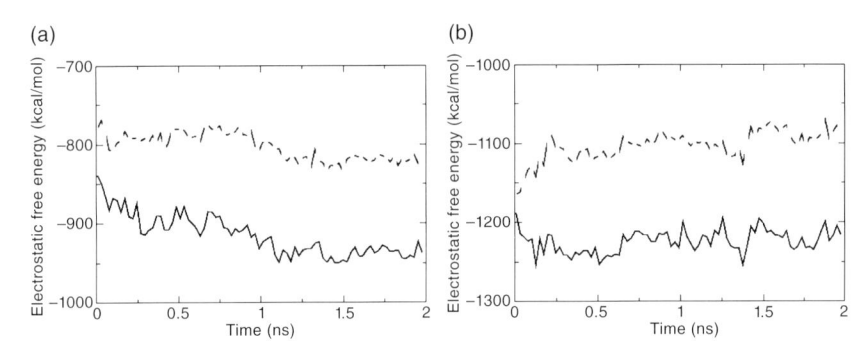

Figure 11.5 Time evolutions of the electrostatic solvation free energy calculated with the PPP model (black lines) and the Poisson–Boltzmann equation (gray lines) for (a) 1PGB and (b) 1BPI proteins. For clarity and convenience, all the plots were translated into the same energy range.

helical structure at position 25–27, have also been inferred from NMR studies [64, 67].

As emphasized in the theory section 11.2.4, the PPP solvent provides an efficient way for estimating "on the fly" the electrostatic solvation free energies during molecular dynamics simulations under the condition that the "instantaneous orientational" free energies are averaged over the translational degrees of freedom of the solvent particles. To test the PPP model solvation energies against accurate calculations based on the Poisson–Boltzmann equation, we proceeded as follows. Over time windows of about 10 ps, an averaged molecular structure is calculated and used to solve the Poisson–Boltzmann equation. The resulting electrostatic free energy can then be compared to the PPP solvation energy averaged over the same time window. Figure 11.5 displays the time evolution of the electrostatic free energies of the two studied proteins, evaluated with the PPP model and the APBS algorithm [68]. The plots indicate that the variations of the PPP electrostatic free energy due to the solute conformational changes correlate quite well with the

Poisson–Boltzmann based calculations. Despite a systematic shift between the two curves, the correlation coefficients between the two methods are 0.86 and 0.76, respectively, for the 1PGB and 1BPI simulations.

11.3.3
Hydration Properties of Nucleic Acids

The highly polar aqueous solvent particularly stabilizes the three-dimensional structure of the strongly charged nucleic acid molecules. Here it is shown that molecular dynamics simulations of oligonucleotides in the PPP solvent are able to reproduce some of their main structural and energetic properties. The four tested molecules are the 17-base anticodon hairpin of Asp-tRNA (TRNA05), the decamer d(CCGCCGGCGG) in both A (ADJ109) and B forms (BD0015), and the d(CGCGAATTCGCG) dodecamer (BDL001), which contains the EcoRI restriction site. These negatively charged solutes were neutralized with the appropriate number of sodium counterions initially placed close to backbone phosphate groups.

It can be seen from Figure 11.6 that the RMSD from the initial conformations of the nucleic acids in the PPP solvent seem to stabilize at low values during the 4 ns trajectories. The highest RMSD is observed around 2.7 Å for the dodecamer BDL001, a value very close to the one seen in a simulation with the TIP3 water model [69]. The quite small 1.7 Å RMSD of the tRNA loop has also been obtained from simulations in SPC solvent [70] and reflects smaller deformations of the RNA than the DNA molecules.

Various helicoidal parameters can be examined to check the fine structure of the nucleic acids [35, 71]. Regarding the sugar puckering phase (Figure 11.7), the distribution peaks of the two studied B-DNA fragments are very close to the canonical B-value associated with the C2'-endo conformation, and the probability distribution for the tRNA is centered around the canonical A conformation C3'-endo. In contrast, the sugar puckering probability of the A-DNA clearly has a

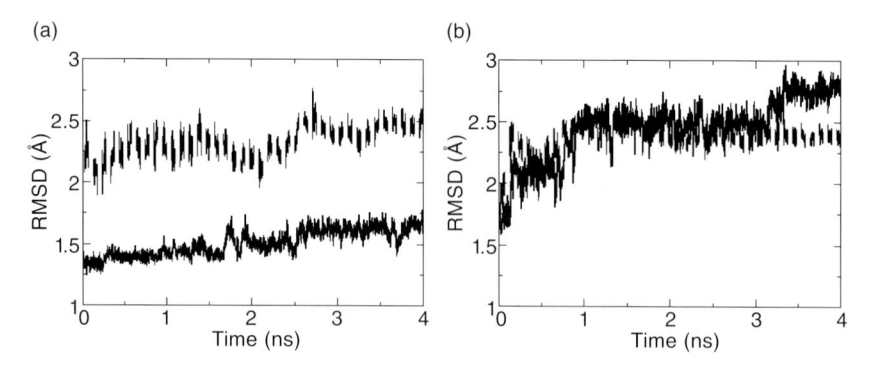

Figure 11.6 Time evolutions of the RMSD from their experimental structure for (a) TRNA05 (black line) and ADJ109 (gray line), and (b) BDL001 (black line) and BD0015 (gray line) simulations.

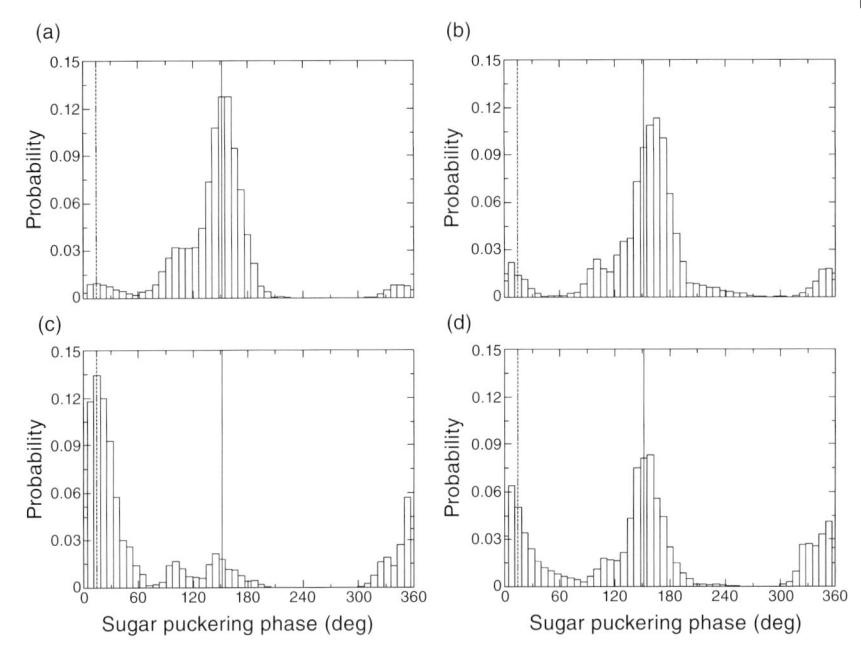

11.3 Applications: Solvation of All-Atom Models of Biomolecules

Figure 11.7 Probability distributions of the sugar puckering phases for the (a) BDL001, (b) BD0015, (c) TRNA05, and (d) ADJ109 oligonucleotides. The canonical A-value and B-value, respectively, are shown with dotted and solid gray lines.

bimodal distribution that reflects a partial transition toward an intermediate structure between an A and B form. This A to B partial transition is also observed by many simulations in explicit solvent [72, 73]. The PPP model is hence overall able to reproduce the structural stability of different kinds of nucleic acids, comparable to simulations using explicit water.

As for the proteins, the PPP model allows the extraction of electrostatic solvation free energy variations during oligonucleotide conformational dynamics. Figure 11.8 compares these energies obtained by the PPP solvent with Poisson–Boltzmann based calculations. It can be seen that, despite a systematic shift between the plots, the two methods yield correlated variations of the electrostatic free energy, the correlation coefficients ranging from 0.83 to 0.89 for the four nucleic acids. This trend is particularly noteworthy for the A-DNA molecule, which undergoes a conformational transition toward an A/B mixed form.

In contrast to purely implicit models, the PPP approach accounts for the molecular aspects of the solvent. It therefore allows the analysis of highly localized water molecules on the surface of biomolecules, which are assumed to have a significant structural influence on their associations, as emphasized by numerous experimental and theoretical studies [4, 74]. To illustrate this possibility, we calculated the maximum residence time (defined in percentage of the trajectory length) of a solvent particle in the major and minor grooves of the studied d(CCGCCGGCGG) decamer in both A and B forms. Figure 11.9 displays this quantity, which is similar

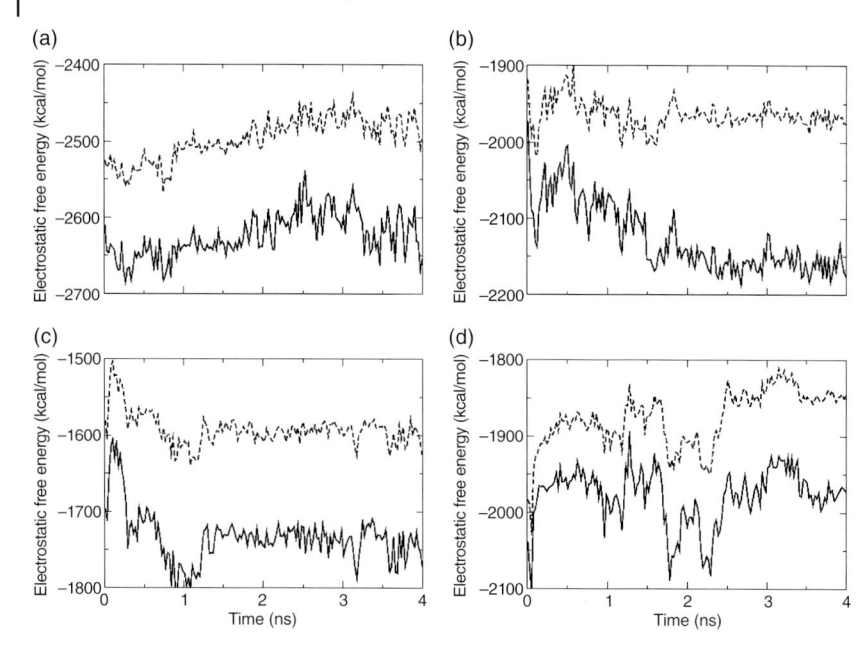

Figure 11.8 Time evolutions of the electrostatic solvation free energy calculated with the PPP model (black lines) and the Poisson–Boltzmann equation (gray lines), for the (a) BDL001, (b) BD0015, (c) TRNA05, and (d) ADJ109 nucleic acids. For clarity and convenience, all the plots were translated into the same energy range.

to the "water residence time" usually measured or calculated from experiments and simulations. Since counterions can also occupy hydration sites, we added to the "water residence time" the maximum residence time of sodium near the same sites. The results obtained from our simulations reproduce the well-known hydration patterns of DNA. The major groove of A-DNA is more hydrated by the solvent particles and sodium ions than is its minor groove, whereas the water molecules and counterions with the longest residence time are found in the B-DNA minor groove rather than in its major one [74]. It appears that, despite the absence of explicit hydrogen bonds in the model, the PPP solvent can nevertheless provide meaningful information about hydration sites, which are not accessible by implicit models.

11.4
Conclusion and Prospects

Molecular dynamics simulations of the tested polypeptides and oligonucleotides indicate that the PPP solvent model provides a reasonably correct electrostatic medium for the solute charge–charge interactions, and allows the modeling of their three-dimensional structures and dynamics in overall agreement with experimental observations. Although the model has no explicit hydrogen bonds, the

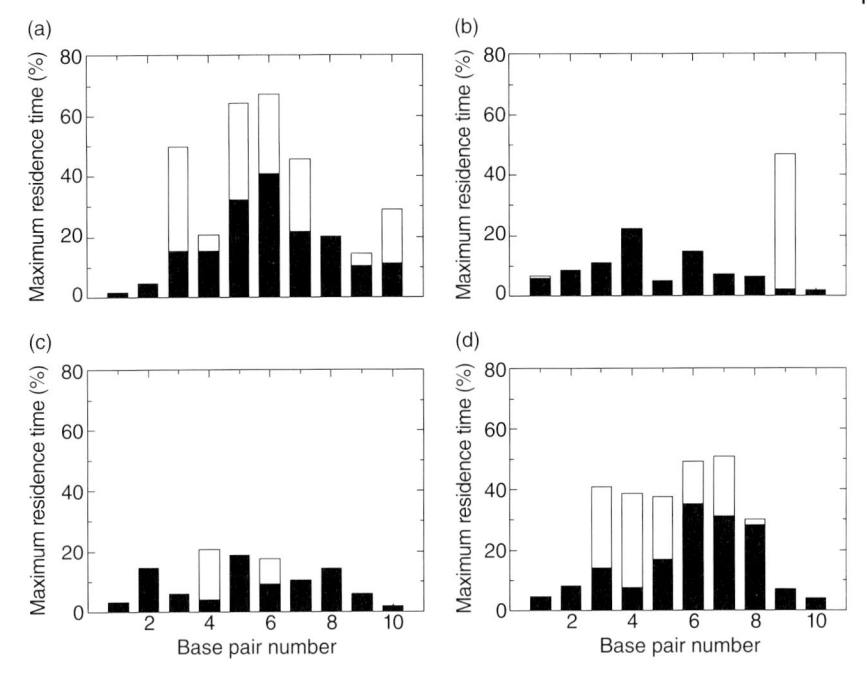

Figure 11.9 Maximum residence times (expressed as a percentage of the simulation length) of solvent particles (black bars) and sodium ions (additional white bars) close to the (a, b) major and (c, d) minor groove of the (a, c) A-DNA and (b, d) B-DNA decamer.

molecular nature of the PPP solvent allows straightforward analysis of the solvation structure around solutes [54]. The polarizable solvent particles can overall correctly estimate the structure-dependent solvation free energy of flexible solutes by means of an affordable equilibration of their translational degrees of freedom. The capability of the PPP model to evaluate the electrostatic solvation free energies of biomolecules is summarized in Figure 11.10, which displays the averaged PPP energies versus the Poisson–Boltzmann ones. The plot reveals a very good correlation between the two approaches for the whole set of studied molecules, the correlation coefficient and slope of the linear relationship being equal to 0.99 and 1.07, respectively.

The PPP model has recently been combined with a so-called "second-generation" force field of proteins including explicit atomic electronic polarizabilities [54] in a natural and consistent way since the solvent particles are themselves polarizable. Another recent coarse-grained version has recently been proposed that is compatible with coarse-grained protein representations such as that in Ref. [37]. The model, called "polarizable coarse-grained solvent" (PCGS) [36], is just a scaled-up version of the PPP model, with pseudo-particles of diameter 4.58 Å, rather than 2.88 Å, thus collecting roughly three water molecules instead of one. Furthermore, the van der Waals interactions are described by softer potentials than the Lennard-Jones ones and compatible with the coarse-grained protein force field of Ref. [37].

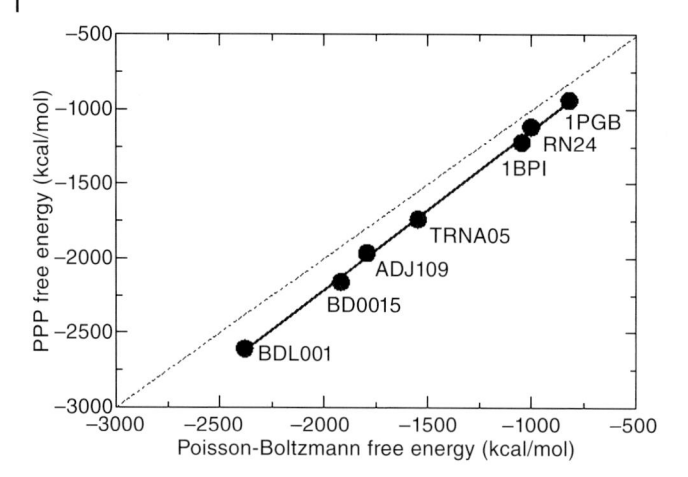

Figure 11.10 Comparison between the PPP and Poisson–Boltzmann electrostatic solvation energies for all the studied biomolecules.

The PCGS particle parameters are similar, in reduced units, to the PPP ones, but an internal dielectric constant $\varepsilon_i = 2$ for the solute is necessary to account for electronic polarization effects. The ability of the PCGS model to yield correct electrostatic solvation free energies has been tested on 17 coarse-grained peptides and proteins. The observed agreement between electrostatic solvation free energies estimated "on the fly" using short simulations and Poisson–Boltzmann calculations appears of the same quality as in Figure 11.10 [36].

All of these results, obtained at either the atomistic or coarse-grained scale, open promising avenues for molecular dynamics studies of the influence of solvent on the protein–protein and DNA–protein association specificity, for which both the solute deformation and dehydration processes are important.

References

1 Goldbaum, F., Schwarz, F., Einsenstein, E., Cauerhff, A., Mariuzza, R., and Poljak, R. (1996) The effect of water activity on the association constant and the enthalpy of reaction between lysozyme and the specific antibodies d1.3 and d44.1. *J. Mol. Recogn.*, **9**, 6–12.

2 Robinson, C. and Sligar, S. (1998) Changes in solvation during DNA binding and cleavage are critical to altered specificity of the EcoRI endonuclease. *Proc. Natl. Acad. Sci. U.S.A.*, **95**, 2186–2195.

3 Lynch, T. and Sligar, S. (2002) Experimental and theoretical high pressure strategies for investigating protein–nucleic acid assemblies. *Biochim. Biophys. Acta*, **1595**, 277–282.

4 Schwabe, J. (1997) The role of water in protein–DNA interactions. *Curr. Opin. Struct. Biol.*, **7**, 126–134.

5 Chalikian, T. and Breslauer, K. (1998) Thermodynamic analysis of biomolecules: a volumetric approach. *Curr. Opin. Struct. Biol.*, **8**, 657–664.

6 Clore, G. and Gronenborn, A. (1992) Localization of bound water in the solution structure of the immunoglobulin binding domain of streptococcal protein G. Evidence for solvent-induced helical

distortion in solution. *J. Mol. Biol.*, **223**, 853–856.

7 Eisenstein, M. and Shakked, Z. (1995) Hydration patterns and intermolecular interactions in A-DNA crystal structures. Implications for DNA recognition. *J. Mol. Biol.*, **248**, 662–678.

8 Sunnerhagen, M., Denisov, V., Venu, K., Bonvin, A., Carey, J., Halle, B., and Otting, G. (1998) Water molecules in DNA recognition I: Hydration lifetimes of *trp* operator DNA in solution measured by NMR spectroscopy. *J. Mol. Biol.*, **282**, 847–858.

9 Makarov, V., Andrews, B., Smith, P., and Pettitt, B. (2000) Residence times of water molecules in the hydration sites of myoglobin. *Biophys. J.*, **79**, 2966–2974.

10 Fields, B., Goldbaum, F., Dall'Acqua, W., Malchiodi, E., Cauerhff, A., Schwarz, F., Ysern, X., Poljak, R., and Mariuzza, R. (1996) Hydrogen bonding and solvent structure in an antigen–antibody interface. Crystal structures and thermodynamic characterization of three fv mutants complexed with lysozyme. *Biochemistry*, **35**, 15494–15503.

11 Wang, Y., Freedberg, D., Grzesiek, S., Torchia, D., Wingfield, P., Kaufman, J., Stahl, S., Chang, C., and Hodge, C.N. (1996) Mapping hydration water molecules in the hiv-1 protease/dmp323 complex in solution by NMR spectroscopy. *Biochemistry*, **35**, 12694–12704.

12 Labeots, L. and Weiss, M. (1997) Electrostatics and hydration at the homeodomain–DNA interface: chemical probes of an interfacial water cavity. *J. Mol. Biol.*, **269**, 113–128.

13 Jorgensen, W., Chandrasekhar, J., Madura, J., and Klein, M. (1983) Comparison of simple potential functions for simulating liquid water. *J. Chem. Phys.*, **79**, 926–935.

14 Berendsen, H., Grigera, J., and Straatsma, T. (1987) The missing term in effective pair potentials. *J. Phys. Chem.*, **91**, 6269–6271.

15 Davis, M. and McCammon, J. (1989) Solving the finite difference linearized Poisson–Boltzmann equation: a comparison of relaxation and conjugate gradient methods. *J. Comput. Chem.*, **10**, 386–391.

16 Nichols, A. and Honig, B. (1991) Rapid finite difference algorithm utilizing successive over-relaxation to solve the Poisson–Boltzmann equation. *J. Comput. Chem.*, **12**, 435–445.

17 Rashin, A. (1990) Hydration phenomena, classical electrostatics, and the boundary element method. *J. Phys. Chem.*, **94**, 1725–1733.

18 York, D. and Karplus, M. (1999) A smooth solvation potential based on the conductor-like screening model. *J. Phys. Chem. A*, **103**, 11060–11079.

19 You, T. and Harvey, S. (1993) Finite element approach to the electrostatics of macromolecules with arbitrary geometries. *J. Comput. Chem.*, **14**, 484–501.

20 Coalson, R. and Beck, T. (1998) *Encyclopedia of Computational Chemistry*, vol. 3, John Wiley & Sons, Inc., New York.

21 Marchi, M., Borgis, D., Levy, N., and Ballone, P. (2001) A dielectric continuum molecular dynamics method. *J. Chem. Phys.*, **114**, 4377–4385.

22 Hassan, S., Guarnieri, F., and Mehler, E. (2000) A general treatment of solvent effects based on screened Coulomb potentials. *J. Phys. Chem. B*, **104**, 6478–6489.

23 Qiu, D., Shenkin, P., Hollinger, F., and Still, W. (1997) The GB/SA continuum model for solvation. A fast analytical method for the calculation of approximate Born radii. *J. Phys. Chem. A*, **101**, 3005–3014.

24 Bashford, D. and Case, D. (2000) Generalized Born models of macromolecular solvation effects. *Annu. Rev. Phys. Chem.*, **51**, 129–152.

25 Hansen, J. and McDonald, I. (1989) *Theory of Simple Liquids*, Academic Press, London.

26 Pettit, B., Karplus, M., and Rossky, P. (1986) Integral equation model for aqueous solvation of polyatomic solutes: application to the determination of the free energy surface for the internal motion in biomolecules. *J. Phys. Chem.*, **90**, 6335–6345.

27 Beglov, D. and Roux, B. (1996) Solvation of complex molecules in a polar liquid: an integral equation theory. *J. Chem. Phys.*, **104**, 8678–8689.

28 Florian, J. and Warshel, A. (1997) Langevin dipoles model for *ab initio* calculations of chemical processes in solution: parametrization and application to hydration free energies of neutral and ionic solutes and conformational analysis in aqueous solution. *J. Phys. Chem. B*, **101**, 5583–5595.

29 Papazyan, A. and Warshel, A. (1997) Continuum and dipole–lattice models of solvation. *J. Phys. Chem. B*, **101**, 11254–11264.

30 Coalson, R. and Duncan, A. (1996) Statistical mechanics of a multipolar gas: a lattice field theory approach. *J. Phys. Chem. B*, **100**, 2612–2620.

31 Chandler, D., McCoy, J., and Singer, S.J. (1986) Density functional theory of nonuniform polyatomic systems. I. General formulation. *J. Chem. Phys.*, **85**, 5971–5982.

32 Evans, R. (1992) Density functionals in the theory of nonuniform fluids, in *Fundamental of Inhomogeneous Fluids* (ed. D. Henderson), Marcel Dekker, New York, pp. 85–176.

33 Lowen, H., Hansen, J., and Madden, P. (1993) Nonlinear counterion screening in colloidal suspensions. *J. Chem. Phys.*, **98**, 3275–3289.

34 Basdevant, N., Borgis, D., and Ha-Duong, T. (2004) A semi-implicit solvent model for the simulation of peptides and proteins. *J. Comput. Chem.*, **25**, 1015–1029.

35 Basdevant, N., Ha-Duong, T., and Borgis, D. (2006) Particle-based implicit solvent model for biosimulations: application to proteins and nucleic acids hydration. *J. Chem. Theory Comput.*, **2**, 1646–1656.

36 Ha-Duong, T., Basdevant, N., and Borgis, D. (2009) A polarizable coarse-grained water model for coarse-grained proteins simulations. *Chem. Phys. Lett.*, **468**, 79–82.

37 Basdevant, N., Borgis, D., and Ha-Duong, T. (2007) A coarse-grained protein–protein potential derived from an all-atom force field. *J. Phys. Chem. B*, **111**, 9390–9399.

38 Ha-Duong, T., Phan, S., Marchi, M., and Borgis, D. (2002) Electrostatics on particles: phenomenological and orientational density functional theory approach. *J. Chem. Phys.*, **117**, 541–556.

39 Marcus, R. (1956) Electrostatic free energy and other properties of states having nonequilibrium polarization. I. *J. Chem. Phys.*, **24**, 979–989.

40 Feynman, R. (1964) *The Feynman Lectures on Physics: Mainly Electromagnetism and Matter*, Addison-Wesley, Reading, MA.

41 Jackson, W. (1999) *Classical Electrodynamics*, John Wiley & Sons, Inc., New York.

42 Kharkats, Y., Kornyshev, A., and Vorotyntsev, M. (1976) Electrostatic models in the theory of solutions. *J. Chem. Soc., Faraday Trans. 2*, **72**, 361–371.

43 Calef, D. and Wolynes, P. (1983) Classical solvent dynamics and electron transfer. 1. Continuum theory. *J. Phys. Chem.*, **87**, 3387–3400.

44 Lee, S. and Hynes, J. (1988) Solution reaction path hamiltonian for reactions in polar solvents. *J. Chem. Phys.*, **88**, 6853–6862.

45 Schaefer, M. and Froemmel, C. (1990) A precise analytical method for calculating the electrostatic energy of macromolecules in aqueous solution. *J. Mol. Biol.*, **216**, 1045–1066.

46 Schaefer, M. and Karplus, M. (1996) A comprehensive analytical treatment of continuum electrostatics. *J. Phys. Chem.*, **100**, 1578–1599.

47 Kirkwood, J. (1934) Theory of solutions of molecules containing widely separated charges with special application to zwitterions. *J. Chem. Phys.*, **2**, 351–361.

48 Cornell, W., Cieplak, P., Bayly, C., Gould, I., Merz, K. Jr, Ferguson, D., Spellmeyer, D., Fox, T., Caldwell, J., and Kollman, P. (1995) A second generation force field for the simulation of proteins, nucleic acids, and organic molecules. *J. Am. Chem. Soc.*, **117**, 5179–5197.

49 MacKerell, A. Jr, Bashford, D., Bellott, M., Dunbrack, R. Jr, Evanseck, J., Field, M., Fischer, S., Gao, J., Guo, H., Ha, S., Joseph-McCarthy, D., Kuchnir, L., Kuczera, K., Lau, F., Mattos, C., Michnick, S., Ngo, T., Nguyen, D., Prodhom, B.,

Reiher, W. III, Roux, B., Schlenkrich, M., Smith, J., Stote, R., Straub, J., Watanabe, M., Wiorkiewicz-Kuczera, J., Yin, D., and Karplus, M. (1998) All-atom empirical potential for molecular modeling and dynamics studies of proteins. *J. Phys. Chem. B*, **102**, 3586–3616.

50 Warshel, A. and Levitt, M. (1976) Theoretical studies of enzymic reactions: dielectric, electrostatic and steric stabilization of the carbonium ion in the reaction of lysozyme. *J. Mol. Biol.*, **103**, 227–249.

51 Warshel, A. and Russell, S. (1984) Calculations of electrostatic interactions in biological systems and in solutions. *Q. Rev. Biophys.*, **17**, 283–422.

52 Pollock, E. and Alder, B. (1980) Static dielectric properties of Stockmayer fluids. *Physica*, **102A**, 1–21.

53 Hansen, J. and Verlet, L. (1969) Phase transitions of the Lennard-Jones system. *Phys. Rev.*, **184**, 151–161.

54 Masella, M., Borgis, D., and Cuniasse, P. (2008) Combining a polarizable force-field and a coarse-grained polarizable solvent model: application to long dynamics simulations of bovine pancreatic trypsin inhibitor. *J. Comput. Chem.*, **29**, 1707–1724.

55 Born, M. (1920) Volumen und Hydratationswarme der Ionen. *Z. Phys.*, **1**, 45–48.

56 Procacci, P., Darden, T., Paci, E., and Marchi, M. (1997) ORAC: a molecular dynamics program to simulate complex molecular systems with realistic electrostatic interactions. *J. Comput. Chem.*, **18**, 1848–1862.

57 Humphreys, D., Friesner, R., and Berne, B. (1994) A multiple-time-step molecular dynamics algorithm for macromolecules. *J. Phys. Chem.*, **98**, 6885–6892.

58 Nose, S. (1984) A unified formulation of the constant temperature molecular dynamics methods. *J. Chem. Phys.*, **81**, 511–519.

59 Rickaert, J., Ciccotti, G., and Berendsen, H. (1977) Numerical integration of the cartesian equations of motion of a system with constraints: molecular dynamics of n-alkanes. *J. Comput. Phys.*, **23**, 327–341.

60 Shoemaker, K., Kim, P., York, E., Stewart, J., and Baldwin, R. (1987) Tests of the helix dipole model for stabilization of alpha-helices. *Nature*, **326**, 563–567.

61 Osterhout, J.J. Jr, Baldwin, R.L., York, E.J., Stewart, J.M., Dyson, H.J., and Wright, P.E. (1989) Proton NMR studies of the solution conformations of an analog of the C-peptide of ribonuclease A. *Biochemistry*, **28**, 7059–7064.

62 Barshi, J., Grasberger, B., Gronenborn, A., and Clore, G. (1994) Investigation of the backbone dynamics of the IGG-binding domain of streptococcal protein G by heteronuclear two-dimensional ^{1}H–^{15}N nuclear magnetic resonance spectroscopy. *Protein Sci.*, **3**, 15–21.

63 Smith, P.E., van Schaik, R.C., Szyperski, T., Wüthrich, K., and van Gunsteren, W. (1995) Internal mobility of the basic pancreatic trypsin inhibitor in solution: a comparison of NMR spin relaxation measurements and molecular dynamics simulations. *J. Mol. Biol.*, **246**, 356–365.

64 Pan, H., Barbar, E., Barany, G., and Woodward, C. (1995) Extensive nonrandom structure in reduced and unfold bovine pancreatic trypsin inhibitor. *Biochemistry*, **34**, 13974–13981.

65 Frank, M., Clore, G., and Gronenbron, A. (1995) Structural and dynamics characterization of the urea denatured state of the immunoglobulin binding domain of streptococcal protein G by multidimensional heteronuclear NMR spectroscopy. *Protein Sci.*, **4**, 2605–2615.

66 Kobayashi, N., Honda, S., Yoshii, H., and Munekata, E. (2000) Role of sidechains in the cooperative beta-hairpin folding of the short C-terminal fragment derived from streptococcal protein G. *Biochemistry*, **39**, 6564–6571.

67 Carulla, N., Woodward, C., and Barany, G. (2000) Synthesis and characterization of a beta-hairpin peptide that represents a core module of bovine pancreatic trypsin inhibitor (BPTI). *Biochemistry*, **39**, 7927–7937.

68 Baker, N., Sept, D., Joseph, S., Holst, M., and McCammon, J. (2001) Electrostatics of nanosystems: application to microtubules and the ribosome. *Proc. Natl. Acad. Sci. U.S.A.*, **98**, 10037–10041.

69 Young, M., Ravishanker, G., and Beveridge, D. (1997) A 5-nanosecond molecular dynamics trajectory for B-DNA: analysis of structure, motions and solvation. *Biophys. J.*, **73**, 2313–2336.

70 Auffinger, P. and Westhof, E. (1996) H-bond stability in the tRNA(Asp) anticodon hairpin: 3 ns of multiple molecular dynamics simulations. *Biophys. J.*, **71**, 940–954.

71 Lavery, R. and Sklenar, H. (1988) The definition of generalized helicoidal parameters and of axis curvature for irregular nucleic acids. *J. Biomol. Struct. Dynam.*, **6**, 63–91.

72 Cheatham, T. III and Kollman, P. (1996) Observation of the A-DNA to B-DNA transition during unrestrained molecular dynamics in aqueous solution. *J. Mol. Biol.*, **259**, 434–444.

73 Trantirek, L., Stefl, R., Vorlickova, M., Koca, J., Sklenar, V., and Kypr, J. (2000) An A-type double helix of DNA having B-type puckering of the deoxyribose rings. *J. Mol. Biol.*, **297**, 907–922.

74 Makarov, V., Pettitt, B., and Feig, M. (2002) Solvation and hydration of proteins and nucleic acids: a theoretical view of simulation and experiment. *Accts. Chem. Res.*, **35**, 376–384.

12
Modeling Electrostatic Polarization in Biological Solvents
Sandeep Patel

12.1
Introduction

Force-field-based molecular simulation methods such as molecular dynamics (MD) are important tools that are available to computational and experimental scientists pursuing research spanning a wide range of biological and physico-chemical systems. Classical statistical mechanics approaches have made their mark within the realm of biophysics, providing atomic-level information relevant to physiological processes that would otherwise be unobtainable with standard experimental approaches, and thus invariably complementing the interpretation of such observations and guiding further experiment [1, 2]. Though classical force-field-based methods are an invaluable tool for studying rather large biomacromolecular systems over a range of time scales, modern approaches rely on empirical force fields that effectively attempt to model the subtle microscopic interactions defining the underlying quantum interaction energy surface. In such methods, the electrostatic interaction is defined by Coulomb interactions between *static* charges assigned to individual sites of molecules. Thus, there is no effective mechanism for allowing an explicit, dynamic electronic response to changes in the local chemical environment. In strongly anisotropic environments (that is, interfacial systems, protein–solvent interface, ion conduction from bulk solution through integral membrane protein channels), one can argue that, at the atomistic level, a fixed-charge representation cannot faithfully model the underlying physics. One must explicitly account for variations in electronic charge distributions (changes in classical charges as condensed representations of the molecular electron density).

Currently, several approaches for first-generation polarizable force fields for biological molecules (proteins, DNA/RNA, ions) are actively being pursued. These include point dipole polarizable models [3–5], shell (Drude) models [6–9], charge equilibration (electronegativity equalization, chemical potential equalization) models [10–21], fluctuating dipole model [22], charge-on-spring (COS) models [23, 24], the sum of interactions between fragments *ab initio* computed (SIBFA) models [25–28] based on molecular polarizability as introduced by Thole, and recent atomic multipole methods such as the AMOEBA [29–31] force field (implemented within the TINKER [32] modeling package). The point dipole polarizable models generally ascribe atomic polarizabilities to various sites within the molecular construct, along with fixed charges, and self-consistently evaluate the induced dipole moments arising from the local electric field generated by the nearby charge density and configuration. The shell (Drude) models introduce finite-mass charged sites generally coupled to more polar nuclei via a harmonic spring with a force constant that intimately determines the local polarizability (the "atomic" polarizability in such models goes as the inverse of the force constant, as one would intuitively presume); effectively, one models, locally, a system with two charges whose separation oscillates with time. Thole-type models introduce intra- and intermolecular dipole–dipole interactions by assigning atomic polarizabilities determined from empirical fits to experimental molecular polarizability tensors [33]. Thole introduced a damping function in order to attenuate the intramolecular dipole–dipole interactions at short distances. The AMOEBA force field introduces permanent atomic monopole, dipole, and quadrupole moments, and explicitly treats polarization by allowing mutual induction of dipoles at prescribed sites with contributions from both permanent multipoles and induced dipoles; in this respect, an iterative approach to self-consistency is required in order to determine the instantaneous moments. Permanent multipoles interact through a multipole interaction matrix. Atomic polarizabilities are fitted to reproduce experimental molecular polarizabilities in the spirit of Thole. These values are used in conjunction with a damping scheme introduced to attenuate the intramolecular dipole–dipole interactions in order to avoid polarization catastrophes at small separations. Though each approach demonstrates unique advantages and shortcomings, it currently remains to be seen if any one approach is superior to the others, much as the current situation with fixed-charge empirical force fields for biomacromolecular modeling.

In the present chapter, we describe efforts in developing polarizable force fields for ultimate application to force-field-based molecular modeling approaches of biologically relevant systems. We will take the approach of presenting the material as a review of recent work in this area, discussing our methodology for developing polarizable models for biological molecules as well as some recent results arising from applications to relevant biological and small-molecule systems. We aim to discuss the nature of polarizable models based on the charge equilibration formalism in particular. In Section 12.2, we discuss several current methods for polarization in classical simulations. In Section 12.3 we describe charge equilibration to incorporating explicit polarizability in molecular simulations of biological systems, in particular describing the protocol within the charge equilibration, or equivalently electronegativity equalization, formalism. Section 12.4 discusses results from application of polarizable models for aqueous and non-aqueous systems (specifically water and liquid alkanes). Section 12.5 will discuss recent work on the application of our charge equilibration force fields to fully hydrated lipid bilayer systems as a prelude to ongoing work exploring integral membrane proteins. Section 12.6 offers concluding remarks.

12.2
Current Approaches for Modeling Electrostatic Polarization in Classical Force Fields

Before discussing in detail the charge equilibration methods (the classical implementations that are of most relevance for the purpose at hand), we discuss several of the current state-of-the-art approaches for electrostatic polarization. As with the status of various fixed-charge force fields (CHARMM, GROMOS, AMBER) [34–36], it is apparent that currently several fundamentally unique approaches are being actively pursued. We reiterate that, despite the decades of development of polarizable force fields, beginning with the early work of Warshel, Lifson, and Hagler, polarizable force fields do not enjoy the popularity that fixed-charge models do. Though it is acknowledged that polarizable models also have not enjoyed the years of refinement and fine-tuning that non-polarizable force fields have experienced, nonetheless the current generation of polarizable force fields can be applied to the study of various classes of biological systems.

The "atomic multipole optimized energetics for biomolecular applications" (AMOEBA) approach integrates distributed *atomic* multipoles (up to quadrupole) along with classical atomic point dipole polarizabilities to account for polarization [29, 31]. Atomic dipole polarizabilities are determined from fits to molecular properties (that is, molecular polarizability) derived from *ab initio* calculations. A Thole-type [33] screening function between diffuse charge distributions, represented as

$$\rho = \frac{3a}{4\pi} e^{-au^3} \tag{12.1}$$

with u as an effective separation determined as a function of the atomic polarizabilities and a an effective width parameter for the diffuse charge density, prevents polarization catastrophes for induction interactions at short distance. The α-component of the induced dipole moment on a particular site i is determined by the electric fields from the static multipoles and induced dipoles in a self-consistent manner:

$$\mu_{i,\alpha}^{\text{induced}} = \alpha_i \left(\sum_{\{j\}} T_\alpha^{ij} M_j \right) + \sum_{\{j'\}} T_{\alpha\beta}^{ij'} \mu_{j',\beta}^{\text{induced}} \quad \alpha, \beta = 1, 2, 3 \tag{12.2}$$

The multipole interaction matrix, T, is defined as follows:

$$\begin{aligned} T &= \frac{1}{R} \\ T_\alpha &= \nabla_\alpha T = -\frac{R_\alpha}{R^3} \\ T_{\alpha\beta} &= \nabla_\alpha T_\beta \\ T_{\alpha\beta\gamma} &= \nabla_\alpha T_{\beta\gamma} \\ \alpha, \beta, \gamma, \ldots &= 1, 2, 3 \end{aligned} \tag{12.3}$$

The reader is directed to the literature for further details of the method and its parameterization for water, organic solutes, and other biomolecular systems [29–31].

The classical Drude oscillator approach incorporates electrostatic polarizability by including pairs of electro-neutral, mobile particles (Drude oscillators with charge $\pm q^{\text{Drude}}$), with one of the pair anchored to a particular nucleus of the molecular construct, and the second a vector displacement $\boldsymbol{d}$ away from the nucleus [37, 38]. An effective "spring," to which is ascribed a force constant, connects the two oscillators. This models the electronic polarization of a given atom by establishing a local dipole that varies dynamically throughout a molecular simulation in response to instantaneous variations in local electric field. The electrostatic component of the Drude force field thus consists of a self-energy for the Drude particles, and the usual Coulomb interactions between two real particles, two Drude particles, and a real and a Drude particle [37, 38]:

$$U(q) = U_{\text{self}} + U_{\text{Coulomb}}$$
$$U_{\text{self}} = \boldsymbol{d} \cdot \boldsymbol{K}^{(D)} \cdot \boldsymbol{d} = \left[K_{11}^{(D)}\right] d_1^2 + \left[K_{22}^{(D)}\right] d_2^2 + \left[K_{33}^{(D)}\right] d_3^2 \quad (12.4)$$

In Equation 12.4, d_1, d_2, and d_3 are the projections of the Drude displacement vector $\boldsymbol{d}$ onto orthogonal axes of a local intramolecular reference frame. Under the Born–Oppenheimer condition, the atomic polarizability tensor determined for each Drude oscillator is given as

$$\boldsymbol{\alpha} = -\left(q_i^D\right)^2 \left(\boldsymbol{K}^{(D)}\right)^{-1} \quad (12.5)$$

From Equation 12.5, it is evident that the polarizability scales inversely as the force constant. Finally, Drude oscillators on atom pairs 1–2 and 1–3 (bond and angle atoms) interact with a screened potential modulated by the function

$$S_{ij}(r_{ij}) = 1 - \left[1 + \frac{(a_i + a_j) r_{ij}}{2(\alpha_i \alpha_j)^{1/6}}\right] \exp\left[-\frac{(a_i + a_j) r_{ij}}{(\alpha_i \alpha_j)^{1/6}}\right] \quad (12.6)$$

where the a are Thole damping parameters and α is the trace of the atomic polarizability tensor [37, 38]. We note that the COS models developed by van Gunsteren, Yu, and coworkers also follow in the spirit of Drude oscillator models [23, 24].

12.2.1
Charge Equilibration Models: Electronic Polarization and Charge Dynamics

Charge equilibration models have been applied to various systems over the past decade [10–17, 20, 21, 39, 40]. The method is based on the density functional theory (DFT) of atoms in molecules as formulated by Yang and Parr [41]. More fundamentally, the method is founded on Sanderson's idea of electronegativity equalization [42, 43]. In the density functional sense, electronegativity equalization amounts to the equalization of the chemical potential in space. In a molecule, this translates to the redistribution of charge among constituent atoms so as to equalize the electronegativity (electrochemical potential) at each point.

The electrostatic energy of a system of M molecules containing N atoms per molecule is

$$E_{\text{electrostatic}} = \sum_{k=1}^{M} \sum_{i=1}^{N} \chi_{ik} Q_{ik} + \frac{1}{2} \sum_{l=1}^{M} \sum_{\alpha=1}^{N} \sum_{\beta=1}^{N} \eta_{\alpha l,\beta l} Q_{\alpha l} Q_{\beta l} + \frac{1}{2} \sum_{i=1}^{MN}{}' \sum_{j=1}^{MN}{}' \frac{Q_i Q_j}{r_{ij}} \quad (12.7)$$

where the χ are atom electronegativities and the η are the atomic hardnesses. The former quantity gives rise to a directionality of electron flow, while the latter represents a resistance, or hardness, to electron flow to or from the atom. The last term in Equation 12.7 is a standard Coulomb interaction between *intramolecular* sites not involved in dihedral, angle, and bonded interactions with each other, as well as between *intermolecular* sites. The second term represents the local charge transfer interaction generally restricted to within a molecule (no charge transfer) or some appropriate charge normalization unit. In the case where charge is constrained over an *entire molecule*, a total molecular charge constraint must be included via a Lagrange multiplier λ_j for each molecule. This leads to an electrostatic energy:

$$E_{\text{electrostatic}} = \sum_{k=1}^{M}\sum_{i=1}^{N}\chi_{ik}Q_{ik} + \frac{1}{2}\sum_{l=1}^{M}\sum_{\alpha=1}^{N}\sum_{\beta=1}^{N}\eta_{\alpha l,\beta l}Q_{\alpha l}Q_{\beta l} + \frac{1}{2}\sum_{i=1}^{MN}{}'\sum_{j=1}^{MN}{}'\frac{Q_iQ_j}{r_{ij}}$$
$$+ \sum_{j=1}^{M}\lambda_j\left(\sum_{i=1}^{N}Q_{ji} - Q_j^{\text{total}}\right) \quad (12.8)$$

More will be said concerning multiple charge normalization units in Section 12.3.2 in the context of describing molecular polarizability within the charge equilibration formalism.

We note that, although the electronegativity and hardness follow exactly from the definitions of the electron affinity and ionization potential [41], they are considered here as empirical parameters to be determined through fitting procedures to match a spectrum of experimental and quantum-mechanical data. Homogeneous hardness values (for each atom type, η_{ii}) are parameterized as discussed in Ref. [10]. Heterogeneous elements (interaction elements between different atom types, η_{ij}) are derived from the individual atom type values based on the combining rule [44]:

$$\eta_{ij}(R_{ij},\eta_i,\eta_j) = \frac{\frac{1}{2}(\eta_i+\eta_j)}{\left[1.0 + \frac{1}{4}(\eta_i+\eta_j)^2 R_{ij}^2\right]^{1/2}} \quad (12.9)$$

where R_{ij} is the separation between atoms (or more generally *sites*) i and j, and the η are the atomic hardness parameters. This local screened Coulomb potential has the correct limiting behavior as $1/r$ for separations greater than about 2.5 Å. This interaction is computed for sites 1–2, 1–3, and 1–4 (sites included in bonds, angles, and dihedrals). Sites in a molecule separated by five or more sites interact via a Coulomb interaction; in the case of interacting molecules, the interaction between sites on different molecules is again of the Coulomb form.

With respect to molecular polarizability, the charge equilibration formalism is in fact a polarizable model, since one can derive the molecular dipole polarizability as [45]

$$\alpha_{\gamma\beta} = \mathbf{R}_\beta^\dagger \boldsymbol{\eta}'^{-1} \mathbf{R}_\gamma \quad (12.10)$$

where $\boldsymbol{\eta}'$ denotes the molecular hardness matrix augmented with extra rows and columns as needed to incorporate total charge constraints for each individual normalization unit, $\mathbf{R}_\beta$ and $\mathbf{R}_\gamma$ represent the Cartesian β and γ coordinates of the atomic position vector, and $\alpha_{\gamma\beta}$ is the $\gamma\beta$ element of the polarizability tensor. Furthermore, the model, being an all-atom representation with partial charges assigned to all atomic species, contains all higher-order electrostatic multipole moments. This is in contrast to point dipole polarizable models [46, 47] and Drude oscillator models [8, 9]. As such, the charge equilibration models incorporate higher-order electrostatic interactions naturally. With respect to the molecular dipole polarizability and its calculation using multiple charge normalization (or charge conservation) units, further discussion is deferred to Section 12.3.2.

In molecular dynamics simulations, the charge degrees of freedom are propagated via an extended Lagrangian formulation imposing a molecular charge neutrality constraint, thus strictly providing for electronegativity equalization at each dynamics step. The system Lagrangian is

$$L = \sum_{i=1}^{M}\sum_{\alpha=1}^{N_i} \frac{1}{2} m_{i\alpha} \dot{r}_{i\alpha}^2 + \sum_{i=1}^{M}\sum_{\alpha=1}^{N_i} \frac{1}{2} m_{Q,i\alpha} \dot{Q}_{i\alpha}^2 - E(Q,r) - \sum_{i=1}^{M} \lambda_i \sum_{\alpha=1}^{N_i} Q_{i\alpha} \quad (12.11)$$

where the first two terms represent the nuclear and charge kinetic energies, the third term is the total potential energy, and the fourth term is the molecular charge neutrality constraint, with λ_i the Lagrange multiplier for each molecule i. The fictitious charge dynamics, analogous to the fictitious wavefunction dynamics in Car–Parrinello (CP) type methods [48], is determined with a fictitious charge "mass" (adiabaticity parameter in CP dynamics). The units for this mass are (energy × time^2/charge^2). The charges are thus propagated based on forces arising from the difference between the average electronegativity of the molecule and the instantaneous electronegativity at an atomic site.

Finally, we comment on the underlying physics of charge equilibration models in general. Though presented in the preceding discussion in an atomic representation of charges, charge equilibration formalisms can be cast in terms of bond charge increments (BCI) [40, 49, 50]. In this representation, where "pseudo-bonds" allowing charge transfer between all atomic entities arises naturally, a molecular entity is considered as a *fully conducting species*, with charge transfer possible between all sites. Though seemingly unrealistic, this effect can be modulated naturally with the bond charge representation [40, 51] as well as in the atomic representation [45, 52]. The latter approach in the context of the present work affords an effective and chemically intuitive mechanism to modulate charge transfer (and equivalently molecular polarizability), as will be discussed further below.

12.3
Parameterization of Charge Equilibration Models

12.3.1
Establishing Atomic Hardness and Electronegativity Parameters

We reiterate that the present approach to implementing the charge equilibration model treats the electronegativities and hardnesses as empirical parameters. There is no attempt to associate the electronegativities to Pauling values (or even to relate the present parameters with each other via some electronegativity scale). One should bear in mind, also, that the lone atom electronegativities as defined in this formalism are not true isolated-atom values, that is, they are not strictly determined for single atoms in the gas phase from the atomic ionization potential and electron affinity, though one could plausibly seed parameter searches with the rigorously defined values.

In order to separate the fitting of the atom type electronegativities and hardnesses, the charge equilibration model is reformulated in terms of a linear response model [53–55]. Equation 12.7 gives the electrostatic energy of a molecule in the presence of some external potential. In the absence of any perturbing external field, the last term vanishes to give

$$E^0_{electrostatic} = \sum_{i=1}^{N} \chi_i^0 Q_i^0 + \frac{1}{2}\sum_{i=1}^{N}\sum_{j=1}^{N} \eta_{ij} Q_i^0 Q_j^0 \quad (12.12)$$

where the Q_i^0 are site charges in the absence of an external field. In each case, at equilibrium, the condition that

$$\frac{\partial E_{electrostatic}}{\partial Q_i} = 0, \quad i = 1,\ldots, N \quad (12.13)$$

and

$$\frac{\partial E_{electrostatic}}{\partial Q_i^0} = 0, \quad i = 1,\ldots, N \quad (12.14)$$

must be satisfied leads to the following expressions for the cases with and without an external potential:

$$\bar{\bar{\eta}}\bar{Q} = -(\bar{\chi}+\bar{\phi}) \quad (12.15)$$

and

$$\bar{\bar{\eta}}\bar{Q}^0 = -\bar{\chi} \quad (12.16)$$

Taking the difference of Equations 12.15 and 12.16 yields an expression for the response of the molecular charge distribution to the external probing field:

$$\bar{\bar{\eta}}\Delta\bar{Q} = -\bar{\phi} \implies \Delta\bar{Q} = -\bar{\bar{\eta}}^{-1}\bar{\phi} \quad (12.17)$$

We now have a relationship between the atomic hardness and the charge response due to the external potential. The fitting thus involves matching the charge responses computed using DFT by variation of the hardnesses.

12.3.2
Polarizability

The electrostatic energy expression (Equation 12.7, *excluding* the charge conservation constraint for the moment) of a *single* molecule j containing N atoms, when differentiated with respect to charge, produces a set of N simultaneous equations:

$$\frac{\partial E}{\partial Q_i} = \chi_i^0 + \eta_i^0 Q_i + \sum_{j \neq i}^{N} \eta_{ij} Q_j = 0 \tag{12.18}$$

Each of these equations can be set equal to zero to solve for a set of charges minimizing the energy restricted to the condition that the elements of the second derivative Hessian matrix are positive definite. Expressed in matrix form, the equations can be recast as

$$\boldsymbol{\eta} \mathbf{Q} = -\boldsymbol{\chi} \tag{12.19}$$

where $\boldsymbol{\eta}$ is the atomic hardness matrix, $\mathbf{Q}$ is the atomic charge vector, and $\boldsymbol{\chi}$ is the atomic electronegativity vector. Explicitly, the matrix representation is

$$\begin{pmatrix} \eta_1 & \eta_{12} & \cdots & \eta_{1N} \\ \eta_{21} & \eta_2 & \cdots & \eta_{2N} \\ \vdots & \vdots & \ddots & \vdots \\ \eta_{N1} & \eta_{N2} & \cdots & \eta_N \end{pmatrix} \begin{pmatrix} Q_1 \\ \vdots \\ \vdots \\ Q_N \end{pmatrix} = -\begin{pmatrix} \chi_1 \\ \vdots \\ \vdots \\ \chi_N \end{pmatrix} \tag{12.20}$$

These equations are augmented with a total charge conservation constraint for a single molecule j:

$$\lambda_j \left(\sum_{i=1}^{N} Q_i - Q^{\text{total}} \right) \tag{12.21}$$

Application of such charge constraints results in the modified set of equations

$$\boldsymbol{\eta}' \mathbf{Q} = -\boldsymbol{\chi}' \tag{12.22}$$

that are explicitly represented as:

$$\begin{pmatrix} \eta_1 & \eta_{12} & \cdots & \eta_{1N} & 1 \\ \eta_{21} & \eta_2 & \cdots & \eta_{2N} & 1 \\ \vdots & \vdots & \ddots & \vdots & \vdots \\ \eta_{N1} & \eta_{N2} & \cdots & \eta_N & 1 \\ 1 & 1 & \cdots & 1 & 0 \end{pmatrix} \begin{pmatrix} Q_1 \\ \vdots \\ \vdots \\ Q_N \\ \lambda \end{pmatrix} = -\begin{pmatrix} \chi_1 \\ \vdots \\ \vdots \\ \chi_N \\ Q^{\text{total}} \end{pmatrix} \tag{12.23}$$

Finally, for multiple charge conservation units (that is, 2 as will be used in the present analysis), the matrix representation straightforwardly follows:

$$\begin{pmatrix} \eta_1 & \eta_{12} & \cdots & \cdots & \eta_{1N} & 1 & 0 \\ \eta_{21} & \eta_2 & \cdots & \cdots & \eta_{2N} & 1 & 0 \\ \vdots & \vdots & \ddots & & \vdots & \vdots & \vdots \\ \vdots & \vdots & & \ddots & \vdots & 0 & 1 \\ \eta_{N1} & \eta_{N2} & \cdots & \cdots & \eta_N & 0 & 1 \\ 1 & 1 & \cdots & 0 & 0 & 0 & 0 \\ 0 & 0 & \cdots & 1 & 1 & 0 & 0 \end{pmatrix} \begin{pmatrix} Q_1 \\ Q_2 \\ \vdots \\ \vdots \\ Q_N \\ \lambda_1 \\ \lambda_2 \end{pmatrix} = -\begin{pmatrix} \chi_1 \\ \chi_2 \\ \vdots \\ \vdots \\ \chi_N \\ Q_1^{\text{total}} \\ Q_2^{\text{total}} \end{pmatrix} \tag{12.24}$$

This last expression is compactly written as Equation 12.20, with the understanding that the $\boldsymbol{\eta}'$ matrix is now augmented with the necessary extra rows and columns to account for the multiple charge conservation units. Molecular dipole polarizabilities are then computed using Equation 12.20 with the η' matrix taken to be as defined by the relation of Equation 12.22. The isotropic polarizabilities presented here are taken as one-third the trace of the molecular polarizability tensor, whose elements are computed from Equation 12.10. From the above discussion, it is now appropriate to mention that the difference between the partitioned and non-partitioned normalization schemes lies in the number of charge constraints by which the set of equations is augmented. That is, in the non-partitioned molecule, a single charge constraint is applied such that the sum of charges over the molecule is zero (or some appropriate constant depending on the system). Conversely, there are two charge constraints in the partitioned molecule, namely, the sum of charges in both partitioned units must be some constant.

12.3.3
Fitting and Force-Field Refinement

In addition to considering electrostatic interactions, polarizable force fields continue to require non-bond dispersion interactions. These interactions, generally modeled using Lennard-Jones potentials, in large part help to define condensed-phase liquid structure. The Lennard-Jones interactions have historically been determined by matching gas-phase structural and energetic properties of clusters (solute–water) simultaneously with condensed-phase properties of pure liquids. This approach affords one path to balance solute–water and solute–solute interactions with the solvent–solvent interactions. With the rapid advances in computational hardware, hydration free energies are becoming more integral components of the parameterization scheme to further characterize and balance solute–solvent interactions in the condensed phase. As implied in such an approach, there is an intimate coupling of the force fields for solvent and solute (be it small molecule or larger biomacromolecule). Furthermore, using small model compounds to determine parameters for larger biomolecular systems relies on the strong assumption of parameter transferability – the suitability of parameters (non-bond, electrostatic, bond-stretch, angle-bend, dihedral).

12.4
Applications of Charge Equilibration Models for Biological Solvents

12.4.1
Water

Water is *the* ubiquitous biological solvent, and represents one of a handful of biological solvents whose properties have been (and continue to be) studied in the utmost detail. From a molecular modeling perspective, there can be no accurate description of biological systems in the absence of the effects of aqueous solvent. Explicit atom water models continue to be developed, with various single-molecule and condensed-phase bulk properties targeted. The reader is referred to the vast literature discussing water force fields. Polarizable force fields for water are much more rare [21, 31, 37]. Currently, several polarizable models are available for use in molecular dynamics simulations; these are based on the approaches discussed in Section 12.2. For the present discussion, we will focus on the "transferable intermolecular potential, four-point, with fluctuating charge" (TIP4P-FQ) water model [21]. The TIP4P-FQ is based on the original TIP4P model of Jorgensen [56] and includes a lone-pair site along the H–O–H angle bisector, 0.15 Å away from the oxygen atom site. The TIP4P-FQ water model is described by a rigid geometry having an O–H bond distance of 0.9572 Å, an H–O–H bond angle of 104.52° and a massless, off-atom M-site located 0.15 Å along the H–O–H bisector, which carries the oxygen partial charge. This geometry is congruent with that of the fixed-charge TIP4P [56] water model. Repulsion and dispersion interactions are modeled using a single Lennard-Jones site located on the oxygen center having parameters σ_O = 3.159 Å and ε_O = 0.2862 kcal mol^{-1}. Partial charges on the atomic sites fluctuate during simulation and yield time-averaged charges of q = 0.62e on the hydrogen sites and q = −1.24e for the M-site.

The AMOEBA water force field [31] is a fully flexible, three-site water model incorporating permanent atomic monopole, dipole, and quadrupole moments. The model explicitly treats polarization by allowing for the induction of dipoles at prescribed sites. In this respect, an iterative approach to self-consistency is required in order to determine the instantaneous moments. Permanent multipoles interact through a multipole interaction matrix. Atomic polarizabilities are fitted to reproduce experimental molecular polarizabilities in the spirit of Thole. These values are used in conjunction with a damping scheme designed to attenuate the intramolecular dipole–dipole interactions in order to avoid polarization catastrophes at small separations. The O–H bond and H–O–H bond angle interactions are based on the MM3 force field and include anharmonic fluctuations arising from higher-order terms. The equilibrium bond length is 0.9572 Å and the equilibrium H–O–H angle is taken to be 108.5° (roughly 4° larger than the experimental value). Additional valence terms are added to model the coupling of intramolecular stretching and bending modes; these are of the Urey–Bradley form. The AMOEBA force field models repulsion and dispersion interactions using a buffered 14–7 potential form having atomic parameters of R_O = 3.405 Å and ε = 0.110 kcal mol^{-1} on the oxygen site, and parameters of R_O = 2.655 Å and ε = 0.0135 kcal mol^{-1} on the hydrogen sites. Thus, the AMOEBA treatment of repulsion and dispersion interactions is qualitatively distinct from the other models we consider here.

The Drude oscillator-based "simple water model with four sites and Drude polarizability" (SWM4-DP) [37] is a fixed-geometry model for water with an O–H bond distance of 0.9572 Å and an H–O–H bond angle of 104.52°, identical to that of the TIP4P-FQ model. A massless M-site is situated along the molecular symmetry axis at a fixed distance of 0.23808 Å from the oxygen atom, slightly further from the oxygen atom than the M-site in the TIP4P-FQ model. Permanent fixed charges of +q and −2q are assigned to the hydrogens and the M-site, respectively, where q = 0.5537e. These charges are similar to those of the TIP4P model (q = 0.5e) but smaller in magnitude than the average partial charges carried by the TIP4P-FQ model. Lennard-Jones parameters of $\sigma_O = 3.1803$ Å and $\varepsilon_O = 0.20568$ kcal mol^{-1} are employed on the oxygen atom to model van der Waals interactions. Despite differences between the models, both the TIP4P-FQ and the SWM4-DP models yield identical gas-phase dipole moments of 1.8 D. The polarizability of SWM4-DP is modeled using an oscillating Drude particle of positive charge $1.77185e$ tethered to the oxygen site (having opposite charge) by a harmonic force constant of 1000 kcal mol^{-1} Å^{-2}. Within this formalism, one modulates polarizability by adjusting the stiffness of the Drude spring or the associated Drude charge. In a recent refinement of the water model, the SWM4-NDP [57], the Drude particles are assigned a *negative* charge and the oxygen carries the positive contribution.

The properties described/predicted by the three models have been described in the literature. Tables 12.1 and 12.2 shows a comparison of several experimental properties for pure water computed for these models (from various sources) as well as select fixed-charge force fields. It is clear that the next generation of polarizable, all-atom water models provide more faithful representations of electrostatic properties, particularly the gas-phase dipole moments, condensed-phase polarization effects, dielectric constants, self-diffusion constants, and enthalpy of vaporization (liquid self-energy). Of particular note is the difference in the molecular polarizabilities of practical use for the various approaches. The Drude model requires a scaling of 0.724 from the gas-phase isotropic polarizability. The TIP4P-FQ water model uses a similar scaling as evidenced by the 1.1 Å^3 molecular polarizability (as opposed to the 1.47 Å^3 gas-phase value) required for stable condensed-phase dynamics simulations. Only the AMOEBA model retains a molecular polarizability that is close to the gas-phase value. It is important to bear in mind that there is sufficient evidence based on *ab initio* calculations and theoretical arguments that the intrinsic molecular polarizability should decrease in the condensed phase, particularly for polar compounds. Thus, the current practice of scaling gas-phase molecular polarizabilities, which arose from practical considerations, appears to be physically relevant. A serious implication that naturally arises is the ability to model this variation, particularly in regions of strong anisotropy (such as the protein–water, membrane–water, and air–water interfaces). Classical models that effectively capture this effect have not been introduced in the literature, though the idea has been discussed [15].

12.4.2
Non-Aqueous Solvents: Charge Equilibration Models of Alkanes

Alkanes, and materials containing aliphatic groups in general, are an important class of fluids, implicated in a variety of industrial processes such as petroleum refining and extraction, a wide range of commercial products, and as industrial solvents. From a biophysical perspective, the aliphatic moiety is ubiquitous in organic and biological systems, and thus requires particular attention in terms of an accurate representation within the context of molecular modeling and statistical mechanics based approaches to studying biophysical processes, including protein

Table 12.1 Comparison of select gas-phase properties predicted by several water force fields.

	AMO^{EB}A[1]	SWM4-NDP[2]	TIP4P-FQ[3]	TIP4P[4]	TIP3P[5]	SPC[6]	SPC/E[7]	TIP4P/EW[8]	TIP3P/EW[9]	Expt
$\mu_{monomer}$	1.773	1.85	1.85	2.18	2.35	2.27	2.35	2.32	2.	1.85
$E_{binding}$	4.96	5.15	4.5	6.3	6.5	6.7	7.17	–	6.87	5.44 (0.7)
R_{O-O}	2.892	2.83	2.92	2.75	2.74	2.75	2.73	–	–	2.976
α		0.97825	1.1	0	0	0	0	0	0	1.44
α_{xx}	1.672		0.82	0	0	0	0	0	0	1.528
α_{yy}	1.225		2.55	0	0	0	0	0	0	1.415
α_{zz}	1.328		0	0	0	0	0	0	0	1.468
Q_{xx}	2.5	2.4247	1.882	2.86	2.30	–	2.71	–	–	2.63
Q_{yy}	−2.17	−2.1768	−1.785	−1.60	−1.38	–	−1.36	–	–	−2.5
Q_{zz}	−0.33	−0.2479	−0.098	0.0	0.0	–	0.0	–	–	−0.13

1) AMOEBA (Atomic Multipole Optimized Energetics for Biomolecular Applications).
2) SWM4-NDP (Simple Water Model, 4-Point with Negative Drude Particle).
3) TIP4P-FQ (Transferrable Intermolecular Potential, 4-Point with Fluctuating Charges).
4) TIP4P (Transferrable Intermolecular Potential, 4-Point).
5) TIP3P (Transferrable Intermolecular Potential, 3-Point).
6) SPC (Simple Point Charge).
7) SPC/E (Simple Point Charge / Extended for polarization).
8) TIP4P/EW (Transferrable Intermolecular Potential, 4-Point, parameterized for Ewald).
9) TIP3P/EW (Transferrable Intermolecular Potential, 3-Point, parameterized for Ewald).

Table 12.2 Comparison of select bulk liquid properties ($T = 298$ K) predicted by several water force fields.

	AMOEBA	SWM4-NDP	TIP4P-FQ	TIP4P	TIP3P	SPC	SPC/E	TIP4P/EW	TIP3P/EW	Expt
Density (g cm^{-3})	10.004 (0.0009)	0.9998	1.00	1.001	0.9995	0.98	0.998	0.995	0.997	0.997
ΔH_{vap} (kcal mol^{-1})	10.48 (0.08)	10.51	10.48	10.65	10.41	10.5	11.7	10.58	10.53	10.51
$\Delta U_{liquid-gas}$ (kcal mol^{-1})	−9.89	−9.923	−9.9	−9.99	−9.82	−9.799	−9.92	−11.110	−9.9	−9.9
Dielectric constant, ε	82 (13)	79 (3)	79 (8)	53	92	65	71	62.9	89	78.3
ε_∞	–	1.68	1.592 (0.003)	1.0	1.0	1.0	1.0	1.0	1.0	1.79
Self-diffusion constant, D ($\times 10^{-5}$ cm^2 s^{-1})	2.02 (0.05)	2.33 (0.02)	1.9	3.29	5.19	3.85	2.49	2.4	4.03, 3.98	2.3
Second virial coefficient, B_2 (l mol^{-1})	−1.058 (0.032)	–	−0.64	–	–	–	–	–	–	−1.158
Constant-volume heat capacity, C_V (kcal mol^{-1} K^{-1})	28.4 (2.0)	–	–	–	–	–	–	–	–	17.8

folding, protein–protein and protein–nucleic acid interactions, as well as processes occurring through channels formed by integral membrane proteins [58–60]. Aliphatic groups constitute a non-trivial, low-dielectric bulk-like environment within lipid membranes; it is this environment that intimately interacts with integral membrane proteins and, in this sense, contributes in mediating myriad physiological processes. Aliphatic groups are present in amino acid side chains and contribute significantly to excluded volume and hydrophobic collapse in the context of protein folding [61]. Likewise, the prevalence of this chemical moiety in carbohydrates and nucleic acids stresses the need for accurate models [62]. As a building block for polarizable force fields for lipidic environments considered as biological solvents for proteins, we next discuss charge equilibration models for linear, saturated alkanes.

12.4.2.1 Bulk Liquid Properties

The charge equilibration force fields for alkanes have been parameterized using an approach discussed in the literature, the discussion of which we will forego in the present context. Moreover, the reader is referred to the literature for a discussion of the simulation protocols used in the modeling of bulk liquid properties of linear alkanes ranging from hexane to pentadecane [63]. The following properties were investigated: enthalpy of vaporization ΔH_{vap}, liquid density ρ, and isothermal compressibility β_T.

Enthalpies of vaporization are determined as

$$\Delta H_{vap} = H_{gas} - H_{liquid} \tag{12.25}$$

Invoking the definition of enthalpy, $H = U + PV$, and the assumption that the vapor phase obeys the ideal gas law, Equation 12.25 becomes

$$\Delta H_{vap} = (\langle E_{gas} \rangle + RT) - (\langle E_{liquid} \rangle + P \langle V_{liquid} \rangle) \tag{12.26}$$

where $\langle E_{gas} \rangle$, $\langle E_{liquid} \rangle$, and $\langle V_{liquid} \rangle$ are obtained from the average values of the relevant property in the gas and condensed phases, and P is constant at 1 atm. The PV term accounts for volumetric expansion from liquid to gas; however, it was found to be very small in relation to the other energy terms, so it was neglected in the calculation of the enthalpy of vaporization. Condensed-phase densities were calculated using the average volume from the bulk simulations and the number of molecules. Isothermal compressibilities were calculated from the volume fluctuations of the bulk simulations using the following statistical mechanical expression:

$$\sigma_V^2 = V N_A k_B T \beta_T \tag{12.27}$$

In keeping with the behavior of fixed-charge and alternative polarizable force fields, the "charge equilibration" (CHEQ) model predicts bulk liquid densities to within 1.6% of experimental values (Figure 12.1), vaporization enthalpies to within 1.3% (Figure 12.2) across the series, and compressibilities to within 5.3% (Figure 12.3). These predictions are comparable to the Drude oscillator model for alkanes [6], though for the longer chains the CHEQ model exhibits slightly more drift from experiment, except in the case of compressibilities, which do not appear to exhibit

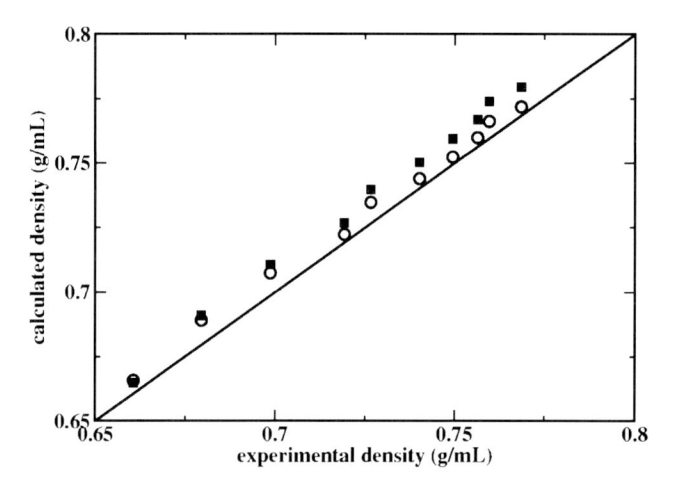

Figure 12.1 Experimental [64] (symbols) and predicted (solid line) alkane bulk liquid density: solid squares, CHEQ model of Davis et al. [63]; open circles, original earlier CHEQ model of Patel and Brooks [13].

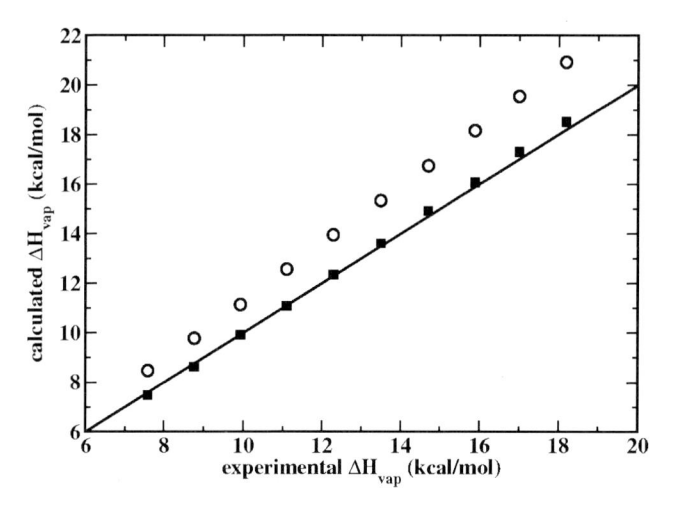

Figure 12.2 Experimental [65] (symbols) and predicted (solid line) vaporization enthalpies for bulk liquid alkanes: solid squares, CHEQ model of Davis et al. [63]; open circles, original earlier CHEQ model of Patel and Brooks [13].

a systematic drift across the series. It must be noted, though, that the Drude model was explicitly fitted to reproduce these properties across the entire homologous series. There is a small, but noticeable, systematic increase in the difference from experimental vaporization enthalpies as the alkane chain length increases. With the charge normalization scheme adopted here (normalization over the entire molecule), the larger alkanes possess a larger polarizability, and hence possess a larger polarization energy contribution to the condensed-phase cohesive energy. This leads to higher vaporization enthalpies relative to experiment. It is also

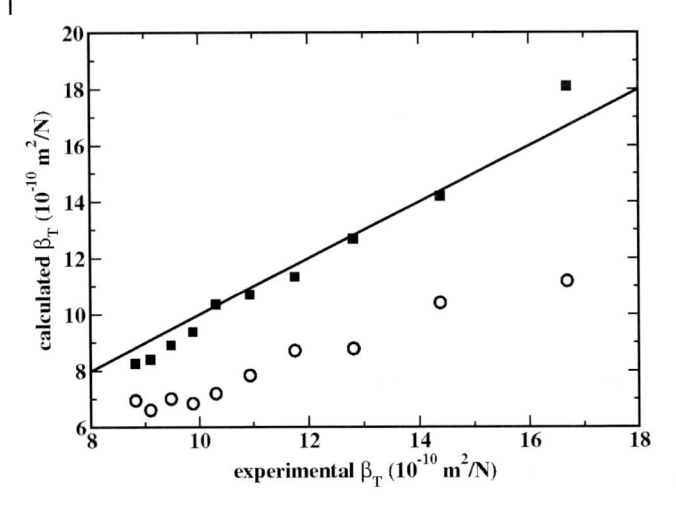

Figure 12.3 Experimental [64] (symbols) and predicted (solid line) isothermal compressibility for bulk liquid alkanes: solid squares, CHEQ model of Davis et al. [63]; open circles, original earlier CHEQ model of Patel and Brooks [13].

important to compare the vaporization enthalpies to those calculated with the previous version of the polarizable force field. That model gave a root-mean-square difference of 14.2% across the series of linear alkanes, compared to 1.3% using the revised parameterization. This can be attributed to the fact that lowering the vaporization enthalpy is equivalent to lowering the difference in energy between the vapor phase and the liquid phase, thus increasing the affinity of the molecule for the vapor phase relative to the previous model.

12.4.2.2 Local Chain Dynamics

Local measures of dynamics along the carbon backbone are related to the ^{13}C nuclear magnetic resonance (NMR) T_1 relaxation time. An accurate representation of these quantities is important in describing the local chain order and dynamics in lipid molecules, particularly in the context of a bilayer environment. Moreover, as one of the ultimate applications of these force fields is to molecular modeling of solvated bilayer and integral membrane protein systems, it is important to characterize the quality of the force field with respect to this property. ^{13}C NMR T_1 relaxation times were calculated using the approach of Klauda et al. [66], in which T_1 is calculated from the reorientation correlation functions of the CH vectors, assuming motional narrowing and an effective bond length of 1.117 Å:

$$\frac{1}{NT_1} = (1.855 \times 10^{10} \text{ s}^{-2})\tau = (1.855 \times 10^{10} \text{ s}^{-2})\int_0^\infty \langle P_2(\hat{\boldsymbol{\mu}}(0)\hat{\boldsymbol{\mu}}(t))\rangle dt \qquad (12.28)$$

Here, N is the number of protons bonded to the relevant carbon, τ is the rotational correlation time, $\hat{\boldsymbol{\mu}}$ is the CH vector, and $P_2(x) = \frac{1}{2}(3x^2 - 1)$, the second Legendre polynomial. Figure 12.4 shows the calculated T_1 values for heptane,

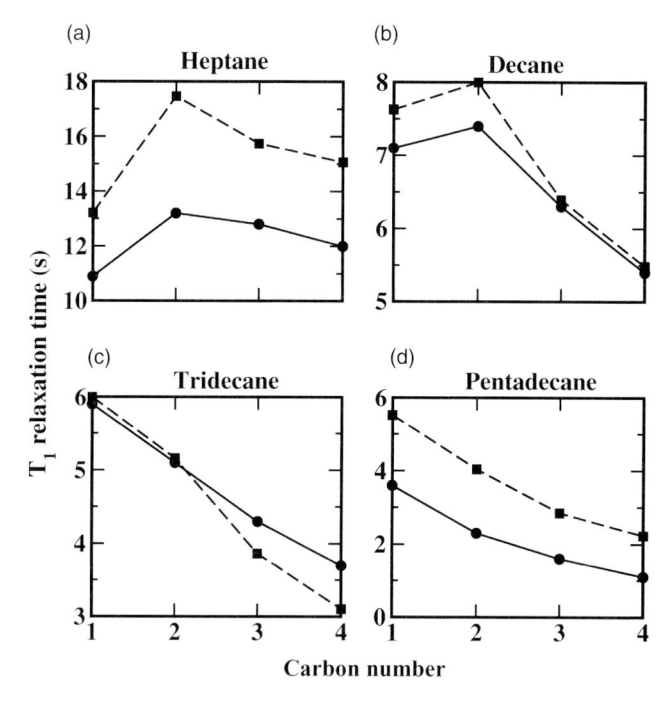

Figure 12.4 The T_1 NMR relaxation times for experiment (solid circles) and predictions based on charge equilibration polarizable force field (solid squares) [63].

decane, tridecane, and pentadecane together with the experimental values [67]. One may note that, since the current torsional component of the alkane force field is commensurate with the CHARMM C27r force field, it is not surprising that the performance with respect to this property is commensurate with that of the nonpolarizable model. This behavior suggests a more detailed investigation into the connection between local chain dynamics and the underlying energy surface in such systems.

12.5
Toward Modeling of Membrane Ion Channel Systems: Molecular Dynamics Simulations of DMPC–Water and DPPC–Water Bilayer Systems

Integral membrane proteins have become an arena of intense fundamental research effort due to the ubiquitous nature of these macromolecules. Comprising roughly one-third of the human genome, they are implicated in myriad physiological function and, unfortunately, dysfunction. For instance, for normal physiological functioning, integral membrane proteins are involved in signaling processes, passive and active transport, and interfacial enzymatic processes [58]. Of course, one cannot speak about integral membrane proteins independent of the lipidic context in which they function. Lipid membranes in their own right have garnered much attention as well, with particular focus on membrane properties and behaviors, including structural deformation (in the presence of small molecules as well as integral membrane proteins), and electrostatic properties, such as the interfacial dipole (or total) potential [68] and dielectric constant variation with location in a bilayer [69]. Of particular recent interest have been protein–lipid interactions, and, specifically, the interactions of charged and/or polar amino acid residues as they pertain to an understanding of the thermodynamics of structural and energetic stability of integral membrane proteins upon desolvation of such systems.

Complementing the enormous volume of experimental effort to understand lipid bilayers and integral membrane proteins, computational approaches based on the analytics of statistical mechanics (molecular dynamics and Monte Carlo simulations) have been an indispensable tool to understanding properties and processes in these systems at the atomic and molecular level, and in providing insights at a resolution inaccessible to current state-of-the-art experimental methods [58–60, 70–78]. Current state-of-the-art all-atom simulation methods employ fixed-charge force fields. The shortcomings of such models, and the plausible importance of explicitly accounting for non-additive electrostatic and charge-transfer effects, have been widely discussed in the literature [79]. The past decade has witnessed an increasing pace of development and application of polarizable force fields for a range of applications, though such models have not yet realized the popularity enjoyed by fixed-charge models. In order to begin to explore the effects of polarization in biological systems, the first step undoubtedly has to be the development of self-consistently parameterized models. In the following discussion, we briefly discuss the parameterization of a polarizable dimyristoylphosphatidylcholine (DMPC) lipid bilayer system and select properties of the fully solvated bilayer using all-atom molecular dynamics simulations.

12.5.1
Fully Polarizable DMPC Bilayers in TIP4P-FQ Solvent: Application of Charge Equilibration Models for Molecular Dynamics Simulations

For constructing a charge equilibration model for DMPC (and related saturated phospholipid molecules), the approach to force-field development constructs a biomacromolecular force field based on smaller, model compound systems. Based on Figure 12.5a, plausible choices for the headgroup regions are the tetramethylammonium and dimethylphosphate ions. The ester groups are modeled by methyl acetate, and the alkyl tails by linear alkanes (as discussed above). As classical force fields attempt to model inter- and intramolecular interactions, we consider interaction energetics and geometries of the model compounds with water as a means to incorporate necessary information into the force-field description. We consider vacuum–water model compound systems as sufficient proxies for developing necessary interaction models; moreover, since water is the solvent of choice, it is necessary to arrive at a reasonably accurate description of this interaction. For the present work, we focus on the phosphorylcholine group atoms (the headgroup

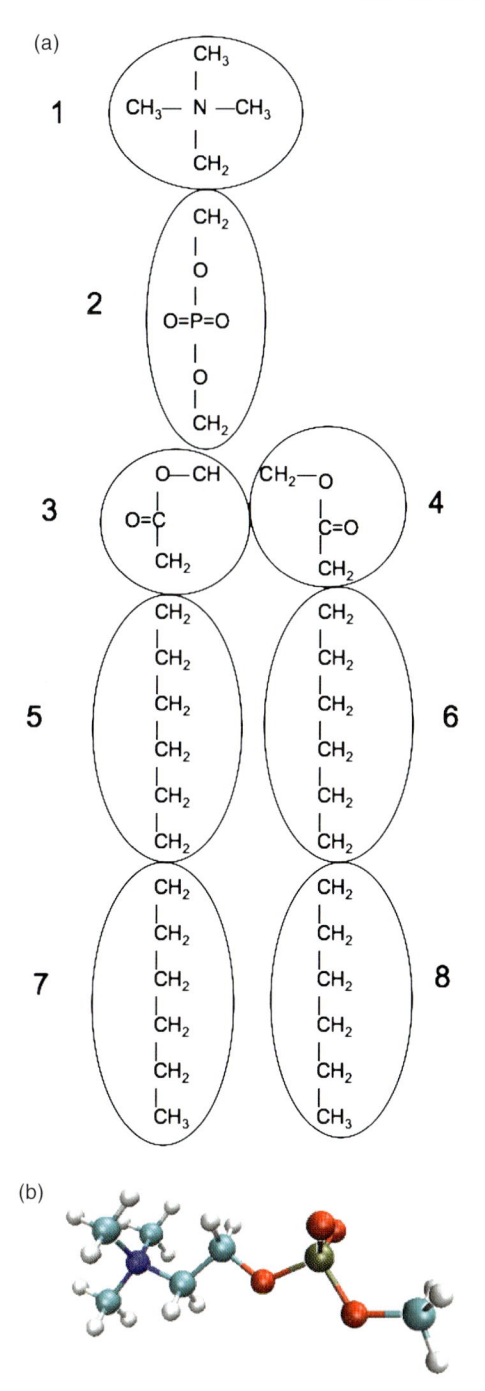

Figure 12.5 (a) Schematic representation of a single DMPC molecule with charge normalization groups explicitly defined within encircled regions. (b) The o-phosphorylcholine model compound for parameterizing DMPC headgroup torsions.

regions) of DMPC while *transferring* force-field parameters for the glycerol group (ester linkage) and associated atoms from earlier work by Patel and Brooks as part of the CHARMM charge equilibration polarizable protein force field [10, 14]. The torsion angles for the *o*-phosphorylcholine model compound are adjusted in the present work in order to reproduce more "global" structural properties of the headgroup region, with particular attention to the distribution of the P–N dipole vector. For the acyl chains, we transfer the force field of Davis *et al.* [63] that was recently revised and tuned to more accurately reproduce a wide range of properties of liquid alkanes, including bulk structural and thermodynamic properties, as well as single-molecule torsional energetics. Finally, intramolecular components of the DMPC force field, including bond and angle stretching and bending, respectively, are transferred from the CHARMM non-polarizable model [80] and the work of Patel *et al.* [10].

The electrostatic parameters of the atoms in the phospholipid headgroup have been adjusted to reproduce the atomic charges and molecular polarizability of model compounds tetramethylammonium and dimethylphosphate. The target data are optimized gas-phase geometries of tetramethylammonium and the *gg* conformer of dimethylphosphate obtained at the MP2/cc-pVTZ level of theory using GAUSSIAN 03 [81]. Atomic charges are determined from fits to the wave-function-based electrostatic potential [82]. We note that the final polarizability for dimethylphosphate (8.057 Å^3) is reduced from the gas-phase *ab initio* value of 8.786 Å^3. The extent of condensed-phase polarizability scaling still remains an empirical matter in the context of polarizable force-field development.

The van der Waals (Lennard-Jones) contribution to the total potential energy is given by the following expression:

$$V_{LJ}(r) = \varepsilon_{ij}\left[\left(\frac{\sigma_{ij}}{r_{ij}}\right)^{12} - 2\left(\frac{\sigma_{ij}}{r_{ij}}\right)^{6}\right] \quad (12.29)$$

The adjustable atomic parameters are the potential well depth ε and the van der Waals radius σ. The parameters of the PL (phosphate phosphorous atom P), OSL (phosphate oxygen singly bonded to P), and O2L (phosphate oxygen doubly bonded to P) atom types were fitted to *ab initio* interaction energies and hydrogen bond distances of dimethylphosphate (DMP) with water. The target data were obtained from calculations performed at the MP2/cc-pVTZ level of theory on two configurations of the DMP–water complex: one in which the water forms hydrogen bonds with both double-bonded oxygens (type O2L) of the phosphate; and one in which the water hydrogen bonds with one bridging oxygen (type OSL) and one double-bonded oxygen. In addition, specific non-bonded interaction terms (NBFIX) were added for interactions of OSL and O2L atoms of dimethylphosphate with the HT (water hydrogen atoms) atoms of TIP4P water in order to better reproduce the *ab initio* energies and geometries. This is in the spirit of the Drude oscillator models for alcohols recently presented by Anisimov *et al.* [8, 9]. These studies revealed the need for special unique interaction parameters between certain atom types in order to match more closely relevant experimental data, in particular hydration free energies of various small-molecule solutes in water. Together, these separate results seem to suggest that, in certain cases, polarizable models may require the addition of further specific combinations of interactions, or at least further atom types. Non-bond interaction parameters are determined by iterating through several values of each parameter. The set of values that resulted in the smallest sum of squared errors was taken to be the final result, which was then modified manually to further increase agreement with the target data.

The refinement of the phospholipid headgroup torsional potential involved optimizing the torsional parameters of the N–C–C–O, C–C–O–P, and C–O–P–O dihedrals. The C–O–P–O torsion was parameterized against *ab initio* conformational energies of dimethylphosphate. Dimethylphosphate has two C–O–P–O torsions that define its conformation; by convention, we refer to the orientation of the two P–O bonds involved in the torsion (φ_1, φ_2) as either *trans* (t), approximately 180°, or *gauche* (g), approximately 60°. The global energy minimum occurs at gg, which may not seem reasonable based on steric considerations, but can be explained by the anomeric or *gauche* effect [83].

In the CHARMM force field, the torsional potential energy of a molecule is represented as a sum of contributions from each dihedral angle in the molecule:

$$V_{\text{dihedral}} = \sum_{\substack{\text{all} \\ \text{dihedral} \\ \text{types}}} \sum_{\varphi} \sum_{j} K_j [1 + \cos(n_j \varphi - \delta_j)] \tag{12.30}$$

The sum is carried out over all parameter sets j and all relevant dihedral angles φ for each dihedral type. The parameterization approach used here is described in more detail elsewhere [63]. The torsional profiles are shown in Figure 12.6. Both illustrate excellent agreement with *ab initio* conformational energies. Also worth noting is the *cis* barrier, which appears to be overestimated by the refined force field. We performed constrained *ab initio* (MP2/cc-pVTZ) geometry optimizations of dimethylphosphate with one dihedral constrained in the *cis* conformation, allowing all other coordinates to minimize. These results suggest an energy barrier of ~8 kcal mol^{-1} for even the lowest-energy *cis* conformer at the MP2/cc-pVTZ level of theory. Using a smaller basis set results in a relative energy of ~6 kcal mol^{-1}, suggested by calculations at HF/6-31G* and MP2/6-31G*. Therefore, our refined model is in good agreement with the MP2 energy barrier despite not having been explicitly fitted to it. It is worth noting that the torsional energetics of existing force fields for the headgroup region reflect the quality of the *ab initio* levels of theory used to originally parameterize them. It appears that the lower levels of theory (HF/6-31G* and MP2/6-31G*) systematically predict lower relative energies for the *gauche–cis* and *trans–cis* conformers relative to the MP2/cc-pVTZ values.

The N–C–C–O and C–C–O–P torsional potentials were fitted simultaneously to the torsional energy surface of o-phosphorylcholine, shown in Figure 12.5b, a compound that has been used previously to model phospholipids [84]. The fit was carried out using the CMAP [85, 86] function in CHARMM. The torsional energy surface for these two coupled torsion angles predicted by the non-polarizable C27r force field [66] is used as the reference. To generate this surface, the two relevant

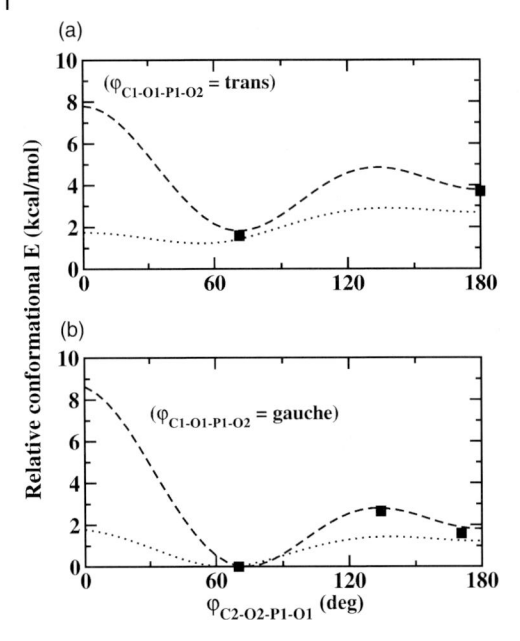

Figure 12.6 Torsional profiles of the C–O–P–O dihedral angle of dimethylphosphate in the (a) *trans* and (b) *gauche* conformations, computed using the non-polarizable CHARMM force-field model (dotted curve), and the CHEQ model (dashed curve). The *ab initio* energies computed at the MP2/cc-pVTZ level of theory using GAUSSIAN 03 are also shown (solid squares).

dihedrals are varied from −180° to +180° in steps of 10°, and energies were calculated relative to (−70°, −60°), a low-energy conformer most likely close to the global minimum. The other dihedrals, analogous to the C–O–P–O torsions in dimethylphosphate, are constrained in the *trans* conformation. Since the energies are relative, contributions from the C–O–P–O torsion cancel out, meaning that this fit is independent of the C–O–P–O torsional parameters. Likewise, the C–O–P–O torsional fitting is independent of the N–C–C–O and C–C–O–P torsional fitting since dimethylphosphate does not involve the latter two dihedral types.

12.5.2
Component Atomic and Electron Density Profiles

One measure of the stability of the bilayer is the number density of various atomic species as a function of distance along the bilayer normal. Figure 12.7 shows the component density profiles for water and atomic species of the lipid (headgroup phosphorus, oxygens, and nitrogen). These are consistent with previous studies [87–89] as well as the non-polarizable CHARMM27 model. There is a slight difference in the extent of water penetration into the membrane interior between the polarizable and non-polarizable force fields. Polarizability of both solvent and lipid would favor, free-energetically, the presence of waters in the low-dielectric environment of the lipid interior. This bears relevance for the thermodynamics associated

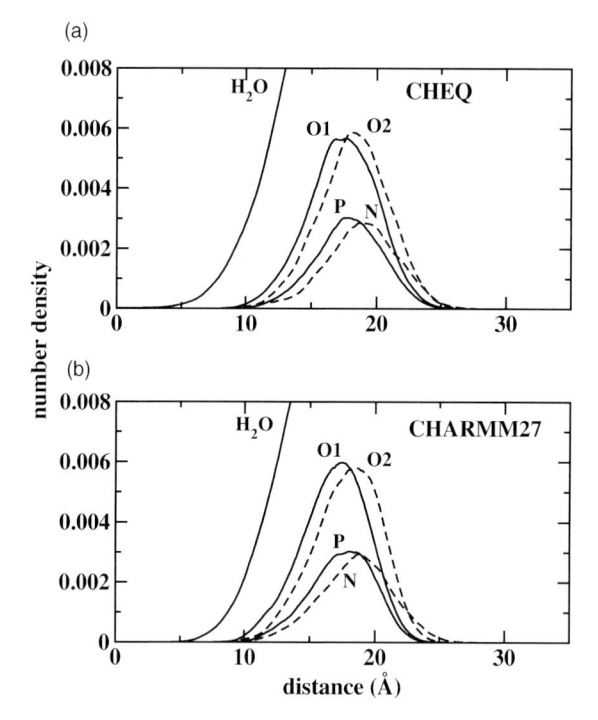

Figure 12.7 Densities of various components as a function of distance along the bilayer normal (z-axis): (a) refined polarizable CHEQ model; (b) non-polarizable CHARMM27 force field. O1 and O2 refer to phosphate oxygens, with O1 being the bridging oxygens.

with integral membrane proteins. Specifically, there has been recent effort to understand the free energetics of integral membrane protein structure and stability associated with desolvation of amino acid side chains of varying degrees of hydrophilicity or hydrophobicity and polarity. This attention has arisen based on the observation of lipid-exposed arginine residues in two recent crystal structures of voltage-gated potassium channels. Based on long-time molecular dynamics simulations, McCallum et al [77]. demonstrated the role of water defects, which form local "solvation cages" around the lipid-exposed arginine, in determining the relative costs for burying charged residues in lipidic environments. The current results, though far from definitive, suggest the role of non-additive electrostatic effects (polarization) in contributing to the stability of lipid-exposed residues of integral membrane proteins.

Figure 12.8 shows the electron density profile for both polarizable and non-polarizable models [90, 91]. Both are consistent with previous simulations of dimyristoylphosphatidylcholine (DMPC) and dipalmitoylphosphatidylcholine (DPPC) [92] as well as with experiment [90]. In order to better compare to experiment, the functional form of the experimental data was generated using models determined by Klauda et al. [91]. Both models [93] are in satisfactory agreement with experiment, with the charge equilibration model displaying a slightly more pronounced dip in the bilayer center.

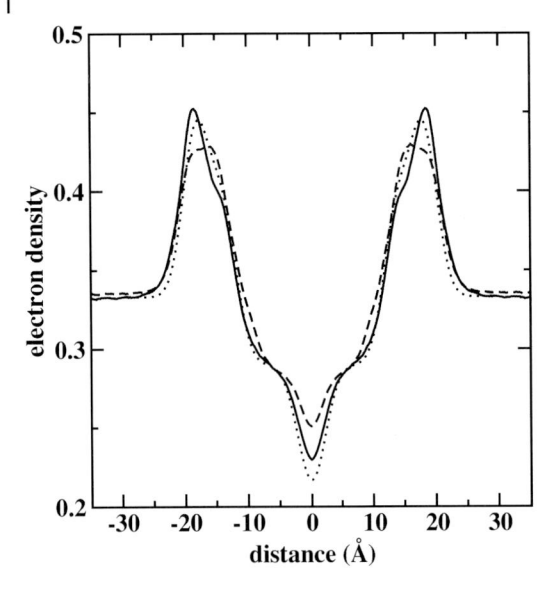

Figure 12.8 Electron density for both polarizable CHEQ (solid line) and CHARMM27 (dashed line) models, as well as experimental values (dotted line) [90, 91].

12.5.3
Lipid Chain Dynamics

The dynamics of the phospholipid tailgroups are probed by the variation of deuterium order parameters with respect to position along the alkyl chain. The deuterium order parameter, S_{CD}, is obtained as

$$S_{CD} = \langle P_2(\cos\theta) \rangle \qquad (12.31)$$

where θ is the angle between a particular CH vector and the bilayer normal, and $P_2(\cos\theta) = \frac{1}{2}\left[3(\cos\theta)^2 - 1\right]$, the second Legendre polynomial. This is equivalently the S_{zz} component of the NMR quadrupolar splitting tensor [94]. Figure 12.9 shows the magnitude of the calculated order parameter as a function of position along the alkyl chain. Also included for comparison are values from the non-polarizable CHARMM force field and experimental values for the sn-2 chain [95]. Both polarizable and non-polarizable models reproduce the experimental trends [95–97]. The polarizable model displays less order in the bilayer center (more closely in agreement with experiment). As discussed earlier, this difference would also contribute to the enhanced water penetration into the bilayer core.

12.5.4
Charge Distributions

The charge equilibration models allow molecular charge distributions to fluctuate over time in response to instantaneous changes in local electrostatic fields. Figure

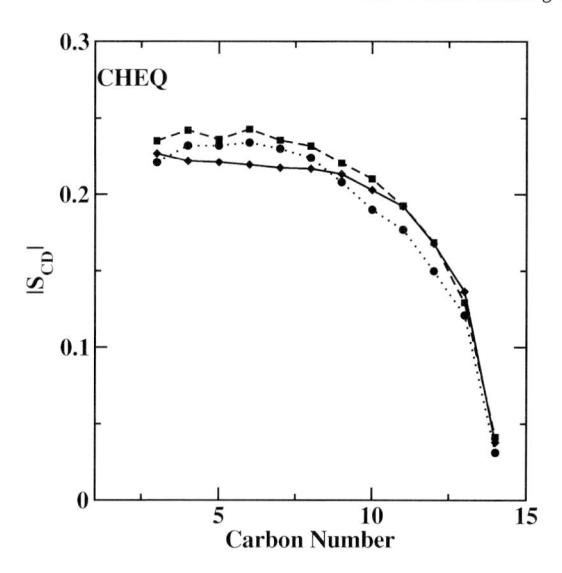

[Figure: Plot labeled CHEQ showing $|S_{CD}|$ (y-axis, 0 to 0.3) vs Carbon Number (x-axis, 0 to 15)]

Figure 12.9 Deuterium order parameters for the tailgroups as a function of position on the hydrocarbon chain: solid squares and dashed line, sn-1; solid diamonds and solid line, sn-2. Experimental data [95] for the sn-2 chain are also shown (solid circles and dotted line).

12.10a–c shows charge distributions for phosphorus, nitrogen, and oxygen atoms of the lipid headgroups. It is worth noting that the two bridging phosphate oxygens (O1 and O2) are both of type OSL, but have different charge distributions. This is a result of differing chemical environments: fixed-charge representations (as in the CHARMM non-polarizable case) ascribe equivalent charges to "like" oxygen atoms (in an average sense, based on gas-phase quantum-mechanical calculations of single or dimer complexes, for instance).

To further quantify this effect, the peak values of each distribution were compared to the average gas-phase minimized charges of all lipid molecules in a snapshot. These values were obtained from a single snapshot in the trajectory by deleting all atoms except for one lipid molecule, allowing the charges to equilibrate without minimizing the atomic coordinates, and averaging the charges over all 72 lipid molecules in the snapshot. This procedure was then repeated using a full coordinate minimization. Table 12.3 shows these charges as well as the peak values of the charge distributions from the trajectories. Comparing these values to the *ab initio* charges of the model compounds used in the parameterization, it is clear that the polarizable model allows the charges to shift in response to the condensed-phase environment. The magnitude of the shift in charge distributions varies with atomic species, but in general is on the order of $0.05e$ to $0.1e$.

Furthermore, Figure 12.10d demonstrates the effect of lipid association on water oxygen atoms. Oxygen atoms on water molecules directly associated with the phosphate groups show the largest shift toward higher values relative to the pure water distribution. This is directly related to the polarization effect on the water hydrogen atoms interacting with the phosphate oxygen atoms. In the case of water

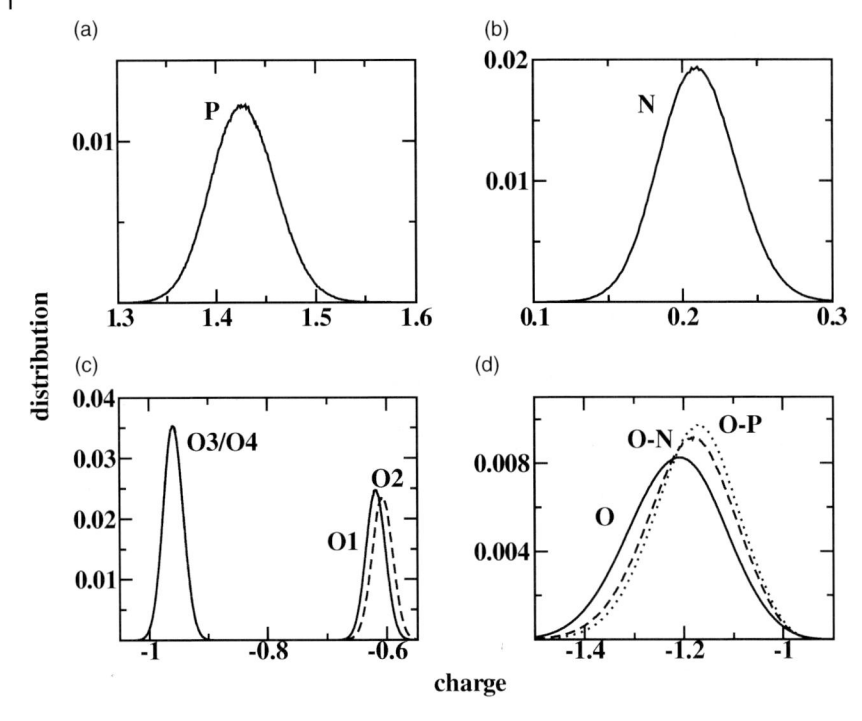

Figure 12.10 Charge distributions for (a) phosphorus, (b) nitrogen, and (c) oxygen atoms of the lipid headgroups and (d) water oxygens over the trajectory. O1 and O2 refer to the bridging phosphate oxygens; O3 and O4 refer to the equivalent double-bonded phosphate oxygens; O–N refers to water oxygens associated with headgroup nitrogen atoms; and O–P refers to water oxygens associated with headgroup phosphorus atoms.

Table 12.3 Peak values of charge distributions over the simulation trajectory compared to average gas-phase minimized charges of all 72 DMPC molecules without (a) and with (b) coordinate minimization.

	Peak	Gas phase[a]	Gas phase[b]
N	0.210	0.151	0.171
P1	1.430	1.514	1.484
O1	−0.619	−0.598	−0.599
O2	−0.608	−0.612	−0.597
O3	−0.960	−0.863	−0.904
O4	−0.960	−0.865	−0.894

a) Without coordinate minimization.
b) With coordinate minimization.

associating with the choline moiety, there is a lesser effect due to the lower polarizability of this group. Nevertheless, there is a shift in oxygen charges due to the larger interaction between the permanent, total choline charge of +1e and the water molecular dipole.

12.5.5
Dipole Moment Variation

Since the polarizable model allows electrostatic response to local chemical environment, one may anticipate a variation in the average dipole moment of water molecules when moving along the interface normal from bulk solution to membrane interior. Figure 12.11 shows the water molecular dipole moment distributions obtained by averaging the dipole moment of water molecules found in slabs of width 0.25 Å along the interface normal. In bulk solution, the water dipole moment plateaus at a value of 2.60 D, 0.8 D above the gas-phase value of 1.85 D; this is consistent with the bulk TIP4P-FQ dipole moment. The profile exhibits a monotonic decrease to a value of approximately 1.9 D within the membrane interior. There are two items of note.

First, the polarizable water model captures the condensed-phase environment effect on the local water molecular electrostatics via an enhanced dipole moment. We note that there is still no single consensus value of the liquid-phase dipole moment of water, with values ranging from 2.5 to 3.0 D [98–100].

Second, we observe that the dipole moment does not fall exactly to the gas-phase value at the center of the membrane. Moving toward the interface from the center,

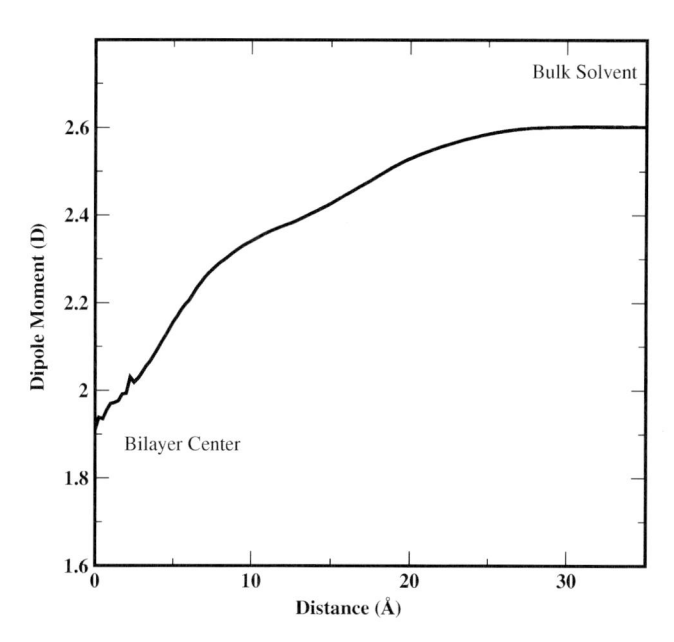

Figure 12.11 Profile of the average molecular dipole moment of water from the center of the bilayer to the bulk solution as a function of distance along the bilayer normal (z-axis).

the average molecular dipole moment increases monotonically. Thus, there is a significant interior region over which the water dipole moment, though not at the gas-phase value, exhibits an enhanced electrostatic moment in the lipidic environment. This suggests a non-trivial dielectric effect exerted by the polarizable membrane, that is, the membrane possesses a dielectric constant different from unity. This arises from the local lipid-chain polarizability as well as from longer-range electric field effects from the polar headgroup region. For the current polarizable lipid model, the alkyl segment contributes approximately 36.3 Å^3 to the overall DMPC molecular polarizability. This is based on the polarizability of tetradecane calculated using two charge constraint units consisting of seven carbons each. Moreover, with a lower water dipole moment in the lipid bilayer center, one can argue that interactions between water molecules not directly associating in the bilayer center are to some degree *reduced* rather than enhanced. This picture contrasts with that expected assuming a constant water dipole moment (hence water dipole–dipole interaction) in the lipid bilayer center.

12.5.6
Dielectric Permittivity Profiles

Electrostatic properties at the solution–water interface play an integral role in mediating transfer processes, interfacial binding and catalysis, and association of small molecules at the solution–bilayer interface. One can ask about the nature of the variation in dielectric constant across this interface. We next consider the application of the approach developed by Stern and Feller [69] for computing the longitudinal profile of the parallel component of the z-dependent dielectric permittivity. We compute profiles of the parallel (in-plane) component of the dielectric permittivity using equations 71 and 26 of Stern and Feller [69] for tin-foil boundary conditions:

$$\varepsilon_{\|} = [4\pi h_{\|}(z)] + 1 \quad (12.32)$$

$$h_{\|}(z) = \frac{1}{2k_B T} \langle \mathbf{P}_{\|}(z) \cdot \mathbf{M}_{\|} \rangle + \langle a_{\|}(z) \rangle \quad (12.33)$$

Here $\mathbf{P}_{\|}(z)$ is the local polarization density and $\mathbf{M}_{\|}$ is the *total* system dipole moment. In their formulation, Stern and Feller [69] decompose the total fluctuation into contributions from fixed charges and/or dipole and explicit polarization (point dipole polarizabilities). However, in the present approach, the polarization component is self-consistently included in the first term of Equation 12.33. We briefly expand on this in the discussion below. The polarization density is computed using a bond-charge approach similar to that outlined in Stern and Feller [69]. Briefly, the charge on a particular atom i is determined from a set of bond-charge increments (BCIs) b_{jk}. The BCIs are defined so that each atom (j or k) associated with a particular BCI b_{jk} receives an amount of charge $\pm b_{jk}$. The total charge on an atom is then a sum over all the contributions from the bond-charge increments to which the atom belongs, as represented by the following mapping:

$$q_i = \sum_{jk} C_{i,jk} b_{jk} \tag{12.34}$$

with

$$C_{i,jk} = \begin{cases} 1 & i=j \\ -1 & i=k \\ 0 & i \neq k, i \neq j \end{cases} \tag{12.35}$$

Given a set of charges, we obtain the bond-charge increments b_{jk} by inverting the C matrix via singular-value decomposition [69], or for well-conditioned matrices via straightforward inversion. The inverse is computed once for a given molecule (using the minimal topology based description of bond-charge increments) and reused for analysis of trajectory snapshots.

The polarization density $P_{\parallel}(z)$ is computed as a sum of the local bond-charge dipole moments in a bin of width dz at position z. As in Stern and Feller [69], we use

$$P(z) = \frac{1}{A} \sum_{jk} \mu_{jk} \delta(z - z_{jk}) \tag{12.36}$$

The bond dipole is determined simply as

$$\mu_{jk} = b_{jk}(r_j - r_k) \tag{12.37}$$

Figure 12.12 shows the z-dependent parallel component of the dielectric constant for the polarizable and non-polarizable models. The bulk water values reflect the properties of the pure solvent models, with the polarizable TIP4P-FQ possessing a value much closer to the experimental value (80 versus 97 for TIP3P). Within the bilayer interior, both models approach a value of unity (the polarizable model being slightly higher due to contributions from polarization), which is the expected value for alkyl-type species via the Kirkwood–Fröhlich formalism. Furthermore, we have neglected to include the infinite-frequency dielectric for the polarizable model, though the contributions from this are on the order of 2 for the membrane interior, and 1.6 to 1.9 for more polar regions (headgroups and bulk solvent). Results are presented as averages of two or three individual trajectories each of length 10–12 ns. It has been shown previously [69] that, for this approach, 10 ns is sufficient to achieve convergence, though more sampling is certain to improve prediction of fluctuation properties.

The DMPC headgroup regions display relatively larger dielectric constants. This has been attributed to the large magnitude of the headgroup dipoles, which experimentally are in the range of 19–25 D (in-plane component) [69]. The qualitative behavior has been observed previously via molecular dynamics simulations of DPPC [69] and palmitoyloleoylphosphatidylcholine (POPC) [101]. The polarizable force field predicts a larger interfacial dielectric arising from the additional fluctuations in the induced dipoles (toward the water phase of the interfacial region). The water dielectric for the polarizable model displays a monotonic decay from a bulk value of 79 to ~1 in the membrane interior (again, these limiting values neglect

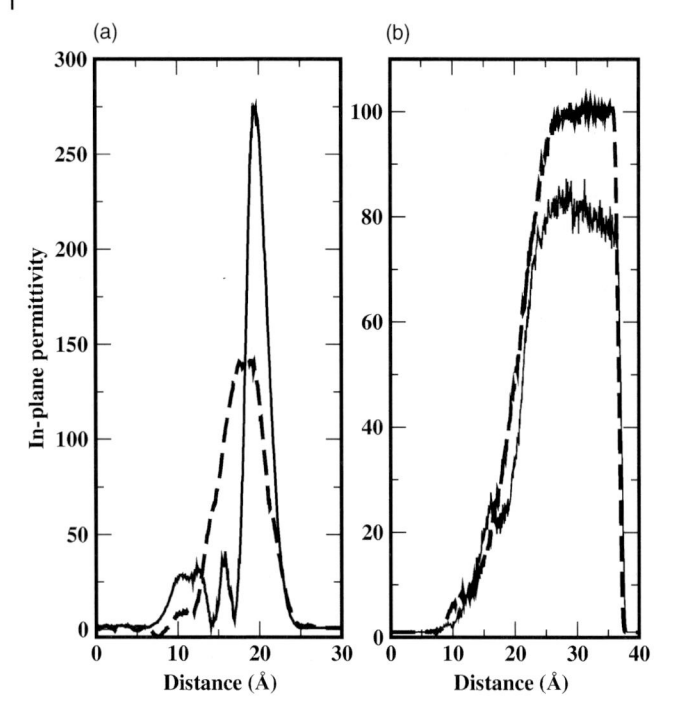

Figure 12.12 Profiles of (a) total and (b) water contributions to the parallel component of the dielectric permittivity as a function of distance along the bilayer normal. Results are shown from both polarizable CHEQ (solid curves) and non-polarizable CHARMM27 (dashed curves) models.

minor contributions from the infinite-frequency dielectric). The non-polarizable TIP3P model similarly displays a monotonic decay, but its bulk value is slightly higher (97); this is a well-accepted property of the TIP3P model [102]. The polarizable model exhibits more fluctuations in the dielectric permittivity in the polar region of the lipid. We attribute this to stronger correlations between static and induced dipoles that are not present in the non-polarizable case.

12.6
Conclusions and Future Directions

We have discussed recent progress in the area of development and application of polarizable force fields for biological solvents (in the most general sense), with focus on charge equilibration models. In particular, we have addressed parameterization issues, as well as the application of charge equilibration force fields to study systems ranging from simple, solvent environments, to large macromolecular assemblies such as lipid bilayers. Moreover, it appears that charge equilibration models afford an efficient means to account classically for charge transfer, even in the case of charged systems (which have proven difficult to model in the past). The addition of a further level of physics has allowed for a novel set of physical behavior described using classical simulations. In the case of small molecules, allowing a simple representation of polarizability allows for a strong condensed-phase effect in terms of enhanced molecule dipole moments, which self-consistently shift to higher values relative to the isolated-molecule case simply with the use of a single parameterization, a physical behavior not described using fixed-charge force fields.

The field of polarizable force-field development and application is now witnessing rapid growth [6–10, 14, 25–29, 37–40, 57, 63]. We believe that detailed characterization of polarizable force fields will be necessary in order to determine the strengths and weaknesses of each of the diverse approaches to polarization now in place. Characterization of these models will require determination of solvation free energetics for a wide array of compounds (representative of the chemical environments in biological systems).

In terms of practical applicability, one of the major issues will be to increase the efficiency of polarizable force fields to be competitive with current state-of-the-art fixed-charge force fields. In the case of charge equilibration models, the bottlenecks arise due to the smaller time steps required in order to propagate the significantly lighter charge degrees of freedom in order to remain on the Born–Oppenheimer surface during dynamics within the extended Lagrangian formulation.

Furthermore, for the charge equilibration models, owing to the requirement for a full accounting of the long-range electrostatic interactions via some variant of an Ewald summation, multiple time-step methods incorporating correct decompositions of the dynamical propagator will be required. Recent work suggests that separating the short-range contributions arising from the reciprocal space sums in Ewald-based methods, which is the correct decomposition of the dynamical propagator, should allow for much larger time steps for the outer loops in multiple time-step methods.

Acknowledgments

The author gratefully acknowledges support from the National Institutes of Health sponsored COBRE (Center of Biomedical Research) Grants, numbers P20-RR017716 and P20-RR015588 at the University of Delaware, Department of Chemistry and Biochemistry, and Department of Chemical Engineering, respectively.

References

1 Brooks, C.L. III, Karplus, M., and Pettitt, B.M. (1988) *Proteins: A Theoretical Perspective of Dynamics, Structure, and Thermodynamics*, in Advances in Chemical Physics, vol. 71 (eds I. Prigogine and S.A. Rice), John Wiley & Sons, Inc., New York.

2 Leach, A.R. (2001) *Molecular Modelling: Principles and Applications*, 2nd edn, Prentice Hall.

3 Xie, W. et al. (2007) Development of a polarizable intermolecular potential function (PIPF) for liquid amides and alkanes. *J. Chem. Theory Comput.*, **3** (6), 1878–1889.

4 Gao, J., Habibollazadeh, D., and Shao, L. (1995) A polarizable intermolecular potential function for simulation of liquid alcohols. *J. Phys. Chem.*, **99**, 16460–16467.

5 Gao, J. and Xia, X. (1992) A priori evaluation of aqueous polarization effects through Monte Carlo QM–MM simulations. *Science*, **258**, 631.

6 Vorobyov, I.V., Anisimov, V.M., and MacKerell, A.D. (2005) Polarizable empirical force field for alkanes based on the classical Drude oscillator model. *J. Phys. Chem. B*, **109**, 18988–18999.

7 Vorobyov, I.V. et al. (2007) Additive and classical Drude polarizable force fields for linear and cyclic ethers. *J. Chem. Theory Comput.*, **3**, 1120–1133.

8 Anisimov, V.M. et al. (2005) Determination of electrostatic parameters for a polarizable force field based on the classical Drude oscillator. *J. Chem. Theory Comput.*, **1**, 153–168.

9 Anisimov, V.M. et al. (2007) Polarizable empirical force field for the primary and secondary alcohol series based on the classical Drude model. *J. Chem. Theory Comput.*, **3**, 1927–1946.

10 Patel, S. and Brooks, C.L. III (2004) CHARMM fluctuating charge force field for proteins: I. Parameterization and application to bulk organic liquid simulations. *J. Comput. Chem.*, **25**, 1–15.

11 Patel, S. and Brooks, C.L. III (2005) A nonadditive methanol force field: bulk liquid and liquid–vapor interfacial properties via molecular dynamics simulations using a fluctuating charge model. *J. Chem. Phys.*, **122**, 024508.

12 Patel, S. and Brooks, C.L. III (2005) Structure, thermodynamics, and liquid–vapor equilibrium of ethanol from molecular dynamics simulations using nonadditive interactions. *J. Chem. Phys.*, **123**, 164502.

13 Patel, S. and Brooks, C.L. III (2006) Revisiting the hexane–water interface via molecular dynamics simulations using non-additive alkane–water potentials. *J. Chem. Phys.*, **124**, 204706.

14 Patel, S., MacKerell, A.D. Jr, and Brooks, C.L. III (2004) CHARMM fluctuating charge force field for proteins: II Protein/solvent properties from molecular dynamics simulations using a non-additive electrostatic model. *J. Comput. Chem.*, **25**, 1504–1514.

15 Patel, S. and Brooks, C.L. III (2006) Fluctuating charge force fields: recent developments and applications from small molecules to macromolecular biological systems. *Mol. Simul.*, **32** (3/4), 231–249.

16 Rick, S.W. (2001) Simulations of ice and liquid water over a range of temperatures using the fluctuating charge model. *J. Chem. Phys.*, **114** (5), 2276.

17 Rick, S.W. and Berne, B.J. (1996) Dynamical fluctuating charge force fields: the aqueous solvation of amides. *J. Am. Chem. Soc.*, **118**, 672–679.

18 Rick, S.W. and Berne, B.J. (1997) Free energy of the hydrophobic interaction from molecular dynamics simulations: the effects of solute and solvent polarizability. *J. Phys. Chem. B*, **101**, 10488–10493.

19 Rick, S.W. and Stuart, S.J. (2002) Potentials and algorithms for incorporating polarizability in computer simulations, in *Reviews in Computational Chemistry*, vol. **18** (eds K.B. Lipkowitz and D.B. Boyd), Wiley-VCH Verlag GmbH, Weinheim, pp. 89–146.

20 Rick, S.W. et al. (1995) Fluctuating charge force fields for aqueous solutions. *J. Mol. Liq.*, **65/66**, 31–40.

21 Rick, S.W., Stuart, S.J., and Berne, B.J. (1994) Dynamical fluctuating charge force fields: application to liquid water. *J. Chem. Phys.*, **101** (7), 6141–6156.

22 Kaminski, G.A. et al. (2004) Development of an accurate and robust polarizable molecular mechanics force field from ab initio quantum chemistry. *J. Phys. Chem. A*, **108**, 621–627.

23 Yu, H. and van Gunsteren, W.F. (2005) Accounting for polarization in molecular simulation. *Comput. Phys. Commun.*, **172**, 69–85.

24 Yu, H. *et al.* (2006) Molecular dynamics simulations of liquid methanol and methanol–water mixtures with polarizable models. *J. Comput. Chem.*, **27**, 1494–1504.

25 Piquemal, J.-P. *et al.* (2007) Key role of polarization anisotropy of water in modeling classical polarizability. *J. Phys. Chem. A*, **111** (33), 8170–8176.

26 Ledecq, M. *et al.* (2003) Modeling of copper(II) complexes with the SIBFA polarizable molecular mechanics procedure. Application to a new class of HIV-1 protease inhibitors. *J. Phys. Chem. B*, **107** (38), 10640–10652.

27 Gresh, N. (1995) Energetics of Zn^{2+} binding to a series of biologically relevant ligands: a molecular mechanics investigation grounded on *ab initio* self-consistent field supermolecule computations. *J. Comput. Chem.*, **16**, 856–882.

28 Gresh, N. and Garmer, D.R. (1996) Comparative binding energetics of Mg^{2+}, Ca^{2+}, Zn^{2+}, and Cd^{2+} to biologically relevant ligands: combined *ab initio* SCF supermolecule and molecular mechanics investigation. *J. Comput. Chem.*, **17**, 1481–1495.

29 Jiao, D. *et al.* (2008) Calculation of protein–ligand binding free energy by using a polarizable potential. *Proc. Natl. Acad. Sci. U.S.A.*, **105** (17), 6290.

30 Ren, P. and Ponder, J.W. (2002) Consistent treatment of inter- and intramolecular polarization in molecular mechanics calculations. *J. Comput. Chem.*, **23** (16), 1497–1506.

31 Ren, P. and Ponder, J.W. (2003) Polarizable atomic multipole water model for molecular mechanics simulation. *J. Phys. Chem. B*, **107**, 5933–5947.

32 Ponder, J.W. (2002) TINKER, Version 4.2, School of Medicine, Washington University.

33 Thole, B. (1981) *Chem. Phys.*, **59**, 341.

34 van Gunsteren, W.F., Dolenc, J., and Mark, A.E. (2008) Molecular simulation as an aid to experimentalists. *Curr. Opin. Struct. Biol.*, **18**, 149–153.

35 Ponder, J. and Case, D.A. (2003) Force fields for protein simulations. *Adv. Protein Chem.*, **66**, 27–85.

36 MacKerell, A.D. Jr *et al.* (1998) CHARMM: the energy function and its parametrization with an overview of the program, in *Encyclopedia of Computational Chemistry* (ed. P.V.R. Schleyer *et al.*), John Wiley & Sons, Ltd, Chichester, pp. 271–277.

37 Lamoureux, G., MacKerell, A.D., and Roux, B. (2003) A simple polarizable model of water based on classical Drude oscillators. *J. Chem. Phys.*, **119** (10), 5185–5197.

38 Lamoureux, G. and Roux, B. (2003) Modeling induced polarization with classical Drude oscillators: theory and molecular dynamics simulation algorithm. *J. Chem. Phys.*, **119** (6), 3025–3039.

39 Chelli, R. and Procacci, P. (2002) A transferable polarizable electrostatic force field for molecular mechanics based on the chemical potential equalization principle. *J. Chem. Phys.*, **117** (20), 9175.

40 Chelli, R. *et al.* (1999) Electrical response in chemical potential equalization schemes. *J. Chem. Phys.*, **111** (18), 8569–8575.

41 Parr, R.G. and Yang, W. (1989) *Density-Functional Theory of Atoms and Molecules*, Oxford University Press, Oxford.

42 Sanderson, R.T. (1951) An interpretation of bond lengths and a classification of bonds. *Science*, **114**, 670.

43 Sanderson, R.T. (1976) *Chemical Bonds and Bond Energy*, 2nd edn, Academic, New York.

44 Nalewajski, R.F., Korchowiec, J., and Zhou, Z. (1988) Molecular hardness and softness parameters and their use in chemistry. *Int. J. Quantum Chem. Quantum Biol. Symp.*, **22** (22), 349–366.

45 Warren, G.L., Davis, J.E., and Patel, S. (2007) Origin and control of super-linear polarizability scaling in chemical potential equalization methods. *J. Chem. Phys.*, **128**, 144110.

46 Caldwell, J.W. and Kollman, P.A. (1995) Structure and properties of neat liquids using nonadditive molecular dynamics: water, methanol, and N-methylacetamide. *J. Phys. Chem.*, **99**, 6208–6219.

47 Caldwell, J.W. and Kollman, P.A. (1995) Cation–pi interactions: nonadditive effects are critical in their accurate representation. *J. Am. Chem. Soc.*, **117**, 4177–4178.

48 Car, R. and Parrinello, M. (1985) Unified approach for molecular dynamics and density-functional theory. *Phys. Rev. Lett.*, **55**, 2471.

49 Banks, J.L. et al. (1999) Parametrizing a polarizable force field from *ab initio* data. I. The fluctuating point charge model. *J. Chem. Phys.*, **110** (2), 741.

50 Kaminski, G.A. et al. (2002) Development of a polarizable force field for proteins via *ab initio* quantum chemistry: first generation model and gas phase tests. *J. Comput. Chem.*, **23**, 1515–1531.

51 Nistor, R.A. et al. (2006) A generalization of the charge equilibration method for nonmetallic materials. *J. Chem. Phys.*, **125**, 094108.

52 Shimizu, K. et al. (2004) Calculation of the dipole moment for polypeptides using the generalized Born–electronegativity equalization method: results in vacuum and continuum dielectric solvent. *J. Phys. Chem. B*, **108**, 4171–4177.

53 Stern, H.A. et al. (2001) Combined fluctuating charge and polarizable dipole models: application to a five-site water potential function. *J. Chem. Phys.*, **115** (5), 2237.

54 Liu, Y.-P. et al. (1998) Constructing *ab initio* force fields for molecular dynamics simulations. *J. Chem. Phys.*, **108** (12), 4739.

55 Stern, H.A. et al. (1999) Fluctuating charge, polarizable dipole, and combined models: parameterization from *ab initio* quantum chemistry. *J. Phys. Chem. B*, **103**, 4730–4737.

56 Jorgensen, W.L. et al. (1983) Comparison of simple potential functions for simulating liquid water. *J. Chem. Phys.*, **79**, 926.

57 Lamoureux, G. et al. (2006) A polarizable model of water for molecular dynamics simulations of biomolecules. *Chem. Phys. Lett.*, **418** (1–3), 245–249.

58 Roux, B. et al. (2004) Theoretical and computational models of ion channels. *Q. Rev. Biophys.*, **37**, 15–103.

59 Li, L.B. et al. (2008) Is arginine charged in a membrane? *Biophys. J.*, **94**, L11–L13.

60 Li, L.B., Vorobyov, I., and Allen, T.W. (2008) Potential of mean force and pK_a profile calculation for a lipid membrane-exposed Arg side chain. *J. Phys. Chem. B*, **112**, 9574–9587.

61 Brylinski, M., Konieczny, L., and Roterman, I. (2006) Hydrophobic collapse in (*in silico*) protein folding. *Comput. Biol. Chem.*, **30** (4), 255–267.

62 Sorin, E., Rhee, Y., and Pande, V.S. (2005) Does water play a structural role in the folding of small nucleic acids? *Biophys. J.*, **88** (4), 2516–2524.

63 Davis, J.E., Warren, G.L., and Patel, S. (2008) Revised charge equilibration potential for liquid alkanes. *J. Phys. Chem. B*, **112**, 8298–8310.

64 Lide, D.R. (ed.) (2003) *CRC Handbook of Chemistry and Physics*, 84th edn, CRC Press, Boca Raton, FL.

65 Lemmon, E.W., McLinden, M.O., and Friend, D.G. (2005) Thermophysical properties of fluid systems, in *NIST Chemistry WebBook, NIST Standard Reference Database Number 69* (ed. P.J. Linstrom and W.G. Mallard), National Institutes of Standards and Technology, Gaithersburg, MD, http://webbook.nist.gov/chemistry/fluid/.

66 Klauda, J.B. et al. (2005) An *ab initio* study of the torsional surface of alkanes and its effect on molecular simulations of alkanes and a DPPC bilayer. *J. Phys. Chem. B*, **109**, 5300–5311.

67 Tofts, P.S. et al. (2000) Test liquids for quantitative MRI measurements of self-diffusion coefficient *in vivo*. *Magn. Reson. Med.*, **43** (3), 368–374.

68 Wang, L., Bose, P.S., and Sigworth, F.J. (2006) Using cryo-EM to measure the dipole potential of a lipid membrane. *Proc. Natl. Acad. Sci. U.S.A.*, **103** (49), 18528.

69 Stern, H. and Feller, S.E. (2003) Calculation of the dielectric permittivity profile for a nonuniform system: application to a lipid bilayer simulation. *J. Chem. Phys.*, **118**, 3401.

70 Pandit, S.A., Bostick, D., and Berkowitz, M.L. (2004) Complexation of phosphatidylcholine lipids with cholesterol. *Biophys. J.*, **86** (3), 1345.

71 Berkowitz, M.L., Bostick, D.L., and Pandit, S.A. (2006) Aqueous solutions next to phospholipid membrane surfaces: insights from simulations. *Chem. Rev.*, **106**, 1527.

72 Tieleman, D.P. (1998) Theoretical studies of membrane models: molecular dynamics of water, lipids and membrane proteins. PhD Thesis. University of Groningen, The Netherlands.

73 Tieleman, D.P. and Berendsen, H.J.C. (1998) A molecular dynamics study of the pores formed by *Escherichia coli* OmpF porin in a fully hydrated palmitoyloleoylphosphatidylcholine bilayer. *Biophys. J.*, **74**, 2786–2801.

74 Tieleman, D.P. et al. (2006) Membrane protein simulations with a united atom lipid and all atom protein model: side chain transfer free energies and model proteins. *J. Phys. Condens. Matter*, **18**, S1221.

75 Aliste, M.P., MacCallum, J.L., and Tieleman, D.P. (2003) Molecular dynamics simulations of pentapeptides at interfaces: salt bridge and cation–pi interactions. *Biochemistry*, **42**, 8976.

76 Aliste, M.P. and Tieleman, D.P. (2005) Computer simulation of partitioning of ten pentapeptides Ace-WLXLL at the cyclohexane/water and phospholipid/water interfaces. *BMC Biochem.*, **6**, 30.

77 MacCallum, J.L., Bennett, W.F.D., and Tieleman, D.P. (2008) Distribution of amino acids in a lipid bilayer from computer simulations. *Biophys. J.*, **94**, 3393–3404.

78 MacCallum, J.L. and Tieleman, D.P. (2006) Computer simulation of the distribution of hexane in a lipid bilayer: spatially resolved free energy, entropy, and enthalpy profiles. *J. Am. Chem. Soc.*, **128**, 125.

79 Bucher, D. et al. (2006) Polarization effects and charge transfer in the KcsA potassium channel. *Biophys. Chem.*, **124**, 292.

80 MacKerell, A.D. Jr et al. (1998) All-atom empirical potential for molecular modeling and dynamics studies of proteins. *J. Phys. Chem. B*, **102** (18), 3586–3616.

81 Frisch, M.J. et al. (2004) Gaussian 03, Gaussian, Inc., Wallingford, CT.

82 Breneman, C.M. and Wiberg, K.B. (1990) Determining atom-centered monopoles from molecular electrostatic potentials. The need for high sampling density in formamide conformational analysis. *J. Comput. Chem.*, **11**, 361.

83 Newton, M.D. (1973) A model conformational study of nucleic acid phosphate ester bonds: the torsional potential of dimethyl phosphate monoanion. *J. Am. Chem. Soc.*, **95**, 256.

84 Woolf, T.B. and Roux, B. (1994) Conformational flexibility of o-phosphorylcholine and o-phosphorylethanolamine: a molecular dynamics study of solvation effects. *J. Am. Chem. Soc.*, **116**, 5916–5926.

85 MacKerell, A.D. Jr, Feig, M., and Brooks, C.L. III (2004) Extending the treatment of backbone energetics in protein force fields: limitations of gas-phase quantum mechanics in reproducing protein conformational distributions in molecular dynamics simulations. *J. Comput. Chem.*, **25**, 1400–1415.

86 Feig, M., MacKerell, J., Alexander, D., Charles, I., and Brooks, L. (2003) Force field influence on the observation of pi-helical protein structures in molecular dynamics simulations. *J. Phys. Chem. B*, **107**, 2831–2836.

87 Pandit, S.A., Bostick, D., and Berkowitz, M.L. (2003) Molecular dynamics simulation of a dipalmitoylphosphatidylcholine bilayer with NaCl. *Biophys. J.*, **84**, 3743–3750.

88 Moore, P.B., Lopez, C.F., and Klein, M.L. (2001) Dynamical properties of a hydrated lipid bilayer from a multinanosecond molecular dynamics simulation. *Biophys. J.*, **81** (5), 2484–2494.

89 Lopez, C.F. et al. (2004) Hydrogen bonding structure and dynamics of water at the dimyristoylphosphatidylcholine lipid bilayer surface from a molecular dynamics simulation. *J. Phys. Chem. B*, **108**, 6603.

90 Kucerka, N. et al. (2005) Structure of fully hydrated fluid phase DMPC and DLPC lipid bilayers using X-ray scattering from oriented multilamellar arrays and from unilamellar vesicles. *Biophys. J.*, **88**, 2626–2637.

91 Klauda, J.B. et al. (2006) Simulation-based methods for interpreting X-ray data from lipid bilayers. *Biophys. J.*, **90**, 2796–2807.
92 Feller, S.E. (2000) Molecular dynamics simulations of lipid bilayers. *Curr. Opin. Colloid Interface Sci.*, **5**, 217–223.
93 Hogberg, C.-J., Nikitin, A.M., and Lyubartsev, A.P. (2008) Modification of the CHARMM force field for DMPC lipid bilayer. *J. Comput. Chem.*, **29**, 2359–2369.
94 Siu, S.W.I. et al. (2008) Molecular simulations of membranes: physical properties from different force fields. *J. Chem. Phys.*, **128**, 125103.
95 Douliez, J.-P., Leonard, A., and Duforc, E.J. (1995) Restatement of order parameters in biomembranes: calculation of C–C bond order parameters from C–D quadrupolar splitting. *Biophys. J.*, **68**, 1727–1739.
96 Trouard, T.P. et al. (1999) Influence of cholesterol on dynamics of dimyristoylphosphatidylcholine bilayers as studied by deuterium NMR relaxation. *J. Chem. Phys.*, **110**, 8802.
97 Otten, D., Brown, M.F., and Beyer, K. (2000) Softening of membrane bilayers by detergents elucidated by deuterium NMR spectroscopy. *J. Phys. Chem. B*, **104**, 12119–12129.
98 Watanabe, K. and Klein, M.L. (1989) Effective pair potentials and the properties of water. *Chem. Phys.*, **131**, 157–167.
99 Chen, B., Xing, J., and Siepmann, J.I. (2000) Development of polarizable water force fields for phase equilibrium calculations. *J. Phys. Chem.*, **104**, 2391–2401.
100 Carnie, S.L. and Patey, G.N. (1982) Fluids of polarizable hard-spheres with dipoles and tetrahedral quadrupoles: integral-equation results with application to liquid water. *Mol. Phys.*, **47** (5), 1129–1151.
101 Nymeyer, H. and Zhou, H.-X. (2008) A method to determine dielectric constants in nonhomogeneous systems: application to biological membranes. *Biophys. J.*, **94** (4), 1185–1193.
102 Price, D.J. and Brooks, I.C.L. (2004) A modified TIP3P water potential for simulation with Ewald summation. *J. Chem. Phys.*, **121** (20), 10096–10103.

Subject Index

a
activity 55ff.
– cosolvent 55
activity coefficient 62, 80
alcohols 18
alkanes
– charge equilibration model 283
AMBER
– all-atom model 256
– force field 275
amino acid
– Lennard-Jones type function 111
– side chains 295
– solvation free energy 172
AMOEBA (atomic multipole optimized energetics for biomolecular applications) force field 274ff.
– water 282
amyloid-β peptide 226
amyloidogenicity profile 215
analytic generalized Born (AGB) 141
– plus non-polar (AGBNP) 141
analytical continuum
– solvation 209ff.
analytical continuum electrostatics (ACE) 141
analytical linearized Poisson–Boltzmann (ALPB) 143
aqueous solution
– forces induced by water 95
– free energy of folding–unfolding process 93
– potential 106
atomic (or self) electrostatic solvation energy 170, 219
atomic fluctuation 224
atomic hardness 278
atomic multipole 275
atomic solvent-accessible surface area 223
atomistic dynamic simulations 96
avian flu virus 145

b
biomolecular force field 56
biomolecular solvation 1
bond charge increment (BCI) 300
bond dipole 301
Born
– approximation 121
– equation 170
– formula 134
– solvation free energy 260
Born–Oppenheimer
– condition 276
– surface 303
tert-butyl alcohol (TBA) 84f.

c
carbon monoxide (CO) 47ff.
Car–Parrinello minimization of the polarization density free energy functional 252
Car–Parrinello (CP) type method 278
cellulose
– water as substrate for enzymatic hydrolysis 45
CG, see coarse-grained
chain dynamics
– local 288
charge
– distribution 72, 135, 275, 296
– dynamics 276
– equilibration 273
– N interacting 115ff.
– single 114
– two interacting charges 115f.
charge equilibration model (CHEQ) 276ff.
– alkanes 283

– molecular dynamics simulation   290
– parameterization   278
charge-on-spring (COS) model   273
CHARMM   137ff., 210ff.
– all-atom model   256
– force field   173ff., 275, 293
– non-polarizable model   292
chemical potential   192
coarse-grained (CG) force field   236ff.
coarse-grained liquid water model
– representability problem   238
– structural property   238
coarse-grained MD simulation   237
coarse-grained potential
– computing   235
– liquid water   235
– state-specific   233ff.
– water   233ff.
coarse-grained solvent   251ff.
– biomolecule   251ff.
– polarizable   251ff.
coarse-grained water   244
computer simulation method   6
– solvation   6
condensed-phase environment   299
cones of hydration   16
configuration integral   32
conformation
– fully extended–folded hairpin   177
conformational dynamics   145
conformational equilibrium
– model peptide   175
conformational sampling   152
constant-pH MD   146
continuum electrostatics   168f.
continuum electrostatics solvation model   173
continuum electrostatics solvent modeling   127ff.
continuum solvent environment   129
cosolute   93ff.
– electrostatic and liquid-structure forces   111
cosolvent   55ff.
– molecule   62ff.
cosolvent activity coefficient   62
Coulomb field   136
Coulomb field approximation (CFA)   136, 218
Coulomb interaction   277
Coulomb interaction energy
– intramolecular   173
Coulomb potential   131, 257
Coulomb term   193

critical micelle concentration (CMC)   86
crystal structure   11
crystal water site   11
crystalline phase   195ff.
– dielectric medium   195

d

Debye–Hückel screening parameter   130, 142
degree of freedom   32
density   109
– fluctuation   34
– liquid   110
– local   110
density pair distribution function   33
density-functional theory (DFT)   252f.
dielectric boundary   138, 172
dielectric constant   31, 256
dielectric continuum theory   253
dielectric dispersion   13
dielectric medium
– crystalline phase   195
dielectric permittivity   114, 300
– profile   300
dielectric response
– water   108
dielectric saturation   110
dielectric system   256
diffusion coefficient   10
– local   10
dimethylphosphate (DMP)   292
dimyristoylphosphatidylcholine (DMPC)   295, 298f.
– molecular polarizability   300
– fully polarizable DMPC bilayers in TIP4P-FQ solvent   290
dipalmitoylphosphatidylcholine (DPPC)   295
– bilayer   144, 289
dipolar saturation   258
dipole moment   299
– variation   299
dipole polarizability   277
Dirac delta function   35
dispersion
– van der Waals type   6
dissolution
– protein in water   38
distribution function   33
– pairwise   3
DMPC–water bilayer system   289
DPPC–water bilayer system   144, 289
DNA, see also nucleic acid
– A-form   17, 266

- B-form 17
- free DNA in solution 145
- hydration 16
- major groove 266
- protein complex 146
Drude force field 276
Drude model 287
Drude oscillator 276
dynamic fluorescence spectroscopy 13

e

effective Born radius 133f., 170, 217
- computing 135
effective energy 211
electric displacement 108
electron density profile 294
electronegativity parameter 278
electrostatic energy 115, 276f.
electrostatic force 114
electrostatic free energy 193, 255f.
electrostatic interaction free energy 120
electrostatic polarization 276
- modeling 275
electrostatic potential 114, 193, 253
electrostatic size
- molecule 143
electrostatic solvation free energy 170, 196, 217ff, 219
electrostatics
- particle 257
enthalpy 286
- vaporization 286
environment 32
Ewald summation technique 7
extended reference interaction-site model (XRISM) 32
extended (β) structure 14

f

FACTS (fast analytical continuum treatment of solvation) approach 141, 209ff.
- application 224
- implicit solvent model 217
- parameterization 223
- scalar and parallel performance 226
- validation 224
femtosecond-resolution fluorescence spectroscopy 13
finite differences 252
finite elements 252
fluctuating dipole model 273
fluorescence spectroscopy 13
folding–unfolding process
- free energy 93

force field (FF) 6, 55ff., 128, 211
- all-atom 260
- biomolecular 56
- first-generation polarizable 273
- modeling electrostatic polarization 275
- optimization 167ff.
- problem 58
- protein 267
- refinement 281
- second-generation 267
force matching (FM) equation 236
force matching method 235f.
free energy 3, 129ff., 169, 210
- electrostatic contribution 114
- folding–unfolding process 93
- hydration 114
- simulation 10

g

Gaussian GB 141
generalized Born (GB) model 131ff., 168
- accuracy–speed tradeoff 147
- application 145
- canonical 133ff.
- limitation 154ff.
- non-aqueous solvent 143
- sphere-based canonical 137
generalized Born surface area (GBSA) theory 183, 210
generalized Born surface area/implicit membrane (GBSA/IM) 143
generalized Born using molecular volume (GBMV) 137, 174, 218
generalized Born with simple switching (GBSW) 139, 173f.
Ginzberg–Landau dynamics 51
glycerol 19, 94
glycine betaine 83f.
graphics processing unit (GPU) 150
Green function 131
GROMOS force field 63ff., 275
Grote–Hynes theory 5
group transfer free energy (GTFE) 81
guanidinium 78

h

Hamiltonian 51, 128, 258
- all-particle-based 257
hard-sphere repulsion 6
Hawkins–Cramer–Truhlar (HCT) model 139
Heaviside step function 37
α-helix 14
Helmholtz free energy 37

Hessian matrix 280
heterogeneous dielectric generalized Born (HDGB) 144
1,1,1,3,3,3-hexafluoropropan-2-ol (HFIP) 18
high interstitial dielectric 141
hybrid explicit/implicit approach 147
hydration
– DNA 16
– free energy 114
– nucleic acid 16, 264
– protein 13
– RNA 16
hydration shell 15ff., 102
hydration structure 41
hydrogen bond (HB) 96
hydrophilic environment 119
hydrophilicity 120
hydrophobic effect 3
hydrophobic environment 119
hydrophobic interaction 86
– urea 80
hydrophobicity 120
hypernetted chain (HNC) 36ff.

i

implicit crystalline phase 195
implicit solvent 169
– modeling protein solubility 191ff.
implicit solvent force-field optimization 167ff.
implicit solvent framework 127ff.
implicit solvent model 210ff.
instantaneous disordered grid 258
integral equation formalisms 4
integral equation theory 5
interaction 1
– bonded 2
– intermolecular 103
– non-bonded 2
– pairwise 3
– solute–water 1
interaction-site model 35
intramolecular correlation function 35
ion
– solvent model 104
ion binding
– selective 42
ionization charge 120f.
iterative Yvon-Born-Green (YBG)
– force field 244
– method 236ff.
– numerical issue 237

k

Kirkwood formula 255
Kirkwood spherical model 137
Kirkwood–Buff (KB)
– based force field (KBFF) 182
– equation 37
– integral 59ff., 182
Kirkwood–Buff derived force field (KBFF) approach 56ff.
– technical aspect 64
– urea model 79
Kirkwood–Buff theory 5, 59ff.
– application 60
Kirkwood–Fröhlich formalism 301
Kronecker delta function 35

l

Lagrange multiplier 121, 245, 277
Langevin dipole 258
Langevin function 259
Langevin piston barostat 240
Laplace equation 131
Lennard-Jones (LJ) coefficient 6
Lennard-Jones energy function 258
Lennard-Jones parameter 176, 259
Lennard-Jones potential 6, 239, 281
Lennard-Jones type function
– amino acid 111
lipid chain dynamics 296
liquid
– bulk liquid property 286
– electrostatic potential 114
– local density 110
– statistical mechanics 31
– structure 33, 116
– structure force 116
local chain dynamics 288
local polarization density free-energy functional 254f
Lorentz–Debye–Sack (LDS) approximation 108ff.
Lorentz–Debye–Sack–Onsager (LDSO) approximation 109

m

macromolecular thermodynamics
– role of water and cosolutes 93
macromolecule
– large-scale motion 145
magnetic relaxation dispersion (MRD) 12
Maxwell–Gauss differential equation 255
mean spherical approximation (MSA) 36

membrane 143
– peptides and proteins in the environment 146
membrane ion channel systems
– modeling 289
microenvironment 121
mixed solution 61
molecular dynamics (MD) 6, 42, 134, 170, 233, 251ff, 289f
– all-atom 235
– constant-pH 146
molecular mechanics (MM) 233
molecular Ornstein–Zernike (MOZ) equation 34
molecular polarizability 300
molecular volume (MV) 138
molecule
– electrostatic size 143
Monte Carlo simulation 6, 290
multi-dimensional optimization 252
multi-scale coarse graining (MS-CG) method 235
multipole interaction matrix 275
myoglobin (Mb) 47ff.
– CO escape pathway 47ff.

n

neutron diffraction and scattering 11
Newtonian dynamics 51
NMR quadrupolar splitting tensor 296
non-aqueous solvent
– charge equilibration model of alkanes 283
non-bonded interaction terms (NBFIX) 292
non-polar solvation 171ff.
nonlinear Poisson–Boltzmann (NLPB) equation 130
normal-mode analysis (NMA) 210
Nosé–Hoover thermostat 240, 260
nuclear magnetic resonance (NMR) technique 12
nuclear Overhauser effect (NOE) 12
nucleic acid
– hydration 16
– hydration property 264
– modeling water effects 112
NVT simulation 245

o

octanol-to-water transfer free energy 193f.
OPLS (optimized potentials for liquid simulations) 65
– urea model 65, 79

optical spectroscopy 13
order parameter 32
orientational free energy 258f.
Ornstein–Zernike
– equation 34
– formalism 111
– integral equation 4
osmolyte
– denaturing 78
– hydrophobic interaction 86
– influence on protein stability 77ff.
– mixed 87
– protecting 83ff.
– water structure 83

p

pair correlation function (PCF) 34
palmitoyloleoylphosphatidylcholine (POPC) 301
partial charge 169ff.
partial molar volume (PMV) 37ff.
– CO escape pathway of myoglobin 49
– protein 38
particle mesh Ewald (PME)
– approximation 149f.
– method 240
partition function 32
PCM optimized structure 48
peptide
– aggregation 213, 226
– conformational equilibrium of model peptide 175
– folding–unfolding simulation 177
– membrane environment 146
– reversible folding of structured peptide 213
peptide conformation
– aqueous solvent 14
Percus trick 36
Percus–Yervick (PY) 36
pH dependence
– protein solubility 197
pK prediction 146
point charge 136, 256f.
point dipole polarizability 300
point dipole polarizable models 273
Poisson equation (PE) 129
Poisson–Boltzmann (PB)
– equation 107, 130, 192, 253ff.
– model 129, 155, 168
polarizability 274ff.
polarizable coarse-grained solvent (PCGS) 267f.
polarizable model 301

polarizable pseudo-particle (PPP)
– solvation energy 263
– solvent 262ff.
– solvent model 253
– stability of small protein 260
polarization
– electronic and orientational 257
polarization charge 254
polarization density 253f., 301
– field 255
polyalanine 14
polyproline II (PPII) 14
potential 97
– aqueous solution 106
– ion 106
potential of mean force (PMF) 168ff.
– pairwise interaction 173
– side-chain dimer 224
pressure constraint 244
pressure denaturation 37
probability density 33
protein
– calculation of pK_a 119
– detecting water molecules trapped inside protein 40
– direct interaction with urea 82
– dissolution in water 38
– DNA complex 146
– hydration 13
– interaction in water-accessible regions 96
– membrane environment 146
– modeling water effect 112
– mutation 201
– partial molar volume (PMV) 38
– second-generation force field 267
– selective ion binding 42
– stability in the PPP solvent 260
– titratable residue (TR) 119
protein folding 14, 77ff., 94, 145
– simulation 179
protein solubility 192ff.
– effects of mutation 201
– modeling 191ff.
– pH dependence 197ff.
protein stability
– osmolyte influence 77ff.
protein structure
– Xe site 47
protein surface
– molecular property of water 14
protein–ligand binding 147
protein–ligand interface
– water 16

protein–protein interface
– water 16

q
quantum mechanics/molecular mechanics (QM/MM) simulation 147
quasi-elastic neutron scattering 12

r
radial distribution function (RDF) 3ff., 33, 56ff., 236
– center-of-mass 241
– pairwise 3
radius
– effective Born 133f., 170, 217
– perfect 134
– perfect effective 136
– van der Waals (VDW) 136
reference interaction-site model (RISM) 4, 32ff., 169
– extended (XRISM) 32
Rekker fragmental hydrophobic constant 120
replica exchange (REX)
– MD simulation 177ff.
– method 176ff.
repulsion
– hard-sphere 6
repulsive interaction
– hard-core-type 38
residue pairwise GB 141
RISM-optimized structure 48
RNA, *see also* nucleic acid
– hydration 16
RNase Sa 197ff.

s
salt
– intermolecular interaction 103
salt effect 141
Savitzky–Golay (SG) algorithm 237
screened Coulomb potentials (continuum) implicit solvent model (SCPISM) 118
self electrostatic solvation energy 170, 219
self-diffusion coefficient 10
self-term 133
shell (Drude) model 273
side-chain dimer
– potential of mean force 224
SIF, *see* solvent-induced forces
simple point charge (SPC) water model 239

simple water model with four sites and Drude polarizability (SWM4-DP)   239ff., 283
– SWM4-NDP   283
simulation
– free energy   10
– solvent structure and dynamics   8
site–site correlation functions
– pairwise   5
small-angle neutron scattering (SANS)   12
small-angle X-ray scattering (SAXS)   12
solubility   192
solute
– electrostatic and liquid-structure forces   111
– uniform dielectric constant   256
solute energy   37
solute–liquid potential   117
solute–solvent dielectric boundary   138
solute–solvent electrostatic force   258
solute–solvent Lennard-Jones interaction   260
solute–solvent van der Waals interaction   142
solute–water interaction   1, 265
solution
– binary mixture   60
– calculation of water forces   113
– microscopic structure   55
solvation   2
– all-atom models of biomolecules   259
– biomolecular   1
– cage   295
– computer simulation method   6
– electrostatic part   142
– energy   134
– environment   168
– equilibrium thermodynamics   2
– experimental method   11
– free energy   3, 36f., 129ff., 168ff., 255
– kinetic effect   5
– model   211
– non-aqueous   18
– non-polar   142, 171
– term   193
solvation free energy   3, 36f., 129ff., 168ff., 260
– Born   260
– small molecule   172
solvent   1
– aqueous   14
– biological   281
– dynamics   8f.
– non-aqueous   283

– structure   8
– volume   139
solvent density   35
– distribution   9
solvent effect   93, 147
– aqueous   93ff.
solvent modeling
– in chemistry and biochemistry   31ff.
– ion   104
– model-free   31ff.
– particle-based   258
solvent-accessible surface area (SASA)   142, 171, 192, 210ff.
solvent-accessible surface area implicit solvent model   210ff.
– limitation   216
solvent-induced forces (SIF)   100ff., 117
SPC/E   7
– water model   63ff.
spine of hydration   16
π-stacking mechanism   78
Stokes–Einstein law   31
structure
– liquid   33
sum of interactions between fragments *ab initio* computed (SIBFA) model   273
surface boundary element   252
symmetry   220

t
thermal volume   15
three-dimensional reference interaction-site model (3D-RISM) theory   32ff.
TIP3P (transferable intermolecular potential, three-point)   7, 173ff., 239ff., 302
TIP4P (transferable intermolecular potential, four-point)   7, 239ff.
TIP4P/FQ (transferable intermolecular potential, four-point, with fluctuating charge)   8, 282ff.
– charge equilibration model for molecular dynamics simulation   290
TIP5P (transferable intermolecular potential, five-point)   8
titratable residue (TR)
– protein   119
torsion parameter   6
torsion potential   177
torsional potential energy   293
total correlation function   34
total electrostatic energy   259
total electrostatic solvation energy   221
total solvation free energy
– FACTS model   223

u

transfer free energy 192
– octanol to water 193
transient-grating (TG) method 47
2,2,2-trifluoroethanol (TFE) 18
trimethylamine N-oxide (TMAO) 83ff.
two-dielectric model 129ff.

u

urea 18, 65
– direct interaction with proteins 82
– hydrophobic interaction 80
– model 65
– OPLS 65
– preferential interaction 68
– van der Waals interaction 82
– water structure 78f.

v

van der Waals (Lennard-Jones) contribution 292
van der Waals energy term 257
van der Waals interaction
– urea 82
van der Waals potential 258
van der Waals (VDW) radius 136ff.
van der Waals volume 220
vibrational spectroscopy 13
viscosity 31
volume 220
– approximation 139
– local 118
– volume term 142
volumetric density distribution 8

w

water 1, 281ff.
– AMOEBA water force field 282
– anisotropic structural property of liquid water 242
– binary solution 65
– bulk water value 301
– bulk-like property 15
– coarse-grained liquid water model 238
– continuum representation 107
– detecting water molecules trapped inside protein 40
– dielectric 301
– dielectric response 108
– dissolution of a protein 38
– DMPC–water and DPPC–water bilayer systems 289
– forces induced in aqueous solution 95
– hydration shell 102
– local properties 93
– macromolecular thermodynamics 93
– model 7, 63ff., 233ff.
– molecular property near protein surfaces 14
– potential 99
– protein–ligand interface 16
– protein–protein interface 16
– solute–water interaction 1
– state-specific coarse-grained potential 233ff.
– structural response 108
– substrate for enzymatic hydrolysis of cellulose 45
water forces
– bulk and non-bulk contribution 99ff.
– calculation in solution 113
water structure 78ff.
– osmolyte 83
– urea 78f.
water residence time 266
weighted histogram analysis method (WHAM) 173

x

X-ray crystallography 11
Xe site 47f.
Xe-free structure 48
Xe-rich structure 48
XRISM, see extended reference interaction-site model

y

Yvon–Born–Green (YBG) equation 236